Sound and Robotics

Sound in human–robot interaction currently encompasses a wide range of approaches and methodologies not easily classified, analyzed or compared among projects. This edited book covers the state of the art in sound and robotics, aiming to gather existing approaches in a combined volume.

Collecting chapters from world-leading academic and industry authors, *Sound and Robotics: Speech, Non-Verbal Audio and Robotic Musicianship* explores how robots can communicate through speech, non-verbal audio and music. The first set of chapters explores how robots use verbal communication, considering the possibilities of speech for human–robot interaction. The second section shifts to roles of non-verbal communication in HRI, including consequential sound, sonification and audio cues. The third and final section describes current approaches to robotic musicianship and their evaluation.

This book is primarily aimed at HRI researchers, ranging from those who have never used sound to those very experienced with sound. Alongside robotic researchers, this book will present avenues for a diverse range of musicians, composers and sound designers to become introduced to the world of HRI and learn of potential creative directions in robotics.

Richard Savery is a research fellow at Macquarie University, Australia, working at the intersection of sound and robotics. He completed his doctorate in music technology at Georgia Tech, USA, focusing on the use of non-verbal audio for improved human–robot interaction.

Sound and Robotics
Speech, Non-Verbal Audio and Robotic Musicianship

Edited by Richard Savery

CRC Press
Taylor & Francis Group
Boca Raton London New York

CRC Press is an imprint of the
Taylor & Francis Group, an **informa** business

A CHAPMAN & HALL BOOK

ISBN: 978-1-032-34084-5 (hbk)
ISBN: 978-1-032-34083-8 (pbk)
ISBN: 978-1-003-32047-0 (ebk)

DOI: 10.1201/9781003320470

Typeset in CMR10
by KnowledgeWorks Global Ltd.

Publisher's note: This book has been prepared from camera-ready copy provided by the authors.

Contents

III   Robotic Musicianship and Musical Robots   241

Preface

Sound and Robotics: Speech, Non-Verbal Audio and Robotic Musicianship is an interdisciplinary exploration of the under-represented area of sound in robotics. This book was developed with the aim to highlight the potential of sound to reshape and improve how humans interact with robots. It addresses the areas of speech, non-verbal audio, and robotic musicianship, focused on understanding how robots can use these forms of audio to communicate.

The contributors to this book come from a diverse range of academic backgrounds, including computer science, mechatronics, psychology and music. Their expertise and knowledge come together to provide a comprehensive overview of the latest developments in the field of sound and robotics and propose future directions for sound and robotics.

Each chapter in this book delves into a specific topic, from sound localization, to voices in storytelling, to musical performance. Through a combination of theoretical discussions, practical applications and case studies, the authors explore the potential of sound to enhance the capabilities of robots and improve their integration into our daily lives.

List of Contributors

Aimee Allen is currently working toward a PhD in human–robot interaction (HRI) within the Faculty of Engineering at Monash University, Australia. Her thesis research is on consequential sounds, which robots unintentionally produce as they move and operate, with a focus on methods to alter these sounds to enhance HRI. Her other research interests include biomimicry, quadruped robots, UAVs and additive manufacturing. She received a BCom with marketing and mechatronic engineering majors from the University of Sydney in 2009. Prior to graduate studies, she worked for several years in software engineering, data analysis, cyber security, management and IT and business education.

Richard Attfield is a PhD student at Monash University within the Human–Robot Interaction Research Group. His research interests include interaction strategies between humans and autonomous multi-robot teams, and the application of multi-modal communication in robotics. His goals are to implement strategies within multi-robot communication that will improve both the effectiveness and experience of a human's role within a human multi-robot team.

Christian Balkenius is a professor of cognitive science at Lund University Cognitive Science (LUCS). His main research goal is to construct systems-level computational models of functional subsystems in the mammalian brain and their interactions using artificial neural networks. His work focuses on various forms of cognitive processes, including sequential processing, categorization, motivation and action selection as well as spatial learning, conditioning and habituation. He has published some 200 research papers and articles on neural network-based modeling of cognitive processes, robotics, vision and learning theory. Prof. Balkenius is the director of the graduate school within The Wallenberg AI, Autonomous Systems and Software Program – Humanities and Society.

Oliver Bown is an associate professor at the UNSW Faculty of Art and Design. He comes from a diverse academic background spanning adaptive systems, music informatics and interaction design. His current active research areas include musical meta creation, new interfaces for musical expression and multi-agent models of social creativity.

Elizabeth A. Croft is an expert in the field of HRI and Vice-President Academic and Provost at the University of Victoria. With over 30 years of experience in the field, she has made significant contributions to the design and development of intelligent robots that can effectively interact with humans. Her research focuses on understanding how people and robots communicate in various settings, including manufacturing, offices and public spaces.

Emma Frid is a postdoctoral researcher at the STMS Laboratory at IRCAM Institute for Research and Coordination in Acoustics/Music and in the Sound and Music Computing group at KTH Royal Institute of Technology. Her current research revolves around Accessible Digital Musical Instruments (ADMIs) and multimodal interfaces. She holds a PhD in sound and music computing and an MScience in engineering from KTH.

Martim S. Galvão is a Brazilian-American composer, percussionist and multimedia artist. Much of his work is concerned with interfaces and how we interact with them. He is especially interested in exploring ideas related to surveillance, consumer-facing technologies and the web. Dr. Galvão earned his bachelor's from Emory University. In 2014, he graduated from the Integrated Composition, Improvisation, and Technology (ICIT) MFA program at the University of California, Irvine. More recently, he earned an MA and PhD from the Computer Music and Multimedia program at Brown University. He is currently an assistant professor at Cal State San Bernardino, where he teaches courses in music technology.

Dr. Birger Johansson is the director of the Lund University Cognitive Robotics Lab. He combines the latest cognition research with deep knowledge in robotics. He has been working with a range of different robots, and specializes in building cognitive robots to make new, real-world experiments that interest both the academic field as well as the general public. He is heavily involved with the software for, and the building of, the humanoid Epi Robot at the Lund University Robotics Group.

Leelo Keevallik is a professor in language and culture at Linköping University, Sweden. She is an expert in interactional linguistics and has worked extensively on conversational language in everyday settings. Her current research concerns sounding practices, such as expressions of strain, pain, proprioception, gustatory pleasure and disgust, as well as how humans interpret robot sounds. She has shown how activities, such as pilates and dance training, are coordinated through the varied use of articulation and prosody. Recent co-edited volumes include "Sounds on the Margins of Language" with R. Ogden and "Sounding for Others" with E. Hofstetter.

Amandus Krantz is a doctoral researcher in cognitive science at Lund University Cognitive Science, Department of Philosophy, Lund University, Sweden. His main research interest is in how perceptions of robots change based on behaviors and interactions. Particularly, he's interested in how non-verbal signals are used to gauge trustworthiness and how these may be read and used by artificial agents as a way of signaling an appropriate level of trustworthiness. Additionally, he's interested in the preconceived negative feelings that are often harbored toward artificial agents, what Isaac Asimov called the Frankenstein Complex. He received an MSc in computer science from Blekinge Institute of Technology.

Dana Kulić develops autonomous systems that can operate in concert with humans, using natural and intuitive interaction strategies while learning from user feedback to improve and individualize operation over long-term use. In collaboration with Prof. Elizabeth Croft, she pioneered systems to quantify and control safety during HRI based on both robot and human perception. She serves as the Global Innovation Research Visiting Professor at the Tokyo University of Agriculture and Technology, and the August-Wilhelm Scheer Visiting Professor at the Technical University of Munich. Before coming to Monash, Dr. Kulić established the Adaptive Systems Lab at the University of Waterloo, and collaborated with colleagues to establish Waterloo as one of Canada's leading research centres in robotics. She is a co-investigator of the Waterloo Robohub, the largest robotics experimental facility in Canada, and a co-principal investigator of the Natural Sciences and Engineering Research Council (NSERC) Canadian Robotics Network, Canada's only federally funded network in robotics.

Birgit Lugrin has held the professorship for computer science (media informatics) as the Chair of Computer Science IX (HRI) at the Institute of Computer Science since 2015, and heads the Media Informatics Group. Prior to this, she was at the University of Augsburg where she wrote her doctoral thesis on "Cultural Diversity for Virtual Characters" and subsequently worked as a faculty member. She holds a BSc and MSc in computer science.

Cynthia Matuszek joined the Computer Science and Electrical Engineering department at UMBC as an assistant professor in 2014, where she founded the Interactive Robotics and Language lab. Matuszek obtained their PhD at the University of Washington in Seattle, where they were co-advised by Dieter Fox and Luke Zettlemoyer. Dr. Matuszek's group in the Interactive Robotics and Language Lab studies how robots can use this kind of language learning to learn to follow instructions or learn about the world. This not only makes robots more useful, but also demonstrates the power of combining robotics with natural language processing. Their work overlaps with the field of *HRI*, or human-robot interaction.

Trinity Melder is a passionate researcher, holding a master's in cybersecurity, witha background in computer science and mathematics. With a deep-rooted love for artificial intelligence, she is constantly pushing boundaries as a self-learner in the field. Growing up in a music-oriented environment, her unique perspective brings a harmonious blend of creativity and technical expertise to every project undertaken. Committed to sharing the magic of music with the world, she strives to bridge the gap between technology and artistry, leaving an indelible mark on both realms.

Hannah Pelikan is a postdoctoral researcher in language and culture at Linköping University, Sweden. She studies how humans make sense of robot behavior from an ethnomethodology and conversation analysis (EMCA) stance, with a particular interest in robot sound. Her PhD research explored autonomous buses on public roads, robots as teammates and Cozmo robots in family homes. Combining EMCA with interaction design, she develops methods for designing HRI that embrace the situated and sequential nature of human interaction. She currently co-edits the special issue on "Sound in HRI" in the *Transactions on Human–Robot Interaction*.

Frederic Anthony Robinson is a Scientia PhD Candidate at the University of New South Wales' Creative Robotics Lab & National Facility for Human-Robot Interaction Research. His research investigates the potential of spatial sonic interaction design to create trust, affinity and companionship in HRI. He is a member of the HRI Pioneers 2020 cohort.

Nicole Robinson is a lecturer at Monash University in the Faculty of Engineering, Department of Electrical & Computer Systems Engineering. She leads an interdisciplinary research group that specializes in both scientific methods and novel engineering. Her research centers around evaluating and improving the quality and success of human-robot teamwork. This work involves theoretical and applied humanrobot interaction work in domestic and field settings, including for mobile, manipulator, mobile manipulator, humanoid and social robots. Prior to Monash, she was Chief Investigator and Program Leader for Human Interaction in the QUT Robotics Centre.

Amit Rogel is a PhD student at Georgia Tech's robotic musicianship lab. His research is centered around robotic gestures and utilizing ancillary movement to enhance robotic musicians and facilitate collaboration between robots and humans. This includes designing and manufacturing new robotic musicians, such as a guitar-playing robot and a novel string and drum robotic instrument. With a background in mechanical engineering from Rochester Institute of Technology, he also has experience playing trombone and producing music.

Liam Roy is a PhD student at Monash University within the HRI research group. His research focuses on leveraging modern-day machine-learning algorithms to develop autonomous systems capable of generating expressive robot sounds and motion.

Anna Savery is a violinist, composer, improviser and music technologist. She creates works that integrate improvised and composed music with cross-media platforms such as Max for Live, Processing and sensor-enhanced musical instruments in collaboration with multidisciplinary artists. She completed an MFA in integrated composition, improvisation and tTechnology at the University of California, Irvine in 2016. She also holds a bachelor's in jazz performance from the Sydney Conservatorium of Music and a bachelor's in classical performance from the Australian Institute of Music. She is due to complete a PhD at University of Technology Sydney, focusing on developing a minimally invasive wireless audio-visual violin bow interface. She is driven by composing narrative-driven, audio-visual works that explore human vulnerability through themes such as immigration, motherhood and belonging.

Richard Savery is a developer of artificial intelligence and robotics, using music and creativity as a medium to program better interactions, understandings and models. He is currently a Research Fellow at Macquarie University, developing new robotic musicians.
He completed a PhD in music technology (minor in humancomputer interaction), at the Georgia Institute of Technologyin 2021.

Yuji Sone is a performance researcher. He initially trained with the experimental theatre company Banyu-Inryoku in Japan. In Australia, he produced numerous media-based performance works throughout the 1990s, and has performed in international and domestic conferences and festivals including Experimenta, TISEA, Sound Watch (NZ), The Adelaide Festival Artists Week, 25 Years of Performance Art in Australia, Canberra National Sculpture Forum, the Sydney Asian Theatre Festival, Sydney Fringe Festival, San Francisco Sound Culture, The Cleveland Performance Art Festival, The Toronto Festival of Performance Art, Rootless UK, Sydney Carnivale, and IFTR Conference Sydney. His hybrid performances have examined issues of cultural representation. His performance experiments, which explore the interaction between live performance and technological systems, work in tandem with his academic research. His research has focused on the cross-disciplinary conditions of mediated performance and the terms that may be appropriate for analyzing such work, especially from cross-cultural perspectives. He is the author of *Japanese Robot Culture: Performance, Imagination, and Modernity* (Palgrave, 2017).

Sophia C. Steinhaeusser completed her studies in media communication in Würzburg and has already worked as a research assistant in media informatics and humancomputer interaction, where she was involved in the ViLeArn project. With a focus on immersive games, she completed her studies in 2020 with an MS. Currently, she is enrolled for a parallel second degree in games engineering. Since 2019, she has been working as a research assistant at the Chair of Media Informatics, where she investigates storytelling with social robots. She is also involved in the ESF-ZDEX project.

Mari Velonaki is director of the UNSW Creative Robotics Lab and the National Facility for Human-Robot Interaction Research, and lead chief investigator at the UNSW Ageing Futures Institute. She has worked as an artist and researcher in the field of electronic art since 1997, creating intellectually and emotionally engaging humanmachine interfaces through the creation of multisensory experiences. She is a pioneer in the field of social robotics in Australia, which she introduced in 2003.

Gil Weinberg is a professor and the founding director of Georgia Tech Center for Music Technology, where he leads the Robotic Musicianship Group. His research focuses on developing artificial creativity and musical expression for robots and augmented humans. Among his projects are a marimba-playing robot called Shimon that uses machine learning to improvise and sing, and prosthetic robotic arms for amputees that restore and enhance human musical abilities. Dr. Weinberg presented his work in venues such as The Kennedy Center, The World Economic Forum, Ars Electronica, among others. His music was performed with orchestras such as Deutsches Symphonie-Orchester Berlin, the National Irish Symphony Orchestra and the Scottish BBC Symphony, while his research has been disseminated through numerous papers and patents. Dr. Weinberg received his MS and PhD in media arts and sciences from MIT and his BA from the interdisciplinary program for fostering excellence at Tel Aviv University.

Lisa Zahray is currently an autonomy engineer at Aurora Flight Science. She completed her bachelor's and master's in computer science and electrical engineering at the Massachusetts Institute of Technology (MIT), where she first became interested in music technology. Her research interests include music information retrieval and AI for creating new music. She plays the trumpet and enjoys casually singing and playing piano.

1

The Landscape of Sound and Robotics

Richard Savery and Trinity Melder

Sound is arguably the biggest missed opportunity in human–robot interaction. While much attention has been given to the visual and physical aspects of robots, such as their movements and appearance, sound has often been overlooked or deemed less important. However, sound can play a crucial role in how robots are perceived and interacted with by humans, especially in social and emotional contexts.

This book emphasizes the ways in which robots can use sound to communicate with humans. This can be through speech, or non-verbal audio, which encompasses all sounds that are created for communication without semantics. We additionally look at ways robots can act as musicians and how musical processes can inform robot design and interaction. This opening chapter presents a brief overview of existing approaches to sound and robotics and offers a summary of the book.

1.1   Overview of Sound and Robotics

To begin this chapter, we conducted a short survey of the use of sound output in robots. We analyzed IEEE's guide to robots (https://robots.ieee.org/), which

DOI: 10.1201/9781003320470-1

1

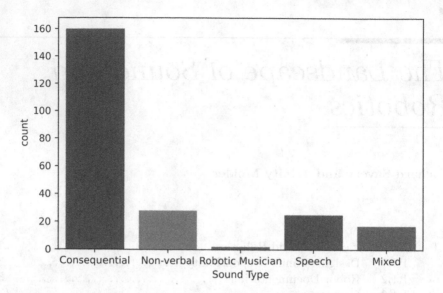

FIGURE 1.1
Sound output from common robot platforms.

includes many common robot platforms and is used in education programs worldwide, making it a valuable resource for our survey. While it is not an exhaustive list, we believe it provides a general overview of the most widely used robots. Our primary goal in the survey was to gain a broad understanding of how sound is currently functioning in robotics, serving as a foundation for the book's focus on leading current research in the use of sound. By examining the robots in the IEEE guide, we were able to identify various broad trends in the use of sound.

Figure 1.1 shows an overview of the types of sound output from the robots in the list. These are broken into speech, non-verbal audio, consequential sound, and mixed. We labeled any robot that primarily used spoken language to communicate as speech, while robots that focused on non-verbal sounds, even if using occasional words, were classified as non-verbal. Mixed robots used speech and non-verbal audio but did not emphasize the use of one or the other. As expected the majority of robots in the list only had consequential sound, which are sounds produced unintentionally as a result of the robot's actions, such as motor or movement sounds. Many of the robots on the list are not focused primarily on human–robot interaction, so a majority focus on consequential sounds is to be expected.

While the survey itself did not show any significant findings, the process made clear many of the existing issues in the research of sound and robotics. The following sections describe some of these key issues and opportunities for future work.

1.1.1 Designers Shift Responsibility

One of the biggest issues in robotics for sound is that many platforms include a speaker, and then leave the responsibility of careful sound design on developers who use the platform. This contrasts the carefully constructed visual and mechanical appearance, which can rarely be changed by end users.

Analyzing the sound between platforms also becomes essentially impossible in these scenarios, as there is no standard sound for a robot, leading to no baseline for comparison. Instead, common robot platforms can be seen with many different sound outputs. While this is a reasonable approach for research, this leaves the field of sound and robotics unable to easily compare results between different robots.

1.1.2 Robot Documentation

Symptomatic of broader robot trends, we had difficulty finding information on some robots that were no longer being produced or discontinued. Often the official website or documentation was unavailable, and we were required to turn to user guides or third-party sources. Likewise, some newer robot platforms described plans for future documentation that were currently unavailable.

This lack of documentation is heightened even further when considering sound and robotics. It was very common to find lengthy descriptions of the design process for the visual appearance, and nothing about the sound. Often robots appearance and motions were demonstrated in videos and publications, while avoiding mention of sound and applying a default speech system.

1.1.3 Categorization

The categories speech, non-verbal audio, and robotic musician are by nature extremely broad. Our original goal included further breakdowns of each category, such as gender for speech, or whether the sounds were emotion-driven, or if they were synthesized or sampled. Ultimately however, it became apparent that it was just not possible to clearly determine how many robots are using sound. This reiterates how much further work is required to understand the role and promote the potential applications of sound for robotics.

1.2 About the Book

Considering the range of sound outputs from robots, this book is split into speech, non-verbal audio, and robotic musicianship. As previously described, the boundary between these categories is not always clear-cut. Nevertheless, the chapters throughout this book are split into sections, with authors primarily addressing one of the forms of communication.

1.2.1 Robot Speech

The first group of chapters in the book delves into the topic of robot speech, exploring the effects of different elements of speech on the user's experience and perception of the robot. Chapter 2 examines the effects of the number of voices and voice type on the storytelling experience and robot perception. The chapter presents findings from an online study comparing human and synthetic voices and the use of single or multiple voices in storytelling. The results suggest that a single voice is preferable for both human and synthetic voices when focusing on the storytelling experience, while the use of different voices is only recommended for synthetic voices when illustrating different characters.

Chapter 3 focuses on how research on vocalizations in human-human interactions can inform the conceptualization of robot sound in human–robot interactions. The chapter highlights three main lessons learned from six examples of human and robot interactions, which include the semantically under-specified nature of both human vocalizations and robot sound, the embodied nature of human sound production, and the need to analyze and design robot sound multi-modally.

Chapter 4 examines the use of speech in social robots for loss-of-trust mitigation. The chapter presents data from two experiments that evaluated the impact of a robot's ability to speak on the user's perception of its trustworthiness, likeability, animacy, and perceived intelligence. The findings suggest that the ability to speak can mitigate the degradation of trust caused by faulty robot behavior, as it increases the perceived intelligence and capability of the robot and makes it appear more like a sentient or living being.

Finally, Chapter 5 explores the topic of grounding spoken language, which involves referencing concepts in the physical world using natural language processing in a robot's environment. The chapter presents a case study of learning grounded language from speech and examines the need for complex perceptual data, the ability to learn to interact directly from speech, and the challenges of learning grounded language from multimodal data. This chapter presents important directions and considerations that are widely overlooked when applying speech to robot systems.

1.2.2 Non-Verbal Audio

The second section addresses the wide range of sound from robots that is non-verbal. The section begins with Chapter 6 which focuses on consequential sounds; the non-intentional sounds from robots such as motor noise. The chapter describes positive and negative effects these sounds can have on interaction, and then highlights the research gap in sound design for consequential sounds and suggests techniques for improving human–robot interaction success.

Chapter 7 explores how a robot's auditory communication can be enhanced by emitting sound across loudspeakers in distributed audio environments. The chapter discusses the design themes for applying interactive sound installation techniques in the context of human–robot interaction, and proposes a generalized design framework for spatial robot sound that can be broken down into three key locations – the robot, objects of interest in the environment, and the space itself.

Chapter 8 examines the concept of sonification and its application in the context of autonomous cars. The chapter presents four different approaches to sonification, discussing the benefits, challenges, and future directions of this emerging field. The goal of the chapter is to describe divergent approaches and cover a range of future possibilities when notifying robot data.

Chapter 9 of this book discusses the use of nonverbal sounds as a means of communication between humans and robots. The authors propose a low-dimensional parameterized communication model based on nonverbal sounds, which is validated using an online survey to investigate the users' experience and shared task performance. The study found that simplified NVS communication could facilitate human–robot collaboration, leading to positive user experience, increased interest in subsequent interactions, and increased collaborative performance.

Chapter 10 describes emotional musical prosody and explores the relationship between personality traits and emotional responses in music-driven emotional interaction in robotics. The authors focus on two of the "Big Five" personality traits, Neuroticism and Extraversion, and investigate how varying the degree of emotional response through sound can be used in robotic systems to demonstrate different personality types. The study found that all human personalities prefer to interact with robots showing low Neuroticism and high Extraversion emotional responses over the short term.

Chapter 11 again uses emotional musical prosody, however focuses on robotic groups. The authors conducted two studies to analyze the impact of emotional musical prosody on trust, likeability, perceived intelligence, and emotional contagion among the participants. The findings suggest that sound and music have broad potential use cases in group human–robot communication and can shape participant responses to robots.

1.2.3 Musical Robots and Robotic Musicians

The final section of the book includes chapters focusing on the intersection of music and robotics. Chapter 12 presents an overview of the field of robotic musicianship, and discusses the evaluation of these systems. This chapter also frames the roles robotic musicians can have in exploring the social and emotional elements of human–robot interaction.

Chapter 13 explores the portrayal of robotic musical emotions through sound and gesture in three different platforms: Stretch by Hello Robot, Panda by Franka Emica, and SeekerBot, an internally developed social robot. The chapter also discusses the use of the framework to create robotic dance based on musical features from a song, where a 7DoF robotic arm was used as the non-humanoid robot.

The final chapter in the book explores the controversy surrounding posthumous holographic performances of deceased singers in the US and Japan. It analyzes how the reception of AI Misora Hibari differed from that of Western counterparts and highlights the importance of voice and culturally specific contexts in robotics research. Importantly it emphasizes broader issues of the role of music and sound in robotic research.

1.3 Conclusion

The chapters in this book provide a diverse and comprehensive exploration of the intersection of sound and robotics. By examining the relationships between sound and robotics from different angles, this book aims to stimulate further research and discussion on the topic, and to foster a deeper understanding of the role and potential of sound in shaping the future of robotics.

Part I

Speech

2

Effects of Number of Voices and Voice Type on Storytelling Experience and Robot Perception

Sophia C. Steinhaeusser and Birgit Lugrin

2.1 Introduction

Our voice is our most important communication medium [48]. This is even true while communication technology is rapidly evolving. For example, thinking of the communicative possibilities provided by smartphones, voice calls are still the most common way of mobile communication [16]. One reason for our preference

DOI: 10.1201/9781003320470-2

on communication by voice is the inherent cues providing information about our interlocuter. For instance, the human voice provides information on a speaker's sex [4], attractiveness [7], personality [4, 38], credibility [55], and emotions [34, 68]. Especially for emotional expression our voice is our primary communication instrument [52]. For example, anger is conveyed by an increase in loudness, fundamental frequency, and intensity [34, 65], whereas sadness is recognized when speaking more slowly with a high pitch and intensity [65].

Thus, it is rather not surprising that humans also want to communicate with machines by naturally and intuitively using their voice [48]. In turn, some technologies have to rely on verbal communication. This is true not only for voice assistants but also for social robots. For effectively supporting humans in a task-related as well as social manner, social robots need to engage on both a cognitive and also on an emotional level [9]. Therefore, a social robot and its behavior should appear comprehensive and human-like [51]. "Robot voice is one of the essential cues for the formulation of robot social attributes" [21, p. 230]. Just like humans, robots can also express emotions through their voices. Doing so fosters a richer social interaction and supports the user's interest in the interaction [8].

One of the many fields in which robots are deployed is the robotic storytelling (see e.g. [1, 58, 61]). Especially within a story context, emotions are crucial to understand a story [45]. Thus, robot storytellers should be provided with expressive voices. For example, Kory Westlund et al. [35] found that when a robotic storyteller's voice included a wide range of intonation and emotion compared to a flat voice without expressiveness, children who listened to the expressive robot showed stronger emotional engagement during the storytelling, greater inclusion of new vocabulary acquired during the storytelling into retelling of the story and greater fidelity to the original story when retelling. In addition to imitating the expressiveness of human voices, social robots as storytellers offer other speech-related capabilities that surpass those of human storytellers. While humans have limited abilities disguising their voices to portray different story characters, robots are able to not only playback every possible voice recording but also generate a sheer endless amount of synthetic voices. Comparing single synthetic voice usage to the adjustment of robotic voice to illustrate different characters, Striepe et al. [60] indicated that adapting a robot's voice to story characters in terms of pitch and speed improves recipients' narrative presence. However, synthetic voices are sometimes negatively remarked by recipients (see e.g. [57]), and human voices are preferred over synthetic voices [17, 24]. Nevertheless, while researchers develop systems automatically producing character-matched synthetic voices for robotic storytelling [42], there is no research on using different human voices to illustrate individual characters in robotic storytelling to the best of our knowledge yet. Therefore, we aim to shed light on this knowledge gap by investigating the effect of number of voices and type of voice – human respectively synthetic – in robotic storytelling on both storytelling experience and robot perception.

2.2 Related Work

Audio books, "typically defined as a recording of a text read aloud by the author, a professional narrator, or a synthetic voice" [30, p. 124], have become one of the most successful and fastest-growing formats of story reception [43]. Especially for this long-form content, the speaker quality and skill are decisive for evaluating the listening experience [13, 15]. For example, narrators can manipulate their voices in terms of volume, pitch, intensity, or pauses to illustrate a story's text [15,65]. Furthermore, giving different characters different voices enhances the listening experience and eases keeping track of a story's characters [40]. In more detail, there are four styles of narration in terms of voicing. Most commonly each character is represented by a different voice acted out by the speaker, called *fully voiced* narration. For instance, reading the books from the *Harry Potter* series by J. K. Rowling out loud, Jim Dale adopts various voices with adapted tones and pitches for all characters in the series [40]. In contrast, in *partially voiced* narration only the primary characters are represented with distinguishable voices, while the other characters share a voice. Even less voices are utilized in *unvoiced* narration in which a story is read straight without acting out distinguishable voices. Last, when multiple narrators are taking part in an audio book production, different speakers can represent different characters, which is called *multivoiced* narration [12]. Since many voice actors are needed to create a multivoiced audio book, the production process can be very expensive and time-consuming. Using synthetic voices is a more affordable way and could be a more feasible approach for audio book producers [47,53]. Yet synthetic voices are already used to produce audio books [30]. While synthesized speech is difficult to understand at first, intelligibility increases within the first five sentences of exposure largely and linearly [64].

2.2.1 Voice Types in Human–Robot Interaction

Synthetic voices as part of voice processing technology are mainly used to transmit messages comprising information from a machine to a human [48]. Synthetic speech is also a key element of human–robot interaction [31]. First of all, the "presence of voice is [a] strong trigger for anthropomorphic perception" [25, p. 204], the attribution of human characteristics to robots. For instance, emotive synthesized speech can improve the attribution of empathy to a robot compared to a flat synthesized voice [32]. Moreover, the choice of voice alone can already affect the interaction with and perception of a robot considerably. For instance, Dou et al. [21] reported that a humanoid robot gained highest ratings on competence when using a male voice, whereas the highest scores on warmth were obtained using a child voice.

Although synthetic voices are common within HRI, researchers also work with pre-recorded human voices played back by the robot (see e.g. [41,61]). Since humans prefer real human voices over synthetic ones [24,44,59] this choice can also positively affect an interaction. Inter alia, human voices are perceived as being more expressive [11], likable [5], truthful, involved [44], credible, pleasant [36,46], and appealing [36]. While some researchers reported no difference between human and synthetic voices regarding persuasion [44,59], Rodero [46] reported that human voices are more persuasive in conveying advertising messages. These positive effects seem to be at least partially transferred to a robot using human voice. Comparing a synthetic and human voice used by the *Alpha* robot, Xu [67] found that when speaking with a human voice Alpha was rated more trustworthy but was as attractive as with a synthetic voice. Choice of voice also influences how people approach a robot. Walters et al. [66] reported that participants' desired comfortable distance to a robot was significantly greater when the robot used a synthetic compared to a human or no voice. Even a mismatch in terms of human-likeness between an agent's voice and motion does not reduce the human voice's positive effect on pleasantness [24]. Thus, Ferstl et al. [24] recommend the highest possible realism of voice for virtual and robot-like agents, even when this produces incongruence between the modalities.

Especially regarding storytelling human voices seem to be preferable due to being considered as more suitable for emotional communication [46,47]. Comparing a storytelling performed by the synthetic voice of Amazon's *Alexa* to a pre-recorded human female voice, Rodero [47] reported stronger emotional responses in the human voice condition, presumably due to a deeper level of processing. In addition, higher physiological levels of attention, arousal, and valence as well as increased self-reported enjoyment, engagement, imagery, and recognition of information were observed in the human voice condition. Similar results were found for robotic storytelling. Examining participants' body language, Costa et al. found exhibited facial expressions indicating emotions and non-verbal arm and head gestures indicating engagement when using a human compared to a synthetic voice during storytelling with the *Aesop* robot [17]. This might be explained by the positive relationship between perceived robot anthropomorphism and engagement as well as narrative presence found by Striepe et al. [60]. According to this finding, human-likeness, e.g. in voice, seems to be an advantage for robotic storytelling. Another but somewhat similar explanation is provided by Rodero and Lucas, who introduce the *human emotional intimacy effect*, proclaiming that people experience closeness and connection when listening to a human voice which in turn leads to a solid and positive emotional response [47]. This theory is in line with Mayer's *voice principle*. This principle states that "people learn more deeply when the words in a multimedia message are spoken in a human voice rather than in a machine voice" [39, p. 345]; therefore, human voices are suggested more effective for teaching than synthesized ones. Following Rodero [47] this effect might be transferred to emotional story content.

However, the predominance of human over synthetic voices seems to vanish with the continuous technical improvement of synthetic voices. Comparing narration in a multimedia learning environment via a modern or older text-to-speech engine or a recorded human voice, Craig and Schroeder reported better perceptions of human voice considering engagement and human-likeness. Nonetheless, there were no significant differences in learning outcomes, credibility, perceptions, or cognitive efficiency to the modern text-to-speech system [19]. Utilizing similar voice conditions with a virtual pedagogical agent the modern text-to-speech engine even had a greater training efficiency and produced more learning on transfer outcomes than the human voice while being rated as equally credible [18]. "This provides consistent evidence against the voice effect" [18, p. 15]. Comparing different synthetic voices newer methods achieve results in likeability closer to human voices than older engines [5]. Even regarding long-form content such as storytelling, several synthetic voices outperform human voices [13].

This might explain newer findings in robotic storytelling. For instance, Goossens et al. [27] indicated no significant difference in terms of engagement and story difficulty between *NAO*'s original synthetic voice and a pre-recorded human voice. Moreover, children who listened to NAO's original voice performed significantly better concerning vocabulary acquisition. In a similar study using the *Pepper* robot, Carolis et al. [20] obtained similar results. Children felt more positive emotions and reported a higher user experience in terms of pleasantness, story understanding, and image clarity when listening to the text-to-speech voice. Thus, new approaches and engines might provide high-quality synthesized speech suitable for robotic storytelling.

2.2.2 Number of Voices in Robotic Storytelling

While voice type used in HRI has been heavily researched, the number of voices used is less explored. Yet, machine learning frameworks are already created to identify synthetic voices matching the story's characters. Based on the idea that common children stories mainly include similar principal characters such as a young, innocent female such as Snow White or Ariel, a young heroic male such as Prince Charming or Robin Hood, and an older villain such as the Evil Queen or Ursula synthetic voices can be clustered following salient attributes such as age and gender but also evilness, intelligence, pitch, and so on [29]. Using Naive Bayes, Greene et al. [29] modeled a relationship between these attributes and synthetic voices to predict appropriate voices for children's stories' characters. However, their approach was not yet evaluated in a perceptual study. Also using an algorithmic approach, Min et al. [42] were able to map synthetic voices to characters in a robotic storytelling outperforming a baseline of random selection. In addition, the authors reported that gender and character differentiation correlate positively with naturalness in the robotic storytelling.

Using several manually pre-shaped versions of NAO's synthetic voice, Ruffin et al. [50] implemented a robotic storytelling of an African tale in which each character was given a distinct voice and LED color. Again, the resulting storytelling sequence was not evaluated. Similarly, Striepe et al. [60] adapted the synthetic voice of the *Reeti* robot in terms of speed and pitch to create a fully voiced robotic storytelling. A user study revealed no significant differences between one voice and character-adjusted voice concerning narrative engagement and anthropomorphism of the robot. However, narrative presence was higher when the robot modified its voice to illustrate the story's characters indicating a positive effect of character illustration via voice.

While the limited body of research on the use of different synthetic voices in robotic storytelling shows positive consequences of character illustration, to the best of our knowledge, no comparable studies have been conducted on the use of different human voices.

2.3   Contribution

In contrast to the attempts of developing systems which automatically produce synthetic voices matching characters in a robotic storytelling [42], there is no research on using different pre-recorded human voices for robotic multivoiced storytelling.

Based on previous work by Striepe et al. [60] and also findings from audio book research [40], the use of multiple character-illustrating voices should improve the storytelling experience. The listener's transportation, "the extent that [they] are absorbed into a story" [28, p. 701], as a key element of narrative engagement [10] is therefore assumed to increase when a robotic storyteller uses distinct voices for different characters compared to an unvoiced narration, regardless of the type of voice used.
H1: Transportation is higher during a multivoiced than during an unvoiced storytelling.

Regarding the type of voice used in HRI mixed results are found. Since positive effects of synthetic voice usage were yet only found whithin children (see e.g. [20, 27]), we form our hypotheses based on the *human emotional intimacy effect* [47], suggesting that a human voice enhances the storytelling experience more than a synthetic voice, regardless of number of voices.
H2: Transportation is higher when a robotic storyteller uses humans compared to synthetic voices.

Concerning robot perception, voice is an important "anthropomorphic ability" [22, p. 183]. Previous studies reported beneficial effects of using human voices in HRI (see e.g. [24, 67]). Especially, human voices carry more emotional cues [47] that are important for perceiving a human-like social entity [23].

H3a: Perceived anthropomorphism is higher when a robotic storyteller uses humans compared to synthetic voices.

H3b: General robot perception is improved when a robotic storyteller uses humans compared to synthetic voices.

H3c: Social robot perception is improved when a robotic storyteller uses humans compared to synthetic voices.

Regarding the number of voices, no suggestions can be made about robot perception due to the limited scope of related work. Thus, the effects of multi-voiced compared to unvoiced robotic storytelling were examined exploratory. Also, the interaction between type and number of voices was investigated exploratory.

RQ1: Does the number of voices used influence the perception of a robotic storyteller?

RQ2: Is there an interaction between type and number of voices used in a robotic storytelling in terms of storytelling experience and robot perception?

2.4 Method

To investigate the effect of number and type of voices used in robotic storytelling on storytelling experience and robot perception, a 2 (one vs. three voices) x 2 (synthetic vs. human voice) between groups design was applied in an online study. The study was approved by the local ethics committee of the Institute for Human–Computer-Media at the University of Würzburg (vote #140222).

2.4.1 Materials

To analyze the influence of robot voice type and number, four settings were implemented using the social robot *NAO V6* [56] and *Choregraphe* version 2.8.6 [2].

We chose the story "Conversation on a Bench in the Park" (available at), which is conceptualized as a dialog between two men moderated by a narrator. The men are talking on a park bench about a murder from the past. At the end of the story, the narrator reveals that possibly one of the men, called Sammy, might have been the murderer himself, while the other man was the responsible police inspector. The story was annotated by four independent raters in terms of basic emotions per sentence. Doing so, a set of six recurring gestures shown in Figure 2.1 resembling the six basic emotions was added to the storytelling. In addition, NAO's line of sight was manipulated to clarify the change of speaker. During the narrator's passages, NAO looked directly into the camera, while its gaze fixed on the men to its left respectively right during their passages. NAO was seated during the whole storytelling.

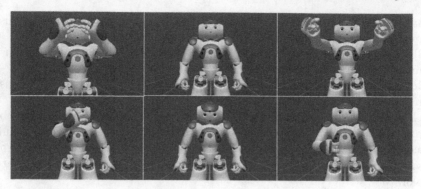

FIGURE 2.1
Emotional gestures expressing fear, disgust, joy, surprise, sadness, and anger
(f.l.t.r.) implemented using Choregraphe.

For the human voice conditions, three male voice actors were recruited,
with the narrator being spoken by a younger man and the two men being
spoken by older men. For the story version using one human voice (*oneHum*
condition), the narrator's voice was recorded reading the whole story out loud
in an unvoiced style, whereas all three voices were recorded in a multivoiced
manner for the version using three human voices (*threeHum* condition). For
the synthetic voice conditions, NAO's internal text-to-speech (TTS) module
was used, which offers a *voice shaping* option that modifies the voice not only
in speed but also tone and thus allows to modificate NAO's voice to generate
new voices of, e.g. different gender and age [2]. For the single synthetic voice
version, we used NAO's standard TTS voice (*oneSyn* condition). The standard
TTS voice was also used for the narrator in the three synthetic voices version
(*threeSyn* condition). For the two old men, the TTS voice's tone was adjusted
to generate two synthetic male voices following the suggestions of Traunmueller
and Eriksson [62].

Combining non-verbal and verbal features described above, four versions of
the story with the same non-verbal behavior but differing in voice usage were
implemented and video-taped[1]. The resulting video stimuli had a length of
approximately 3:50 minutes. Camera angle and picture section are displayed
in Figure 2.2.

2.4.2 Measures

To analyze participants' storytelling experience their *transportation* was measured using the *Transportation Scale Short Form* (TS-SF) [3]. It includes six
items, e.g. "I could picture myself in the scene of the events described in the
narrative", which are answered on a seven-point Likert-Scale anchored by

---

[1]www.soundandrobotics.com/ch2

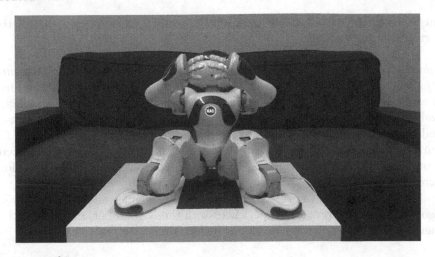

FIGURE 2.2
Freeze frame showing NAO expressing fear when narrator is speaking.

1 – "I totally disagree" and 7 – "I totally agree". Two items concern story characters and thus were adapted to our story, e.g. "As I listened to the story, I could vividly picture Sammy". Appel et al. [3] reported reliability of .80 to .87, whereas Cronbach's alpha of .91 was calculated for the current sample.

General robot perception was operationalized by the *Godspeed* questionnaire series [6], comprising five scales measured on five-point semantic differentials: (1) *Anthropomorphism* comprising five items, e.g. "machinelike" versus "humanlike", (2) *Animacy* including six items, e.g. "mechanical" versus "organic", (3) *Likeability* with five items, e.g. "unfriendly" versus "friendly", (4) *Perceived Intelligence* comprising five items, e.g. "foolish" versus "sensible", and (5) *Perceived Safety* including three items, e.g. "anxious" versus "relaxed". Bartneck et al. [6] reported reliability of .88 to .93 for *Anthropomorphism*, .70 for *Animacy*, .87 to .92 for *Likeability*, and .75 to .77 for *Perceived Intelligence*. Reliability was not reported for the *Perceived Safety* scale. For the current sample, Cronbach's alpha of .76 was calculated for *Anthropomorphism*, .80 for *Animacy*, .86 for *Likeability*, .83 for *Perceived Intelligence*, and .76 for *Perceived Safety*.

To get deeper insights into NAO's perceived *anthropomorphism*, the *multidimensional questionnaires to assess perceived robot morphology* (RoMo) by Roesler et al. [49] was applied. The questionnaire includes four scales targeting (1) *Appearance*, (2) *Movement*, (3) *Communication*, and (4) *Context*. Since we did not manipulate the robot's appearance and context and due to our focus on the robot's speech, we only used the *Communication* scale which comprises ten items on verbal and non-verbal expression such as "How human-like is the speech rhythm of the robot?". The items were answered using a slider anchored by 0% – "not at all" and 100% – "fully". Cronbach's alpha for the current sample was .89.

Social robot perception was operationalized using the *Robotic Social Attributes Scale* [14]. The questionnaire includes the three factors (1) *Warmth*, (2) *Competence*, and (3) *Discomfort*, each comprising six items in the form of adjectives, for example "emotional", "reliable", and "strange", which are evaluated on a 9-point Likert-scale anchored by 1 – "definitely not associated" and 9 – "definitely associated". The authors reported reliability of .92 for warmth, .95 for competence, and .90 for discomfort, while for the current sample values of .87 for warmth, .91 for competence, and .78 for discomfort were calculated.

Last, participants were asked to provide their gender and age. They were also asked on their previous experiences with the NAO robot, namely whether they had seen it in pictures or videos or had already interacted with it.

2.4.3   Study Procedure

The online survey was hosted using *LimeSurvey* version 444 [37]. When accessing the website, individuals first gave informed consent to take part in the study. After being randomly assigned to one of the four conditions, they watched the respective video described in section 2.4.1. Afterward, they filled in the questionnaires on transportation, general robot perception, perceived anthropomorphism, and social robot perception. Last, participants provided demographic data and were thanked and debriefed.

Participation in the study took about 15 minutes. We recruited our participants from the students enrolled at the University of Würzburg using the internal online-recruitment system. For their participation they received credits mandatory for obtaining their final degree.

2.4.4   Participants

Overall, 145 persons with a mean age of 21.32 ($SD = 2.23$) years took part in the study. While 28 participants identified as male (age: $M = 22.00$, $SD = 2.09$), the majority of 117 participants self-reported being female (age: $M = 21.16$, $SD = 2.24$). No one self-indicated as a diverse gender. Only 21 participants stated to have never seen the NAO robot before, whereas 124 already saw it in pictures or videos. In contrast, only 46 participants had already interacted with the NAO robot in person.

Being randomly assigned to one of the four conditions, 36 persons watched the story told by NAO using one human voice ($n_{male} = 9$, $n_{female} = 27$; age: $M = 21.39$, $SD = 2.33$), whereas 41 participants watched the story told using one synthetic voice ($n_{male} = 8$, $n_{female} = 33$; age: $M = 21.34$, $SD = 2.25$). In each case, 34 people watched the video, with three humans ($n_{male} = 5$, $n_{female} = 29$; age: $M = 21.32$, $SD = 2.48$) respectively three synthetic ($n_{male} = 6$, $n_{female} = 28$; age: $M = 21.34$, $SD = 1.88$) voices.

2.5 Results

All analyses were conducted using *JASP* version 0.16.0.0 [33]. An alpha-level of .05 was applied for all statistical tests. Descriptive data are presented in Table 2.1. Calculated Levene's tests indicated homogeneity of variances for all scales, $ps > .05$.

2.5.1 Transportation

A two-way ANOVA was used to analyze the effects of number of voices (H1) and type of voice (H2) on recipients' *Transportation* into the story told. Calculated results indicate a significant main effect of number of voices ($F(1, 141) = 4.33$, $p = .039$, $\omega^2 = .02$), whereas no significant main effect of type of voice was shown, $F(1, 141) = 2.39$, $p = .148$, $\omega^2 = .01$. Also, no significant interaction effect was revealed, $F(1, 141) = 0.11$, $p = .739$, $\omega^2 = .00$. Bonferroni-adjusted post-hoc comparisons again revealed no significant interaction ($ps > .05$), but a significant difference comparing number of voices when averaging over the levels of type of voice with higher values for one compared to three voices as displayed in Figure 2.3.

TABLE 2.1
Descriptive data per condition.

	oneHum		oneSyn		threeHum		threeSyn	
	M	*SD*	*M*	*SD*	*M*	*SD*	*M*	*SD*
Transportation[a]	2.73	1.23	2.42	1.06	2.31	0.97	2.11	0.96
Anthro.[b]	1.99	0.58	1.66	0.57	1.43	0.43	1.62	0.57
Animacy[b]	2.72	0.54	2.35	0.57	2.44	0.57	2.50	0.71
Likeability[b]	3.54	0.68	3.21	0.77	3.11	0.59	3.32	0.84
P. Intelligence[b]	3.41	0.60	3.29	0.76	3.04	0.57	3.12	0.84
P. Safety[b]	3.06	0.89	3.34	0.81	3.05	0.89	3.07	0.85
Anthro. Comm.[c]	38.92	19.32	23.18	15.80	30.79	13.80	27.10	17.44
Warmth[d]	3.93	1.48	3.88	1.62	3.64	1.55	3.88	1.69
Competence[d]	5.08	1.81	5.33	1.83	4.69	1.63	4.72	1.69
Discomfort[d]	2.62	1.38	3.03	1.12	3.26	1.68	3.31	1.57

P. = Perceived, Anthro. = Anthropomorphism, Anthro. Comm. = Anthropomorphism in Communication.
a. Calculated values from 1 to 7.
b. Calculated values from 1 to 5.
c. Calculated values from 0% to 100%.
d. Calculated values from 1 to 9.

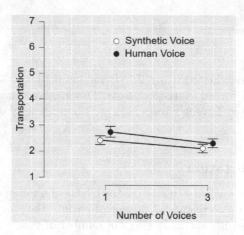

FIGURE 2.3
Descriptive plot for transportation. Error bars display standard error.

2.5.2   General Robot Perception

First, both anthropomorphism-related scales were analyzed using two-way
ANOVAs (H3a & RQ1). Considering *Anthropomorphism* a significant main
effect of number of voices ($F(1, 141) = 3.97$, $p = .001$, $\omega^2 = .06$) was indicated,
while no significant main effect of type of voice was observed, $F(1, 141)$
$= 0.68$, $p = .450$, $\omega^2 = .00$. Further, a significant interaction was found
between number and type of voices, $F(1, 141) = 2.49$, $p = .004$, $\omega^2 = .05$.
Bonferroni-adjusted post-hoc comparisons indicate significantly higher values
of anthropomorphism for *oneHum* compared to *threeHum* ($p < .001$) as well
as for *oneHum* compared to *threeSyn*, $p = .029$. In addition, the difference
of higher vaues in *oneHum* compared to the *oneSyn* condition just missed
significance, $p = .050$. This disordinal interaction can be seen in Figure 2.4. In
contrast, for *Anthropomorphism in Communication* no significant main effect
of number of voices ($F(1, 141) = 0.57$, $p = .450$, $\omega^2 = .00$) was shown, while
the main effect of type of voice was significant, $F(1, 141) = 12.18$, $p < .001$, ω^2
$= .07$. Also, the interaction effect was significant, $F(1, 141) = 4.68$, $p = .032$,
$\omega^2 = .02$. Bonferroni-corrected post-hoc comparisons reveal significantly higher
values for *oneHum* compared to *oneSyn* ($p < .001$) as well as for *oneHum*
compared to *threeSyn*, $p = .022$. This disordinal interaction is displayed in
Figure 2.4.

   Further, to analyze general robot perception again two-way ANOVAs were
calculated (H3b & RQ1). Regarding *Animacy*, neither main effect of number of
voices ($F(1, 141) = 0.37$, $p = .544$, $\omega^2 = .00$) nor main effect of type of voice
($F(1, 141) = 2.35$, $p = .128$, $\omega^2 = .01$) were significant. However, ANOVA
calculation revealed a significant interaction between number and type of
voices, $F(1, 141) = 4.70$, $p = .032$, $\omega^2 = .03$. Bonferroni-adjusted post-hoc

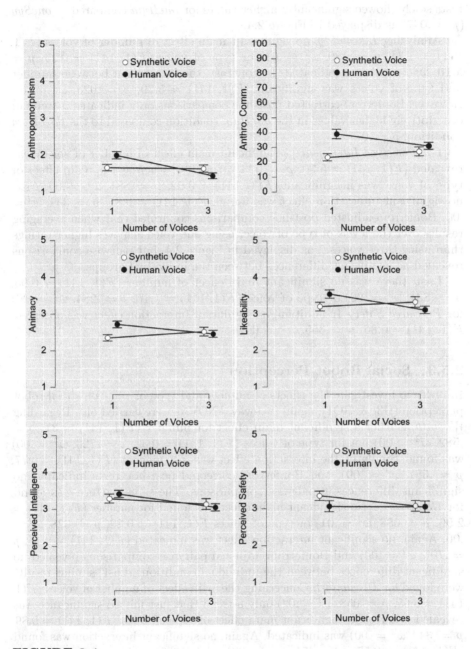

FIGURE 2.4

Descriptive plots for general robot perception. Error bars display standard error. Anthro. Comm. = Anthropomorphism in Communication.

tests solely showed significantly higher values for *oneHum* compared to *oneSyn* ($p = .047$) as displayed in Figure 2.4.

Analyzing *Likeability*, no significant main effect of number of voices ($F(1, 141) = 1.74$, $p = .189$, $\omega^2 = .01$) or type of voice ($F(1, 141) = 0.25$, $p = .616$, $\omega^2 = .00$) was indicated. In contrast, the interaction between number and type of voices was significant, $F(1, 141) = 5.29$, $p = .023$, $\omega^2 = .03$. However, Bonferroni-corrected post-hoc comparisons only indicated a trend of descriptively higher values in the *oneHum* condition compared to the *threeHum* condition, $p = .077$.

For *Perceived Intelligence* a significant main effect of number of voices was revealed ($F(1, 141) = 5.30$, $p = .023$, $\omega^2 = .03$), whereas the main effect of type of voice was insignificant, $F(1, 141) = 0.02$, $p = .881$, $\omega^2 = .00$. Also, no significant interaction effect was found, $F(1, 141) = 0.81$, $p = .369$, $\omega^2 = .00$. Bonferroni-adjusted post-hoc comparisons confirmed that when averaging results over the levels of type of voice, using one voice achieves higher ratings than using three voices. , as displayed in Figure 2.4, but pairwise comparisons revealed no significant differences between the individual groups.

Last, there was no significant main effect of number ($F(1, 141) = 0.95$, $p = .331$, $\omega^2 = .00$) or type of voice ($F(1, 141) = 1.16$, $p = .284$, $\omega^2 = .00$) for *Perceived Safety*. In addition, no significant interaction effect was observed, $F(1, 141) = 0.86$, $p = .355$, $\omega^2 = .00$.

2.5.3   Social Robot Perception

In order to investigate the effects of number and type of voice on social robot perception (H3c & RQ1), again two-way ANOVAs were carried out. Regarding *Warmth*, neither a significant main effect for number ($F(1, 141) = 31$, $p = .582$, $\omega^2 = .00$) nor for type of voices ($F(1, 141) = 0.32$, $p = .723$, $\omega^2 = .00$) was found. Similarly, the interaction effect was insignificant ($F(1, 141) = 0.77$, $p = .581$, $\omega^2 = .00$), and Bonferroni-corrected post-hoc tests indicated no significant differences in pairwise comparisons. The same pattern was found for *Competence*. No significant main effect was found for number ($F(1, 141) = 2.96$, $p = .088$, $\omega^2 = .01$) or type of voices $F(1, 141) = 0.22$, $p = .628$, $\omega^2 = .00$. Again, no significant interaction effect was indicated ($F(1, 141) = 0.14$, $p = .707$, $\omega^2 = .00$) and Bonferroni-adjusted pairwise comparisons revealed no significant differences between the individual conditions. Last, similar results were found for *Discomfort*. Concerning the main effect of number of voices ($F(1, 141) = 3.65$, $p = .058$, $\omega^2 = .02$) only a trend that just missed significance was revealed, whereas no significant main effect of type of voice ($F(1, 141) = 0.89$, $p = .348$, $\omega^2 = .00$) was indicated. Again, no significant interaction was found ($F(1, 141) = 0.57$, $p = .451$, $\omega^2 = .00$), and Bonferroni-corrected post-hoc comparisons indicated no significant differences between the conditions.

2.6 Discussion

To investigate the effects of type of voice (human vs. synthetic) and number of voices (one vs. three) in a robotic storytelling scenario on both storytelling experience and robot perception an online study was carried out.

Participants' transportation into the story told by the robot was significantly higher for the unvoiced compared to the multivoiced narration for both synthetic and human voices. Thus, **H1** is rejected. This finding is not only in contrast to our presumption but also to findings from audio book research [40] and related research on robotic storytelling using fully voiced narration [60]. One possible explanation might be that humans are looking for consistency. If we listen to a multivoiced audio book we imagine multiple voice actors. Consequently, watching a single robot using different voices is inconsistent in terms of expected number of speakers. This conflict might have impeded participants' transportation into the story. In contrast, this finding could also potentially be due to a poor distinguishability of the voices used. We used only male voices in the human voice conditions and also male-sounding voices in the synthetic voice conditions. Although the narrator was way younger than the other two speaker the voices may have been too similar. Additionally, Min et al. [42] report a positive effect not only of character but also gender distinction. A wider variety of voices should be used in future studies to improve character and gender distinction and elicit naturalness as reported by Min et al. [42].

Further, no significant difference was found between human versus synthetic voice(s). Thus, **H2** is rejected, too. Also, no interaction between number and type of voice was indicated (**RQ2**). Overall, transportation was relatively low in all conditions. Reasons might be the story itself and the NAO robot. The story comprises a dialog between two men having a conversation about a murder from the past. While the story's genre *crime* is one of the most popular book genres today [54] and should therefore meet most of the participants' interest, the story's structure is relatively uncommon and thus interfering – although the dialog structure was suiting our research aim. The isochronous narration includes rather individual information about the murder stringed together than a continuous plot and thus might have been hard to follow. Vaughn et al. [63] suggest a positive relationship between transportation and fluency or easiness of processing a story. In turn, the missing fluency in our story and the mental effort required to follow the storytelling and grasp all relevant information may have hindered participants' transportation into the story. Another impeding factor might have been the mechanical sounds from the robot's motors. Frid et al. [26] reported that "certain mechanical sounds produced by the NAO robot can communicate other affective states than those originally intended to be expressed" [p. 8]. This could have led to confusion among our participants, which in turn worsens fluency of the storytelling and increases mental demand. Moreover, NAO's motor sounds were generally found to be disturbing [26]. Therefore, in future studies these motor sounds should be completely masked as suggested by Frid et al. [26] or the use of other robots should be considered. Additional attention should be paid to the fact that the story used is easy to process and follows familiar structures.

Effects of choice of voice type and number on general robot perception were mixed. Especially for anthropomorphism conflicting results were found. While anthropomorphism in communication was significantly higher when using humans compared to synthetic voices in both unvoiced and multivoiced conditions, no such finding was revealed for anthropomorphism measured using the *Godspeed* scale focusing on robot appearance, i.e. "moving rigidly" or "machinelike" [6]. However, for both measures the *oneHum* condition yielded the highest scores, so that **H3a** can be partially accepted. Human voices seem to be preferable for triggering anthropomorphism. Regarding number of voices (**RQ1**), following results from our pairwise comparisons unvoiced narration should be preferred when utilizing human voices, whereas when using synthetic voices both unvoiced and multivoiced narration are acceptable in terms of perceived anthropomorphism.

For animacy, likeability, perceived intelligence, and perceived safety no significant differences were found when comparing human and synthetic voices. Therefore, **H3b** is rejected. This finding is in contrast to the *human emotional intimacy effect*. Even though the synthetic voices used in our study were perceived less human-like as shown above, this lack of human-likeness did not affect general robot perception. Our findings support the claim that modern synthetic voice engines achieve results close to human speech [5, 13] and may have reached a point where they can deliver narration as credible as human voices [19].

Regarding **RQ1**, no difference between unvoiced and multivoiced narration was found for animacy, likeability, and perceived safety. In contrast, unvoiced narration scored higher on perceived intelligence compared to using three voices independently from voice type. While no interaction (**RQ2**) between number and type of voice was found for perceived intelligence and safety, both factors interact in terms of animacy and likeability. While human voice should be preferred for unvoiced narration in terms of animacy, no differences were obtained between human and synthetic voice for multivoiced narration. For improving likeability, either one human or multiple synthetic voices should be used. However, this is only a trend in the data. Overall, the *oneHum* condition scored highest on all scales except for perceived safety. Thus, unvoiced narration using a human voice is suggested to improve general robot perception but also synthetic voices are acceptable.

Regarding social robot perception, no differences in warmth, competence, and discomfort were obtained between human and synthetic voice usage, thus **H3c** must be rejected. Similarly, no differences were indicated comparing unvoiced and multivoiced narration (**RQ1**). Also, no interaction between both factors was observed (**RQ2**). Neither type nor umber of voice seems to affect social robot perception.

2.7 Practical Implications for Choice of Voice

First of all, the choice of voice for robotic storytelling can be made independently from a robot's perceived sociality in terms of warmth, and more important discomfort and competence. None of the voice conditions tested made our participants feel uncomfortable. In terms of storytelling experience unvoiced narration is suggested, while type of voice can be freely chosen. This is also true if a robot's perceived intelligence shall be improved. The decision between human and synthetic voice becomes relevant only for scenarios in which anthropomorphism is crucial. If high levels of anthropomorphism are desired, human unvoiced narration should be preferred. Otherwise, human and synthetic voices performed almost the same, so that human voices are partially recommended for unvoiced narration, whereas when using synthetic voices suit both unvoiced and multivoiced narration.

2.8 Conclusion

An online study was carried out to shed light on the choice of voice for robotic storytelling not only in terms of type of voice, namely human or synthetic but also in terms of number of voices used, leading to unvoiced or multivoiced narration. While audio book research reports positive effects of multivoiced narration on the storytelling experience our results suggest a preference for unvoiced narration potentially due to the robot's physical embodiment. Regarding type of voice, our findings support the assumption that modern synthetic voice engines achieve results close to human speech and may have reached a point where they can deliver narration as good as human voices. At this point, mixed productions of synthetic and natural signals [53] might be a next step to be tested in future work.

Acknowledgments

The authors would like to thank Alisa Dianov, Angela Ast, Annika Büttner, Alisa Ebner, and Jana Luksch for preparing the stimulus material.

Bibliography

[1] AHN LE, Q., D'ALESSANDRO, C., DEROO, O., DOUKHAN, D., GELIN, R., MARTIN, J.-C., PELACHAUD, C., RILLIARD, A., AND ROSSET, S.

Towards a storytelling humanoid robot. In *2010 AAAI Fall Symposium Series* (2010), Association for the Advancement of Artificial, Ed.

[2] ALDEBARAN ROBOTICS. Choregraphe, 2016.

[3] APPEL, M., GNAMBS, T., RICHTER, T., AND GREEN, M. C. The transportation scale–short form (ts–sf). *Media Psychology 18*, 2 (2015), 243–266.

[4] ARONOVITCH, C. D. The voice of personality: stereotyped judgments and their relation to voice quality and sex of speaker. *The Journal of Social Psychology 99*, 2 (1976), 207–220.

[5] BAIRD, A., PARADA-CABALEIRO, E., HANTKE, S., BURKHARDT, F., CUMMINS, N., AND SCHULLER, B. The perception and analysis of the likeability and human likeness of synthesized speech. In *Interspeech 2018* (ISCA, 2018), ISCA, pp. 2863–2867.

[6] BARTNECK, C., KULIĆ, D., CROFT, E., AND ZOGHBI, S. Measurement instruments for the anthropomorphism, animacy, likeability, perceived intelligence, and perceived safety of robots. *International Journal of Social Robotics 1*, 1 (2009), 71–81.

[7] BERRY, D. S. Vocal types and stereotypes: Joint effects of vocal attractiveness and vocal maturity on person perception. *Journal of Nonverbal Behavior 16*, 1 (1992), 41–54.

[8] BREAZEAL, C. Emotive qualities in robot speech. In *Proceedings 2001 IEEE/RSJ International Conference on Intelligent Robots and Systems. Expanding the Societal Role of Robotics in the Next Millennium (Cat. No.01CH37180)* (2001), IEEE, pp. 1388–1394.

[9] BREAZEAL, C., DAUTENHAHN, K., AND KANDA, T. Social robotics. In *Springer Handbook of Robotics*, B. Siciliano and O. Khatib, Eds. Springer International Publishing, Cham, 2016, pp. 1935–1972.

[10] BUSSELLE, R., AND BILANDZIC, H. Measuring narrative engagement. *Media Psychology 12*, 4 (2009), 321–347.

[11] CABRAL, J. P., COWAN, B. R., ZIBREK, K., AND MCDONNELL, R. The influence of synthetic voice on the evaluation of a virtual character. In *Interspeech 2017* (ISCA, 2017), ISCA, pp. 229–233.

[12] CAHILL, M., AND RICHEY, J. What sound does an odyssey make? content analysis of award-winning audiobooks. *The Library Quarterly 85*, 4 (2015), 371–385.

[13] CAMBRE, J., COLNAGO, J., MADDOCK, J., TSAI, J., AND KAYE, J. Choice of voices: A large-scale evaluation of text-to-speech voice quality for long-form content. In *Proceedings of the 2020 CHI Conference on Human Factors in Computing Systems* (New York, NY, USA, 2020), R. Bernhaupt, F. F. Mueller, D. Verweij, J. Andres, J. McGrenere, A. Cockburn, I. Avellino, A. Goguey, P. Bjørn, S. Zhao, B. P. Samson, and R. Kocielnik, Eds., ACM, pp. 1–13.

[14] CARPINELLA, C. M., WYMAN, A. B., PEREZ, M. A., AND STROESSNER, S. J. The robotic social attributes scale (rosas). In *Proceedings of the 2017 ACM/IEEE International Conference on Human-Robot Interaction - HRI '17* (New York, New York, USA, 2017), B. Mutlu, M. Tscheligi, A. Weiss, and J. E. Young, Eds., ACM Press, pp. 254–262.

[15] CLARK, L. Speaking pictures: the role of sound and orality in audio presentations of children's picture books. *New Review of Children's Literature and Librarianship 9*, 1 (2003), 1–19.

[16] COOP ITALIA. Most common ways to communicate with a smartphone in italy in 2019, by type [graph]. https://www.statista.com/statistics/1085033/most-common-ways-to-communicate-with-smartphones-in-italy/, 2019.

[17] COSTA, S., BRUNETE, A., BAE, B.-C., AND MAVRIDIS, N. Emotional storytelling using virtual and robotic agents. *International Journal of Humanoid Robotics 15*, 03 (2018), 1850006.

[18] CRAIG, S. D., AND SCHROEDER, N. L. Reconsidering the voice effect when learning from a virtual human. *Computers & Education 114* (2017), 193–205.

[19] CRAIG, S. D., AND SCHROEDER, N. L. Text-to-speech software and learning: Investigating the relevancy of the voice effect. *Journal of Educational Computing Research 57*, 6 (2019), 1534–1548.

[20] DE CAROLIS, B., D'ERRICO, F., AND ROSSANO, V. Pepper as a storyteller: Exploring the effect of human vs. robot voice on children's emotional experience. In *Human-Computer Interaction – INTERACT 2021*, C. Ardito, R. Lanzilotti, A. Malizia, H. Petrie, A. Piccinno, G. Desolda, and K. Inkpen, Eds., vol. 12933 of *Lecture Notes in Computer Science*. Springer International Publishing, Cham, 2021, pp. 471–480.

[21] DOU, X., WU, C.-F., NIU, J., AND PAN, K.-R. Effect of voice type and head-light color in social robots for different applications. *International Journal of Social Robotics*, 14 (2021), pp. 229–244.

[22] DUFFY, B. R. Anthropomorphism and the social robot. *Robotics and Autonomous Systems 42*, 3-4 (2003), 177–190.

[23] EYSSEL, F., HEGEL, F., HORSTMANN, G., AND WAGNER, C. Anthropo-morphic inferences from emotional nonverbal cues: A case study. In *19th International Symposium in Robot and Human Interactive Communication* (092010), IEEE, pp. 646–651.

[24] FERSTL, Y., THOMAS, S., GUIARD, C., ENNIS, C., AND MCDONNELL, R. Human or robot? In *Proceedings of the 21th ACM International Conference on Intelligent Virtual Agents* (New York, NY, USA, 2021), ACM, pp. 76–83.

[25] FINK, J. Anthropomorphism and human likeness in the design of robots and human-robot interaction. In *Social Robotics*, D. Hutchison, T. Kanade, J. Kittler, J. M. Kleinberg, F. Mattern, J. C. Mitchell, M. Naor, O. Nier-strasz, C. Pandu Rangan, B. Steffen, M. Sudan, D. Terzopoulos, D. Tygar, M. Y. Vardi, G. Weikum, S. S. Ge, O. Khatib, J.-J. Cabibihan, R. Sim-mons, and M.-A. Williams, Eds., vol. 7621 of *Lecture Notes in Computer Science*. Springer Berlin Heidelberg, Berlin, Heidelberg, 2012, pp. 199–208.

[26] FRID, E., BRESIN, R., AND ALEXANDERSON, S. Perception of mechanical sounds inherent to expressive gestures of a nao robot-implications for movement sonification of humanoids. *Sound and Music Computing* (2018).

[27] GOOSSENS, N., AARTS, R., AND VOGT, P. Storytelling with a social robot. *Robots for Learning R4L* (2019).

[28] GREEN, M. C., AND BROCK, T. C. The role of transportation in the persuasiveness of public narratives. *Journal of Personality and Social Psychology 79*, 5 (2000), 701–721.

[29] GREENE, E., MISHRA, T., HAFFNER, P., AND CONKIE, A. Predicting character-appropriate voices for a tts-based storyteller system. In *Thirteenth Annual Conference of the International Speech Communication Association* (2012).

[30] HAVE, I., AND PEDERSEN, B. S. Sonic mediazation of the book: Affor-dances of the audiobook. *MedieKultur — Journal of Media and Commu-nication Research 54* (2013), 123–140.

[31] HENNIG, S., AND CHELLALI, R. Expressive synthetic voices: Consid-erations for human robot interaction. In *2012 IEEE RO-MAN: The 21st IEEE International Symposium on Robot and Human Interactive Communication* (2012), IEEE, pp. 589–595.

[32] JAMES, J., BALAMURALI, B. T., WATSON, C. I., AND MACDONALD, B. Empathetic speech synthesis and testing for healthcare robots. *Interna-tional Journal of Social Robotics 13*, 8 (2021), 2119–2137.

[33] JASP TEAM. Jasp, 2021.

[34] JOHNSTONE, T., AND SCHERER, K. R. Vocal communication of emotion. *Handbook of Emotions 2* (2000), 220–235.

[35] KORY WESTLUND, J. M., JEONG, S., PARK, H. W., RONFARD, S., ADHIKARI, A., HARRIS, P. L., DESTENO, D., AND BREAZEAL, C. L. Flat vs. expressive storytelling: Young children's learning and retention of a social robot's narrative. *Frontiers in Human Neuroscience 11* (2017), 295.

[36] KÜHNE, K., FISCHER, M. H., AND ZHOU, Y. The human takes it all: Humanlike synthesized voices are perceived as less eerie and more likable. evidence from a subjective ratings study. *Frontiers in Neurorobotics 14* (2020), 593732.

[37] LIMESURVEY GMBH. Limesurvey, 2022.

[38] MALLORY, E. B., AND MILLER, V. R. A possible basis for the association of voice characteristics and personality traits. *Speech Monographs 25*, 4 (1958), 255–260.

[39] MAYER, R. E. Principles based on social cues in multimedia learning: Personalization, voice, image, and embodiment principles. In *The Cambridge Handbook of Multimedia Learning*, R. E. Mayer, Ed., Cambridge handbooks in psychology. Cambridge University Press, New York, NY, 2014, pp. 345–370.

[40] MEDIATORE, K. Reading with your ears: Readers' advisory and audio books. *Reference & User Services Quarterly 42*, 4 (2003), 318–323.

[41] MEGHDARI, A., SHARIATI, A., ALEMI, M., VOSSOUGHI, G. R., EYDI, A., AHMADI, E., MOZAFARI, B., AMOOZANDEH NOBAVEH, A., AND TAHAMI, R. Arash: A social robot buddy to support children with cancer in a hospital environment. *Proceedings of the Institution of Mechanical Engineers, Part H: Journal of Engineering in Medicine 232*, 6 (2018), 605–618.

[42] MIN, H.-J., KIM, S.-C., KIM, J., CHUNG, J.-W., AND PARK, J. C. Speaker-tts voice mapping towards natural and characteristic robot storytelling. In *2013 IEEE RO-MAN* (26.08.2013 - 29.08.2013), IEEE, pp. 793–800.

[43] MOYER, J. E. Audiobooks and e-books: A literature review. *Reference & User Services Quarterly 51*, 4 (2012), 340–354.

[44] MULLENNIX, J. W., STERN, S. E., WILSON, S. J., AND DYSON, C.-L. Social perception of male and female computer synthesized speech. *Computers in Human Behavior 19*, 4 (2003), 407–424.

[45] PARK, S.-H., BAE, B.-C., AND CHEONG, Y.-G. Emotion recognition from text stories using an emotion embedding model. In *2020 IEEE International Conference on Big Data and Smart Computing (BigComp)* (2020), IEEE, pp. 579–583.

[46] RODERO, E. Effectiveness, attention, and recall of human and artificial voices in an advertising story. prosody influence and functions of voices. *Computers in Human Behavior 77* (2017), 336–346.

[47] RODERO, E., AND LUCAS, I. Synthetic versus human voices in audiobooks: The human emotional intimacy effect. *New Media & Society* (2021), 146144482110241.

[48] ROE, D. B., Ed. *Voice Communication Between Humans and Machines.* National Academy Press, Washington, DC, 1994.

[49] ROESLER, E., ZUR KAMMER, K., AND ONNASCH, L. *Multidimensionale Fragebögen zur Erfassung der wahrgenommenen Robotermorphologie (RoMo) in der Mensch-Roboter-Interaktion.* 13, 2023.

Roesler, E., zur Kammer, K., and Onnasch, L. Multidimensionale Fragebögen zur Erfassung der wahrgenommenen Robotermorphologie (RoMo) in der Mensch-Roboter-Interaktion [Submitted manuscript]. 2023. - Comment on page 31 at [60]: Volume 13.

[50] RUFFIN, M., NIAS, J., TAYLOR, K., SINGLETON, G., AND SYLVAIN, A. Character development to facilitate retention in a storytelling robot. In *Proceedings of the 2020 ACM Southeast Conference* (New York, NY, USA, 2020), M. Chang, D. Lo, and E. Gamess, Eds., ACM, pp. 276–279.

[51] SALEM, M., EYSSEL, F., ROHLFING, K., KOPP, S., AND JOUBLIN, F. Effects of gesture on the perception of psychological anthropomorphism: A case study with a humanoid robot. In *Social Robotics*, B. Mutlu, C. Bartneck, J. Ham, V. Evers, and T. Kanda, Eds., vol. 7072 of *Lecture Notes in Computer Science*. Springer Berlin Heidelberg, Berlin, Heidelberg, 2011, pp. 31–41.

[52] SCHERER, K. R. Expression of emotion in voice and music. *Journal of Voice: Official Journal of the Voice Foundation 9*, 3 (1995), 235–248.

[53] SHAMSI, M., BARBOT, N., LOLIVE, D., AND CHEVELU, J. Mixing synthetic and recorded signals for audio-book generation. In *Speech and Computer*, A. Karpov and R. Potapova, Eds., vol. 12335 of *Lecture Notes in Computer Science*. Springer International Publishing, Cham, 2020, pp. 479–489.

[54] SIMON-KUCHER & PARTNERS. Welche genres lesen sie unabhängig vom format?, 2020.

[55] SIMONDS, B. K., MEYER, K. R., QUINLAN, M. M., AND HUNT, S. K. Effects of instructor speech rate on student affective learning, recall, and perceptions of nonverbal immediacy, credibility, and clarity. *Communication Research Reports 23*, 3 (2006), 187–197.

[56] SOFTBANK ROBOTICS. Nao: V6, 2018.

[57] STEINHAEUSSER, S. C., GABEL, J. J., AND LUGRIN, B. Your new friend nao vs. robot no. 783 - effects of personal or impersonal framing in a robotic storytelling use case. In *Companion of the 2021 ACM/IEEE International Conference on Human-Robot Interaction* (New York, NY, USA, 03082021), C. Bethel, A. Paiva, E. Broadbent, D. Feil-Seifer, and D. Szafir, Eds., ACM, pp. 334–338.

[58] STEINHAEUSSER, S. C., SCHAPER, P., AND LUGRIN, B. Comparing a robotic storyteller versus audio book with integration of sound effects and background music. In *Companion of the 2021 ACM/IEEE International Conference on Human-Robot Interaction* (New York, NY, USA, 03082021), C. Bethel, A. Paiva, E. Broadbent, D. Feil-Seifer, and D. Szafir, Eds., ACM, pp. 328–333.

[59] STERN, S. E., MULLENNIX, J. W., DYSON, C., AND WILSON, S. J. The persuasiveness of synthetic speech versus human speech. *Human Factors 41*, 4 (1999), 588–595.

[60] STRIEPE, H., DONNERMANN, M., LEIN, M., AND LUGRIN, B. Modeling and evaluating emotion, contextual head movement and voices for a social robot storyteller. *International Journal of Social Robotics* (2019), 1–17.

[61] STRIEPE, H., AND LUGRIN, B. There once was a robot storyteller: Measuring the effects of emotion and non-verbal behaviour. In *Social Robotics*, A. Kheddar, E. Yoshida, S. S. Ge, K. Suzuki, J.-J. Cabibihan, F. Eyssel, and H. He, Eds., vol. 10652 of *Lecture Notes in Computer Science*. Springer International Publishing, Cham, 2017, pp. 126–136.

[62] TRAUNMÜLLER, H., AND ERIKSSON, A. The frequency range of the voice fundamental in the speech of male and female adults. *Unpublished manuscript 11* (1995).

[63] VAUGHN, L. A., HESSE, S. J., PETKOVA, Z., AND TRUDEAU, L. "this story is right on": The impact of regulatory fit on narrative engagement and persuasion. *European Journal of Social Psychology 39*, 3 (2009), 447–456.

[64] VENKATAGIRI, H. Effect of sentence length and exposure on the intelligibility of synthesized speech. *Augmentative and Alternative Communication 10*, 2 (1994), 96–104.

[65] VERMA, R., SARKAR, P., AND RAO, K. S. Conversion of neutral speech to storytelling style speech. In *2015 Eighth International Conference on Advances in Pattern Recognition (ICAPR)* (04.01.2015 - 07.01.2015), IEEE, pp. 1–6.

[66] WALTERS, M. L., SYRDAL, D. S., KOAY, K. L., DAUTENHAHN, K., AND TE BOEKHORST, R. Human approach distances to a mechanical-looking robot with different robot voice styles. In *RO-MAN 2008 - The 17th IEEE International Symposium on Robot and Human Interactive Communication* (01.08.2008 - 03.08.2008), IEEE, pp. 707–712.

[67] XU, K. First encounter with robot alpha: How individual differences interact with vocal and kinetic cues in users' social responses. *New Media & Society 21*, 11-12 (2019), 2522–2547.

[68] YOGO, Y., ANDO, M., HASHI, A., TSUTSUI, S., AND YAMADA, N. Judgments of emotion by nurses and students given double-bind information on a patient's tone of voice and message content. *Perceptual and Motor Skills 90*, 3 Pt 1 (2000), 855–863.

3

Learning from Humans: How Research on Vocalizations can Inform the Conceptualization of Robot Sound

Hannah Pelikan and Leelo Keevallik

3.1 Introduction

Beeps and whirrs are just some examples of sounds that robots produce. Such sounds are not exclusive to robots: non-lexical vocalizations such as *ouch*, *wohoo*, and *tadaa* have recently been shown to be an important and effective element of human-human communication: people consistently make sense of these sounds when interacting [36]. Taking an interactional perspective, this chapter provides examples of how insights on human vocalizations and prosody can inform the analysis and design of robot sound.

Human vocalizations feature special prosodies and convey important information on the state of its producer [34] and some of them are understood universally [9]. They function at the margins of human language, provide affordances for cross-cultural understanding, and crucially facilitate interaction. In our chapter we discuss how vocalizations and prosodies that index emotional states such as sighs and moans indicating disappointment [23, 25] or sensorial and proprioceptive experiences such as the vocalization of gustatory pleasure with a *mmm* [84] or the sonification of body movement [32] are intuitively understood and acted on by humans in concrete activity contexts. Even though semantically less specific than lexical words, vocalizations are an important interactional resource precisely because they are sufficiently flexible to be adaptable to a variety of contexts. In this chapter we argue that their interactional function and special properties can inspire design of robot sounds for interaction with humans.

In robots, similar sounds have been glossed as "semantic-free utterances" [86], including gibberish speech, paralinguistic aspects such as backchannels, voice quality or pitch, and "non-linguistic" sonifications such as beeps. The majority of studies have evaluated them through questionnaires, whereas knowledge of how people interpret them in real-time interaction is still lacking. While semantic-free utterances may stand out as deliberately designed for communicational purposes, even "consequential sounds", originating from the physical embodiment of the robot contribute to how a robot is perceived [46] and may be interpreted in interaction. Both of these categories feature in our work, as we are interested in designing recognizable robot behavior that displays interactional affordances in an intuitive and implicit way [30, 48]. Our aim is to enrich the sounding opportunities for robots by taking inspiration from recent findings regarding the use of vocalizations in human-to-human interaction.

We will discuss how insights on various aspects of human vocal behavior can be used as a resource for designing recognizable robot behavior that makes robots more expressive and natural to interact with. First, we demonstrate how one and the same vocalization or sound can be interpreted differently in different contexts: it gains situated meanings depending on the exact interactional context that it occurs in. This problematizes the design of sound all too rigidly for specific goals but also points to opportunities for sounds to be used for more flexible outcomes depending on when exactly they are produced, i.e. their sequential context in a particular activity. Second, we discuss how sounds are tightly connected to embodiment and movement, thus not to be designed as an independent mode. Third, we highlight some qualitative differences between sounds, such as single versus repeated sounds, and how they can achieve coordination between several agents in interaction. We end the chapter by briefly presenting a method for designing and prototyping the timing of non-lexical sound based on video recordings of concrete interactions. Drawing on three transcribed video recordings from human interaction and three from human–robot interaction, we contribute lessons that could inform the methods of sound design for robots.

3.2 Robot Sound Design

Evaluating users' interpretations is a central concern for robot sound design, usually with the goal of ensuring that they are in line with the designer's intentions [5, 13, 46, 64, 69, 73, 78]. This is typically done by playing audio or video recordings to study subjects and asking them to rate the sounds along different scales. Few studies have taken a different approach, testing sounds in interactional contexts. Using different video scenarios, such as a robot being hit or kissed just before a sound was played, Read and Belpaeme [65] demonstrated that the situational context influences how sounds are interpreted. Others have explored how users interpret sound during live interactions with a robot in the lab [54]. In real world encounters the evaluation of robot sound has to deal with practical issues such as whether sound can even be heard in a particular environment, such as on a busy road [45, 59].

While some studies focus on audio as a separate modality, sound is often tightly intertwined with a robot's material presence. Whether sound is suitable or not may depend on the specific embodiment of the robot [41, 66] and people may prefer different robot voices depending on the context in which they are used [80]. Consequential sound and musical sonifications are naturally paired with movement [13, 29, 69, 73, 78]. These can even accomplish interactionally relevant actions such as managing delay, for instance through a combination of cog-inspired sounds and turning away from the human interlocutor [57]. Work on backchannels and affect bursts in robots combines facial expressions with vocalizations [54]. The interplay of different modalities has been given particular attention in the design of emotion displays [79].

Concerning how a sound can be varied, studies have compared the use of beeps versus words [12] and explored how variations in intonation, pitch, and rhythm influence the interaction [11, 63]. While the majority of studies focus on evaluating specific robot sounds, more recent work has started to formulate general design principles, reflecting among others on how sound could be varied and modified throughout longer interactions [68].

In short, audio is typically not designed as a standalone resource but is intertwined with another resource such as a movement or a facial expression – but rarely involves a range of modalities at the same time. While much work has focused on designing a set of particular animations, some studies are exploring how sound can be varied throughout an interaction. Importantly, studies of robot sound in real-time interaction remain rare, a gap which our work tries to address.

3.3   Vocalization in Human Interaction

While human interaction can be markedly centered around language, non-lexical vocalizations provide different affordances from lexical items, as they are, a) underspecified (vague in meaning), b) part of complex multimodal conjectures that may reveal information about the mental or physical state of the body producing them, and c) relatively variable in their form. In addition, prosodic aspects of all vocal delivery contribute specific meanings. All of these aspects may be useful for robot sound design.

While lexical items have meanings that are traditionally captured in dictionaries, vocalizations such as pain cries (*uuuw*), strain grunts, or displays of shivering (*brrr*) do not generally figure there. They are semantically underspecified, and we understand them in the context of someone hitting their head on a microwave door, lifting a spade of manure, or wading into cold water. Indeed, they lack propositional meaning but humans make sense of them in concrete activity contexts where they can also take specific action: parental lipsmacks encourage infants to eat [85], a strain grunt recruits others to rush to help [33], and a pain cry shows that students have understood a self-defence technique [82]. Vocalizations such as clicks may be used to hearably not say anything, thus leaving assessments implicit [50]. This vagueness makes the vocalizations usable in a broad variety of functions, while it is also true that the meaning of any word is determined by its context of use.

As is clear from the brief examples above, vocalizations are necessarily embedded in multimodal trajectories of action. The accomplishment of social action is intimately tied to people's ability to behave in a comprehensible manner and to competently interpret these very behaviors in entirety, not merely separating out a single stream of information, such as contained in the vocal tract sound. If someone sniffs and gazes away, it may make evident that the person will not speak at this point in conversation [24]. When a glass is simultaneously lifted to the sniffer's nose, they may be publicly demonstrating access to a source of a smell, such as in beer tasting sessions [44]. When someone gasps there is reason to check where their gaze is for a spill or potentially broken glasses [2]. A *mmm* with a specific rise-fall intonation after taking a spoonful of food is typically interpreted as expressing gustatory pleasure [84]. In short, from a human interaction perspective, sound is not a standalone resource but gets interpreted in combination with other aspects such as movement, facial expressions and so forth. Many vocalizations also express sensory immediacy, such as just having smelled, tasted, or dropped something. A reaction to pain or discomfort needs to be immediate in order to be deemed visceral [83]. A Finnish *huh huh* (a double heavy outbreath) is uttered at transitions from strenuous activities that have just come to an end [55]. Notably, emotion displays such as "surprise" or "appreciation" feature distinct embodied aspects, as was shown very early by Goodwin and Goodwin [19]. A display of "disappointment" may be performed through a particular interjection in combination with a distinct pitch movement such as the English "oh" [7] or by a visible deflation of the body [6]. In co-present activities in particular, human actions are performed as, and interpreted through, multimodal displays.

Vocalizations can furthermore be adjusted in many ways: repetition, loudness, lengthening, or sound quality. A moan expressing disappointment at a boardgame move can include and combine any back vowel, i.e. a, o, and u and feature variable lengths [25]. Rhythmicity (and arrhythmicity) is a way to exhibit affiliative (or disaffiliative) relations between turns. Vocalized celebrations can be performed in chorus [77]. All of this means that vocalizations can be adjusted to their sequential and action environment and interpreted flexibly [36]. Work in robotics has often glossed parts of this variability under the category of prosody [86], including elements such as loudness and pitch curves. For designing robot sound, the variability and flexibility can be of particular interest, since it means that they need not emulate a very specific human vocalization to be understood as meaningful, and can be adapted to a variety of contexts.

To summarize, research on vocalizations in interaction has resulted in a better understanding of the contextualized methods and resources participants use for making sense of each other. These methods include attention to not only the position of the item in an action sequence but also its indexical aspects, exact timing in relation to current bodily action, the articulatory and prosodic features of the utterance, as well as material, spatial, and other contextual matters in the local environment. In this chapter, we will proceed to use the same method to target robot sounds in interactional settings and show that it can inform new ways of thinking about those.

3.4   Method

We take an Ethnomethodology and Conversation Analysis (EMCA) approach to studying interaction. EMCA originated from sociology, with an initial focus on human spoken interaction in phone calls [72]. Fairly soon it made some of its most impactful advances within anthropology, self-evidently using video rather than merely a tape-recorder to capture human interaction holistically [16, 18] and in linguistics, pioneering a new branch that came to be called interactional linguistics [8, 49]. In these areas, close studies of video-recorded interaction in naturally occurring situations have provided a solid ground for revealing the underlying organization of human collaboration. EMCA has been successfully applied to study human-computer interaction [3, 38, 62, 76] and interaction with robots [58, 61, 81].

In this chapter we draw on video recordings from a variety of settings. Participants in all recordings have given their consent to be videorecorded and we only show data from the persons who agreed to share their videos. The examples from human-human interaction have been previously published [31, 34,85]. The examples from human–robot interaction stem from two previously collected corpora: A Cozmo toy robot in Swedish and German family homes [57,60], and autonomous shuttle buses on public roads, on which we also tested own sound designs [59]. Please see the original publications for details on the video corpora.

EMCA treats video recordings as data, beginning the analysis with a multimodal transcription. Transcription is done manually because it is at a level of detail that cannot (yet) be handled by automated transcription and image analysis software, crucially because all the locally relevant details cannot be predicted. Several transcription conventions are available, and in this article we follow the most established and readability-focused Jeffersonian transcription system for verbal utterances [21] and sounds [56], while we use Mondada's transcription system [43] for tracing embodiment and movements at tenth-of-a-second intervals. A close transcription enables the analyst to unpack how interactions evolve in real time.

EMCA methodology is specialized to find "order at all points" [71, p.22], revealing how people systematically calibrate their behavior to each other and to machines, even though it may look disorderly at first sight. Analytic questions include: How do others respond to what someone (or a robot) just did? What understanding of the prior action do they display in their response? And what opportunities and expectations do they create through their own subsequent action? Drawing on detailed transcripts, the researcher typically looks at each turn in an interaction, trying to identify what is accomplished by it. Our particular study objective is to use this method for both human-to-human and human–robot interaction in order to locate similarities and identify possible sources of inspiration.

3.5   Applying Insights on Human Vocalizations to Robots

In this section we present examples of video recorded interactions to highlight three main aspects of how humans make sense of vocalizations among themselves and how they interpret robot sounds: meaning potentials, the multimodal embeddedness of sounds, and their flexible production. In each section, we highlight how the findings can be leveraged by roboticists to create sounds designs that are more in line with human expectations.

3.5.1 Sounds are Semantically Underspecified

When robot sound is evaluated in user studies, the aim is usually to validate that a specific sound can convey the meaning intended by design, for instance a particular emotion [5, 28]. We would like to highlight that this is important for narrowing down the overall range of potential meanings, but this can never nail down a sound to one fixed interpretation. In everyday interaction, sound is always interpreted in the specific local context that it occurs in, which can be an asset for robot sound design: Sequencing (the way in which behavior by different participants follows one after another) and context are essential for how humans make sense of each other's actions and moves. Importantly, all meaning is negotiated in interaction. Utterances get their meaning specified in a local context: Consider the word "nice", which according to the Merriam Webster Dictionary can mean different things, ranging from "polite, kind", to "pleasing, agreeable", to "great or excessive precision". While a dictionary can provide a number of example phrases, it is impossible to list what exactly it would mean at any moment in interaction [22, p. 143]. Similar arguments can be made for many words, Norén and Linell [47] for instance provide a parallel discussion on the Swedish word *ny* "new". Thus, even though lexical items are seemingly more fixed in their meaning, it is ultimately only a question of degree. It is only possible to establish meaning potentials both for vocalizations and words [39].

Consider the extract in Figure 3.1, where an infant produces a short vocalization *mmh* [85, p.248-250]. It is easy to imagine that a *mmh* by an infant could mean anything from satisfaction to dawning unhappiness, depending on the circumstances, while in this case, taken from a mealtime, it basically emerges as a request. The infant produces a *mmh* sound with rising pitch and attempts to grab a bite of food on a tray (Figure 3.1, line 1). Mum gazes at the infant and interprets it as "wanting" that specific bit (l.3). Making sense of all kinds of sounds is an inherent aspect of human interaction and happens with regard to the activity context, gesture, prosody, etc. A "mmh" in a different context, such as after a question, can easily be interpreted as a positive answer [15].

Similarly, sound designed for robots gets assigned a specific situated meaning when a robot plays the sound at a particular moment in interaction. Consider an animation by the Cozmo robot which the designers intended to mean "happy to meet you" (see Figure 3.2) highlighting success after a relatively lengthy face learning activity: When played at the dinner table in a context when Cozmo has just been offered a sip of beer, the animation may get understood as "Cozmo likes beer". Cozmo is programmed to play this animation at the end of every successful face learning sequence. After saying the user's name twice, the robot launches a sequence of quick sounds, while showing smiley eyes on its display and waving its forklift arms. In our corpus the animation typically gets responded to by smiles and petting the robot. We could thus observe that the sound gets treated as closing the face learning sequence, as designed

```
1 INF   mmh?#
  inf   >>gaze on a piece of food
  img        #img1
2       (0.5)^(0.4)
  mum         ^reaches over to the tray-->>
3 MUM   want that bit over there,
4       (0.3)&(0.9)
  inf          &lifts a handful
```

Img 1. Infant utters 'mmh' and reaches for a bit of food.

FIGURE 3.1
Finger-licks with and without lip-smacks (Lewis002_0515_LSS30 & 31).

for. However, we also recorded an example in which the sound-animation gets interpreted in a quite different way, with participants in a family home formulating their understanding of the animation as "full agreement" to a question whether Cozmo likes beer [60].

The extract in Figure 3.3 provides a transcript of the interaction in which a couple, Ulrich and his wife, meet the robot. The robot has been scanning Ulrich's face when he gets impatient and proposes a different thing. Ulrich proceeds to ask "do you like Giesinger beer?" (l. 01), while grabbing his glass. Cozmo says *Ulrich* (l. 03) while Ulrich is speaking, but it gets ignored by everybody present. In this moment Cozmo plays a *oaaaaow* sound that resembles a "wow" (l. 06), which Ulrich responds to with the German change-of-state token *ah* (l. 07) (this resembles an "oh" in English). Cozmo's second formulation of Ulrich's name drowns in laughter (l. 08-11). When Cozmo then finally plays the happy animation (l. 16), Ulrich interprets it as "oh yes, full approval" (l. 17-19). This example highlights very vividly that designers cannot ultimately define how a robot sound gets interpreted in the specific local context of a dynamic family life. While a verbal utterance such as "happy to meet you" would simply not fit as a response to the question "do you like beer", the sound (as part of a display of happiness) gets interpreted as a fitted, positive response about the beer.

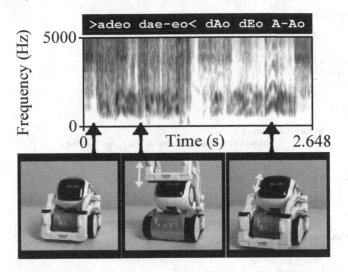

FIGURE 3.2
Cozmo's happy animation at the end of a face learning sequence (adapted from [60]).

The above extracts highlight how humans make sense of vocalizations and sounds, treating them not as pre-defined but instead as meaning different things depending on the specific local context that they are uttered in, and the multimodal aspects that go with them. The examples highlight the importance of repeatedly deploying and testing prototypes in real-world contexts, in order to learn about the range of possible associations by humans. This can also help to identify at what moments narrow meaning potentials are necessary and when interaction would be eased by more interpretative flexibility. Featuring broader meaning potentials, sounds invite broader interaction possibilities and can function efficiently in different interaction contexts.

3.5.2 Sound Production is Embodied

Some of the studies that explore the interplay of robot sound with other modalities focus on identifying which modality is most effective for conveying a specific message or robot state, such as happiness [79]. In contrast, work in sonification and perception generally tends to treat sound as a multisensorial and multimodal phenomenon [4, 14], in which these modalities are not ranked but contribute to an impression that may be more than the sum of its parts. In interactional sense-making, modalities are necessarily intertwined. As we highlighted in the extracts in Figure 3.1 and 3.3, neither the infant's nor Cozmo's sounds stand isolated in defining their meaning. They are interpreted alongside body movements and facial expressions.

In human interaction, vocalizations generally acquire their meaning from their context and ongoing bodily activities. Being essentially embodied, they

```
01 HUS   ma[gst du k- magst du giesin]ger bier?=
         do you like do you like Giesinger beer?
                    ((beer from local brewery))
02 COZ   [u::lrich:::              ]
03 WIF   =[(h)]
04 JON   [k(h)]a
05 RES   e(h)[h]
06 COZ      [o]aaaaow
07 HUS   A[H    ]
         oh
08 RES   [ha↑ha]↓ha[haha]
09 HUS           [haha][ha[haha hahaha]ha  ha ha he]
10 WIF               [haha   (h)(h)(h)   (h)(h)(h)]
11 COZ                    [°°   rich°°]
12 HUS   &[.hhh [hehe [ha]heha .h] (h)] (h)      ]
13 WIF   [(h)    (h)(h)(h) (h)  (h)]
14 RES   [(h)(h)   (h)(h)       (h) (h)]
15 HOS       [↑hi: ↓ha]
16 COZ          [>adeo dae-eo< dAo deo A]-Ao=
17 HUS   =ahJA#
         oh yes
 img          #img1
18       (1.6) ((more joint laughter))
19 HUS   volle zustimmung#
         full approval
 img                          #img2
     ((joint laughter))
```

Image 1. Cozmo just played the "happy" animation.

Image 2. Ulrich translates it as 'full approval'.

FIGURE 3.3
Full approval (A1 [00:38-00:50]).

can instruct movement. An example is shown in the extract in Figure 3.4, which is taken from a Pilates class. The teacher has just demonstrated a new exercise called helicopter that involves moving legs and hands in a circular motion to opposite sides while balancing on the buttocks. When the students try it out, the teacher produces a long nasal sound in combination with a large gesture to accompany them [34]. The pitch trace is shown above the transcript line.

To start with, the teacher times the beginning of the exercise by uttering a slightly lengthened *ja* "and" while simultaneously launching a two-hand gesture that shows the required circular movement of the legs (Figure 3.4, Images 1-2). As she dips into the shape, she produces the strained vocalization that ends in a very high pitch (Figure 3.4, l. 1), marking the limbs' arrival

```
1 TEA   #ja: qnnnn#nnnnnnmmmmuhh  (0.2)#
        and
    img  #img1     #img2              #img3

2       ja teiselepoole,
        and to the other side
```

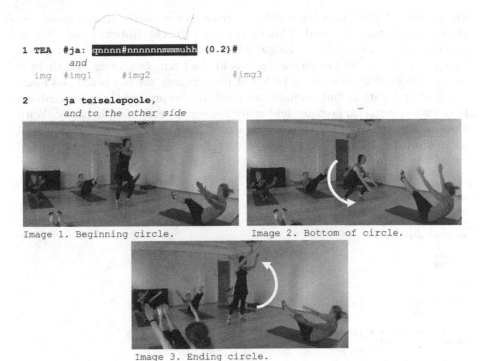

Image 1. Beginning circle.　　　Image 2. Bottom of circle.

Image 3. Ending circle.

FIGURE 3.4
Pilates I.

back at the top position (Figure 3.4, Image 3). The teacher's bodily-vocal performance makes the exercise visually and audibly available, highlighting the shape, trajectory, as well as the ostensible proprioceptive experience of the exercise: the strain and its temporal extension are illustrated through her voice. The students toiling each at their own pace show that they attend to the teacher's production as an instruction.

This instance illustrates how one person's vocalizing in combination with their embodiment can have an immediate impact on other people moving. Sounding practices are usable for scaffolding others' activities (as above) as well as coordinating bodies together, such as a collaborative lift of a heavy object [33], making others move in a certain manner [32], and achieving synchrony in movement [26]. Humans may raise their voice [67] or increase repetition tempo, in order to persuade others to comply [52]. There is thus a wide range of ways how sounds embedded in multimodal action can be used for coordination across participants. Needless to say, a *qnnn* sound might mean very different things in other contexts.

The lack of comparable expressivity for robots is often attributed to the difficulty of designing a convincing coordination of different robot resources, such as movements, facial expressions, and sound. Depending on their morphology, robots may have different resources available than humans, including for instance colored lights or vibration [40, 75]. Robots have different bodies than humans and it is unclear whether they could meaningfully instruct fine-grained physical movement such as in the Pilates example, but there is nevertheless a general point to be learnt here. Sounds are not heard and made sense of in isolation, but humans are used to interpreting them in combination with movement, gestures and other nonverbal, embodied behavior. With Cozmo, an *uuuuuu* sound only becomes meaningful as a greeting when paired with a lifted head and large eyes on its screen-face, as we illustrate in the extract in Figure 3.5.

```
01 COZ   quack +chrrrrrɨrr((motor sound))
   coz          +drives forward-->
   coz                  ±lifts head-->

02 COZ   u#uɨudeo+:
   coz   -->ɨ
   coz          -->+
   img   #img1

03       (0.6)

04 MOM   °j↑a↓a°
         yes

05 COZ   ɨch+r[r +]# ((motor sound))
06 SON       [(h)]
   coz   ±lifts head-->>
   coz      +turn+
   img           #img2

07       (0.6)

08 COZ   +chr + ((motor sound))
   coz   +turn+

09       (0.4)

10 MOM   hej
         hello

11       +(0.3)      +
   coz   +lifts fork+

12 COZ   ±‖uuuuuu‖#
   coz   ±large eyes-->
   img          #img3

13 MOM   H↑Eɨ:J
         hello
   coz   -->ɨ
```

Image 1. Moving forward.   Image 2. Turning.   Image 3. Large eyes.

FIGURE 3.5
Cozmo's sound animation is interpreted as a greeting (FAM6 Day 3 [01:28-01:36]).

Cozmo has just been switched on, has left the charger and played an animation that resembles "waking up". It drives forward and raises its head, playing a gibberish *uuudeo* sound (l. 02), which is acknowledged by mother and son with a *ja* "yes" in Swedish (l. 04) and a laughter particle (l. 06). The robot turns and lifts its head, now facing the mother, who greets the robot by saying *hej* "hello" (l. 10). The robot briefly moves its forklift arms and displays large, "cute" eyes on its display while playing an *uuuuu* sound with rising intonation (l. 12). The mother treats this as a response to her greeting and greets the robot once more, now with increased prosodic marking (l. 13). It is the activity context of Cozmo just waking up that this precise multimodal production is interpreted as a greeting by a human and responded to affectively. Paired with a backwards driving movement and a raised forklift, *uuuuu* could indicate something else, for instance surprise.

In short, sound should be considered as a multimodal and multisensory phenomenon. Having demonstrated how humans tightly intertwine body and voice, we want to encourage designers to pay particular attention to this intertwinement. Studies that focus on how various resource combinations are interpreted in environments of everyday life can be particularly informative for robot design, helping to evaluate for instance how sound draws attention to particular movements.

3.5.3 Sound can be Adapted for Complex Participation

In addition to the sequential interpretation, pitch movement, and embodied nature of human sounding practices, we would like to highlight a further aspect: sounds and words can be repeated with particular prosodies and thereby convey persuasion to follow a contextualized request for coordination. Vocal devices can suggest actions, addressing not only one person at a time but also dealing with complex participation frameworks involving several people [17,20].

We may consider another example from Pilates training in the extract in Figure 3.6, where the teacher is asking the students to roll up and balance on their buttocks [31]. She first provides the instruction to stay up (l. 1-2) and then repeats the word "hold (it)" seven times (l. 3-4), while the students are rolling up each at their own tempo (l. 1-4). The words are uttered at a high speed (indicated with >< in the transcript), with sounds floating together almost to the point of the words not being recognizable. The pitch trace is marked above the transcription line. With her level prosody on the first five items (l. 3) she is indicating that she will continue to provide this instruction, while on the last two repetitions her pitch rise already projects an end (l. 4), which is informative for all the students in the class. Furthermore, she coordinates her final repetition minutely with the last student (marked in a circle) arriving in the balanced position.

This example shows how humans mutually coordinate actions across multiple participants: the teacher instructs while she is also accommodating to the students moves, the students comply with the teacher's instruction and the ones who have arrived early wait for the others as well as the teacher who is to

```
1 Teacher   ni*i, jääge      üles^se, leidke      tasakaal,
           okay, stay up, find (your) balance,
  stuA,B        *stay up
  stuC,D                            ^stay up

2            jääge ülesse#
             stay up
  img                        #1
```

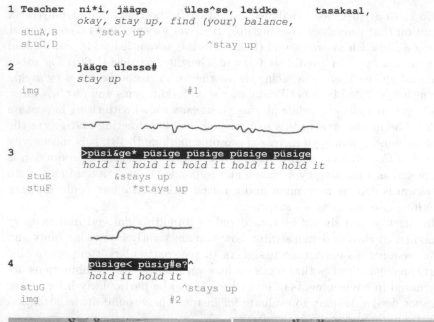

```
3          >püsi&ge* püsige püsige püsige püsige
           hold it hold it hold it hold it hold it
  stuE          &stays up
  stuF            *stays up
```

```
4          püsige< püsig#e?^
           hold it hold it
  stuG                      ^stays up
  img                   #2
```

Image 1. Students at various stages   Image 2. Last student arriving.
         of the exercise.

FIGURE 3.6
Pilates II.

produce a next instruction. Different participants need to do different things to coordinate with others: while some need to stay in position, others have to deploy abs to get into alignment with the class (comp. Images 1 and 2 in Figure 3.6), and the teacher is merely using her voice and gestures to coordinate with students. The teacher's repetition indicates that the required action by others or its quality has not yet been reached [42, 51, 74]. It also highlights that the message is less about the semantic meaning of the word itself than about the action that is supposed to be accomplished and coordinated with others.

Adapting these findings to robots, it is important to notice this mutual reflexivity and continuous adjustment in human-human interaction. While robots may not reach the same level of fine coordination, they can likewise repeat sounds and preferably do so in a manner that is minutely adjusted to other participants' actions. In the following extract we look at an EasyMile EZ10 autonomous shuttle bus for public transport, on which we tested a range of sounds in live traffic. The bus drives on invisible tracks and is often stopped by cyclists and pedestrians that are getting too close, and we were interested in exploring ways that could help the bus to maintain traffic flow by asking other road users to keep a distance [59]. The researcher acts as a Wizard of Oz, playing pre-recorded samples on top of the buse's own soundscape through a Bluetooth speaker. In the extract in Figure 3.7 she is triggering three saxophone rolls with rising intonation while the bus is moving forward.

The bus leaves its designated stop (Figure 3.7, l. 01) and approaches a crossing where it often triggers unnecessary emergency braking when people get too close. Two cyclists are approaching the bus on its left in the same lane, and a pedestrian walks toward the crossing from the right. The wizard triggers a first sound (l. 02), a saxophone riff with rising pitch, inspired by question intonation. Both the cyclists and the pedestrian are still relatively far away but seem to be aware of the approaching bus evident in their gaze and head orientation (l. 03). As the groups are moving closer toward the approaching bus, the wizard triggers another sound (l. 04). While the sound is playing, the right cyclist clearly starts gazing at the bus (l. 04), displaying an orientation to the sound. The left cyclist immediately moves further toward the left, onto a sidewalk lane (l. 04, Image 1). Soon after, the right cyclist also steers away (l. 05, Image 2). Meanwhile, the pedestrian has been slowing down their steps, but keeps approaching the intersection (l. 05). The wizard triggers a third sound (l. 06), drawing the gaze of the right cyclist once more (l. 06), who then moves even further toward the left, onto the sidewalk lane (l. 07). The pedestrian who has slowed down further (Image 3) now also gazes at the bus (l. 07), and finally stops completely, until the bus has passed.

We showed how different parties in interaction mutually adjust to each other, even though they may maintain asymmetrical roles, such as a bus being on invisible tracks or when one person is officially instructing the others. In both examples, repetition of a sound fitted to the current movement worked as a tool for continuous responsivity to others, and it seems to achieve the wanted outcome: mutual attention and adjustment. In the example with humans exercising, each person was completing the exercise at their own speed. In the example with the autonomous bus, everyone needed to act slightly differently, depending on their activity trajectory (walking, cycling, or driving), which results in the whole situation being mutually coordinated. Notably, this reflexive adjustment was in the current instance achieved by a human being and not an automated machine, but it provides an example of where robot sound design could be headed, dynamically adjusting sound to an evolving situation.

```
01      |(1.5)
   bus  |leaves bus stop->>

02 WIZ  da↑deep

03      (2.3)*(1.7)      *(1.5)+(0.3)*(0.4)   +(0.4)
   cycR      *gazes fwd*gaze cycL--*gazes fwd-->
   ped                            +glance to bus+

04 WIZ  da*↑deep%
   cycR -->*gazes at bus-->
   cycL         %turns onto sidewalk-->

05      (0.1)†(0.3)#(0.7)•(0.5)#(0.3)*(0.1)•(0.7)%
   ped      †slows down steps-->
   cycR            *steers away------•
   cycR                       -->*
   cycL                             -->%
   img          #1          #2

06 WIZ  da↑deep*
   cycR       *gazes at bus-->

07      (0.3)+•(0.2)#(0.1)*†(1.2) •
   ped      +gazes at bus->>
   cycR     •moves onto sidewalk•
   cycR        -->*
   ped             -->†stops->>
   img          #3
```

Image 1. The cyclist on the left steers away from the bus.

Image 2. The cyclist on the right steers away from the approaching bus.

Image 3. The pedestrian stops.

FIGURE 3.7

Prototyping sounds for an autonomous bus (EM-f sax round 3 cyclists at Blå Havet).

3.6 Discussion and Implications

By comparing human-human to human–robot interaction, we have demonstrated different ways in which sound can be a constitutive part of performing social actions, showing instances where coordination in real time is relevant. We have highlighted that it is never the sound alone that creates meaning, the message is not exclusively encoded into the sound. Instead, a sound entails a meaning potential that achieves significance in relation to the embodiment of its producer and other details of the local context. In the following, we highlight three main lessons that sound designers and researchers in robotics can take away from this work.

3.6.1 Meaning as Potentials

Reviewing literature and examples on non-lexical vocalizations, prosodies, and sense-making in humans, we highlighted that the term "semantic-free" [86] for robot sound is not entirely accurate. These sounds, like human vocalizations, do not necessarily carry a fixed meaning but they certainly feature potentials to be interpreted in particular ways, depending on the local interactional contexts. More generally, an important lesson for robotics is that human language does not function like math, with pre-defined symbols that always mean the same thing and lead to inevitable outcomes. Instead, meaning is partially flexible and negotiated, even for regular words. One may conceptualize robot audio as a continuum of fixedness, words like for instance "Hello" have relatively narrow meaning potentials, implying mostly a greeting at the beginning of encounters, while sound can have broader potential meanings, such as Cozmo's happy animation, which may be treated as a greeting or as accepting an offer for beer and a range of other things, depending on what exactly has just happened.

Our work highlights how studying human non-lexical vocalizations can inform the goals and questions relevant for robot sound. While it is important to ensure that potential meanings are going in the right direction (such as happy or sad valence in emotion displays), the design should not strive for setting an absolute meaning to them. Rather, we demonstrated that they gain rather specialized meanings once placed in concrete interactional contexts: asking for help, answering a question, instructing a move, greeting, coordinating, etc. For robotics, this kind of flexibility could be an advantage in settings where lexical expressions – especially in a specific language – are inappropriate or inefficient. The above explorations also highlight that it is worth considering and designing meaning potentials carefully: a sound that evokes associations with warnings such as a horn may not be appropriate for inviting people to come closer to a robot (the rolling bus). At the same time, what users and other people who encounter robots ultimately make of these sounds cannot be entirely pre-planned by the designers, leaving space also for creativity on behalf of the users. Accordingly, field studies are essential in working out the meaning of each particular sound in situated contexts.

3.6.2 Sound and Multimodality

We demonstrated how humans skillfully design combinations of vocalizations and gestures and that Cozmo's visual behavior contributes to how its sounds get interpreted. Previous work has explored the interplay between sound and facial displays in experimental settings [40, 79], and we want to highlight the importance of studying sound in combination with embodied behavior like movement, gestures and facial expressions even in interaction. Our work demonstrates how various resource combinations are interpreted in the "messy" environments of everyday life; and we are showing the orderliness and logic in those. In real world interaction the amplifying function of sound [79] may be what draws people's attention to a movement or change in facial expression that would otherwise go unnoticed. Beyond communicating specific things through sound quality, the presence of sound may sometimes be particularly useful in marking a behavior as intentional or in highlighting the character of a movement, such as the Pilates teacher in the extract in Figure 3.4 was doing when accompanying an exercise.

Robot sound design can take inspiration from the growing body of work on human vocalizations, particularly during physical activities [1, 32, 33, 67]. This can be informative for settings in which humans and robots collaborate on physical tasks, such as to minutely coordinate lifts and handovers, or when working alongside large industrial robots. Sonifying parts of movements that are especially difficult for a robot could contribute to making the robot's behavior explainable for humans, rendering the collaboration more rewarding. Such sounds can also be a natural, implicit [30] tool for robots to ask for help (cf. Figure 3.1, the infant example). We do not necessarily envision that robots should copy humans but suggest looking more closely at the interplay of sound and visual behavior, also in the designers' own bodies when brainstorming robot sounds. Are we making a specific facial expression, movement or gesture while vocalizing a suggestion? How can this be translated to robot behavior in a meaningful way? We believe that paying attention to the interplay of these modalities, as argued for instance also in somaesthetic design approaches [27], can be beneficial even for robot sound design.

3.6.3 Variable Form and Reflexive Adaptation to Multiple Participants

Finally, we pointed out how sound can be variably produced in regard to pitch height, length, and repetition in order for it to be reflexively adjusted to ongoing action and the locally emerging context, such as for the moving bus. Our research extends prior work on robot sound that has focused on different intonation curves [11], highlighting that modifying other prosodic and rhythmic elements may be a promising direction for further research. Such variation becomes especially relevant when robots interact in naturalistic settings, where interaction is not limited to one-user-one-robot, but where multiple participants need to coordinate.

When designing sound that can function in a range of contexts [68], repetition is a particularly promising resource. Varying the rhythm and number of repetitions, a single sample such as the saxophone roll used on the autonomous bus in the extract in Figure 3.7, can accomplish a range of different actions. Rather than designing a wide range of complex variations, design of so-called "semantic-free" utterances may benefit from inspecting closely how one sound can accomplish multiple actions in different real-life contexts. With modifications such as lengthening, repetition, and loudness variations, an agent can signal urgency, extension of the activity, as well as the fact that attempts at coordination (at least by some participants) are not yet sufficient. Even though the highest level of reflexivity between the sounding and those moving around a robot was only achieved by a human wizard in our study, the success highlights that there are opportunities for future design.

3.6.4 Designing Sound for Interactional Sequences

When aiming to design robot sound for interaction, we argue that it is crucial to consider its interpretability in the precise context where it will be used. Field methods such as observations and recordings of actual situations of sound use are essential, as is the close documentation of participant action. We hope to have demonstrated how a video-based study of sound in specific settings, through transcription can yield detailed insights of the relative timings of participant behavior and provide an empirical ground for discussion.

Most importantly, robot sound should not be designed in isolation, but prototyped in interactional sequences, in which timing, embodiment, and multimodal aspects of the local context play a crucial role. We are specifically interested in developing methods for adopting an interactional perspective that do not require specialist training in transcribing video. Close observation and video recordings in the setting in which the robot is used, and repeated sound design interventions in concrete interactional contexts, captured on video are key to such an approach. Specifically, we developed a video voice-over technique [59] that extends vocal sketching techniques [53, 70] by sketching over recordings of human-human or human–robot interaction: A short video snippet (about 30 seconds) is played on loop and sound is prototyped by performing voice-overs of how a robot could sound in this situation, either with one's own voice or by playing a sample. Repeatedly testing sounds on top of video recordings of actual interaction with the moving robot helps to get a sense of how they fit the particular embodiment of the robot. More importantly, it enables designers to intuitively produce the most interactionally relevant timing and duration, and potentially this kind of "annotated" data could be further used for teaching robots at what moments in interaction they should produce sound [54]. Further exploring this interactional sound design approach, we also tested sound with Wizard-of-Oz setups in real world environments, as exemplified in the extract in Figure 3.7. This provides a sense of the specific soundscape and the interplay of sound and other multimodal aspects. The setup can reveal different meaning potentials in a range of real-life consequential situations and is particularly well-suited for gaining insights on variations through repetition, pitch modulation, and duration.

3.7   Conclusion

We set out to highlight lessons for robot sound design that can be learnt from studying how humans use non-lexical vocalizations and prosody to accomplish social actions. We demonstrated how a semantically underspecified sound gains meaning locally in concrete interactions, making sound particularly useful for contexts in which verbal utterances risk being too specific. We then provided insights on the multimodal nature of interaction, showing how sound and visual behavior are intertwined, arguing that robot sound designers can gain from scrutiny of their own bodily moves when brainstorming sounds and from paying close attention to how a robot moves in a concrete space and context while playing the designed sounds. Finally, we looked at how sound can be adapted to multiple addressees at the same time by prosody and repetition and provided an example from our own sound design with a Wizard-of-Oz setup on a public road. Overall, we argued that "non-semantic" or "semantic-free" sound is indeed semantically underspecified but not meaningless, and provided examples of how sound can be flexibly adjusted to coordinate actions with multiple participants. Robot sounds should preferably be designed as multimodal displays for interactional sequences.

Acknowledgment

We would like to thank Sally Wiggins, Adrian Kerrison and Emily Hofstetter for their invaluable comments on a previous draft of this chapter. This work is funded by the Swedish Research Council grant "Vocal practices for coordinating human action" (VR2016-00827).

Bibliography

[1] ALBERT, S., AND VOM LEHN, D. Non-lexical vocalizations help novices learn joint embodied actions. *Language & Communication 88* (Jan 2023), 1–13.

[2] BEN MOSHE, Y. Hebrew stance-taking gasps: From bodily response to social communicative resource. *Language & Communication* (in press).

[3] BROWN, B., AND LAURIER, E. The trouble with autopilots: Assisted and autonomous driving on the social road. In *Proceedings of the 2017 CHI Conference on Human Factors in Computing Systems*. Association for Computing Machinery, New York, NY, USA, 2017, pp. 416–429.

[4] CARAMIAUX, B., BEVILACQUA, F., BIANCO, T., SCHNELL, N., HOUIX, O., AND SUSINI, P. The role of sound source perception in gestural sound description. *ACM Transactions on Applied Perception 11*, 1 (Apr 2014), 1–19.

[5] CHAN, L., ZHANG, B. J., AND FITTER, N. T. Designing and validating expressive cozmo behaviors for accurately conveying emotions. In *2021 30th IEEE International Conference on Robot & Human Interactive Communication (RO-MAN)* (Aug 2021), IEEE, pp. 1037–1044.

[6] CLIFT, R. Visible deflation: Embodiment and emotion in interaction. *Research on Language and Social Interaction 47*, 4 (Oct 2014), 380–403.

[7] COUPER-KUHLEN, E. A sequential approach to affect: The case of disappointment. In *Talk in Interaction: Comparative Dimensions*, M. L. Markku Haakana and J. Lindström, Eds. Finnish Literature Society (SKS), Helsinki, 2009, pp. 94–123.

[8] COUPER-KUHLEN, E., AND SELTING, M. *Interactional Linguistics: Studying Language in Social Interaction*. Cambridge University Press, 2017.

[9] DINGEMANSE, M., TORREIRA, F., AND ENFIELD, N. J. Is "huh?" a universal word? conversational infrastructure and the convergent evolution of linguistic items. *PLOS ONE 8*, 11 (2013), 1–10.

[10] EKMAN, I., AND RINOTT, M. Using Vocal Sketching for Designing Sonic Interactions. *Proceedings of the 8th ACM Conference on Designing Interactive Systems*, (2010) 123–131. https://doi.org/10.1145/1858171.1858195

[11] FISCHER, K., JENSEN, L. C., AND BODENHAGEN, L. To beep or not to beep is not the whole question. In *Social Robotics. ICSR 2014. Lecture Notes in Computer Science, vol 8755*, M. Beetz, B. Johnston, and M. Williams, Eds. Springer, Cham, 2014, pp. 156–165.

[12] FISCHER, K., SOTO, B., PANTOFARU, C., AND TAKAYAMA, L. Initiating interactions in order to get help: Effects of social framing on people's responses to robots' requests for assistance. In *The 23rd IEEE International Symposium on Robot and Human Interactive Communication* (Aug 2014), pp. 999–1005.

[13] FRID, E., AND BRESIN, R. Perceptual evaluation of blended sonification of mechanical robot sounds produced by emotionally expressive gestures: Augmenting consequential sounds to improve non-verbal robot communication. *International Journal of Social Robotics 14*, 2 (2022), 357–372.

[14] FRID, E., LINDETORP, H., HANSEN, K. F., ELBLAUS, L., AND BRESIN, R. Sound forest: Evaluation of an accessible multisensory music installation. In *Proceedings of the 2019 CHI Conference on Human Factors in Computing Systems* (Glasgow Scotland UK, May 2019), ACM, pp. 1–12.

[15] GARDNER, R. *When Listeners Talk: Response Tokens and Listener Stance.* John Benjamins, Amsterdam, 2001.

[16] GOODWIN, C. *Conversational Organization: Interaction between Speakers and Hearers.* Academic Press, London, 1981.

[17] GOODWIN, C. Participation, stance and affect in the organization of activities. *Discourse & Society 18*, 1 (2007), 53–73.

[18] GOODWIN, C. *Co-Operative Action.* Learning in Doing: Social, Cognitive and Computational Perspectives. Cambridge University Press, Cambridge, Nov 2018.

[19] GOODWIN, C., AND GOODWIN, M. H. Concurrent operations on talk: Notes on the interactive organization of assessments. *IPrA Papers in Pragmatics 1*, 1 (1987), 1–54.

[20] GOODWIN, C., AND GOODWIN, M. H. Participation. In *A Companion to Linguistic Anthropology*, A. Duranti, Ed. Blackwell, Oxford, 2004, pp. 222–244.

[21] HEPBURN, A., AND BOLDEN, G. B. *Transcribing for Social Research.* SAGE, London, 2017.

[22] HERITAGE, J. *Garfinkel and Ethnomethodology.* Polity Press, 1984.

[23] HOEY, E. M. Sighing in interaction: Somatic, semiotic, and social. *Research on Language and Social Interaction 47*, 2 (Apr 2014), 175–200.

[24] HOEY, E. M. Waiting to inhale: On sniffing in conversation. *Research on Language and Social Interaction 53*, 1 (Jan 2020), 118–139.

[25] HOFSTETTER, E. Nonlexical "moans": Response cries in board game interactions. *Research on Language and Social Interaction 53*, 1 (Jan 2020), 42–65.

[26] HOFSTETTER, E., AND KEEVALLIK, L. Prosody is used for real-time exercising of other bodies. *Language & Communication 88* (Jan 2023), 52–72.

[27] HÖÖK, K. Designing with the Body: Somaesthetic Interaction Design. (2018) MIT Press.

[28] JEE, E.-S., JEONG, Y.-J., KIM, C. H., AND KOBAYASHI, H. Sound design for emotion and intention expression of socially interactive robots. *Intelligent Service Robotics 3*, 3 (2010), 199–206.

[29] JOHANNSEN, G. Auditory displays in human–machine interfaces of mobile robots for non-speech communication with humans. *Journal of Intelligent and Robotic Systems 32*, 2 (2001), 161–169.

[30] JU, W. *The Design of Implicit Interactions*. Springer International Publishing, 2015.

[31] KEEVALLIK, L. Linguistic structures emerging in the synchronization of a pilates class. In *Mobilizing Others: Grammar and Lexis within Larger Activities*, C. Taleghani-Nikazm, E. Betz, and P. Golato, Eds. John Benjamins, Amsterdam/Philadelphia, 2020, pp. 147–173.

[32] KEEVALLIK, L. Vocalizations in dance classes teach body knowledge. *Linguistics Vanguard 7*, s4 (2021), 20200098.

[33] KEEVALLIK, L. Strain grunts and the organization of participation. In *Body, Participation, and the Self: New Perspectives on Goffman in Language and Interaction*, L. Mondada and A. Peräkylä, Eds. Routledge, London, 2023.

[34] KEEVALLIK, L., HOFSTETTER, E., WEATHERALL, A., AND WIGGINS, S. Sounding others' sensations in interaction. *Discourse Processes* (Jan 2023), 1–19.

[35] KERSTIN NORÉN, PER LINELL "Meaning potentials and the interaction between lexis and contexts: an empirical substantiation", *Pragmatics*, 17 (2007), pp. 387–416.

[36] KEEVALLIK, L., AND OGDEN, R. Sounds on the margins of language at the heart of interaction. *Research on Language and Social Interaction 53*, 1 (Jan 2020), 1–18.

[37] KERSTIN NORÉN AND PER LINELL, "Meaning potentials and the interaction between lexis and contexts: an empirical substantiation", *Pragmatics*, vol. 17, no. 3 (2007), pp. 387–416.

[38] LICOPPE, C., LUFF, P. K., HEATH, C., KUZUOKA, H., YAMASHITA, N., AND TUNCER, S. Showing objects: Holding and manipulating artefacts in video-mediated collaborative settings. In *Proceedings of the 2017 CHI Conference on Human Factors in Computing Systems* (New York, NY, USA, 2017), CHI '17, Association for Computing Machinery, pp. 5295–5306.

[39] LINELL, P. *Rethinking Language, Mind, and World Dialogically.* Advances in Cultural Psychology: Constructing Human Development. Information Age Publishing, Charlotte, NC, 2009.

[40] LÖFFLER, D., SCHMIDT, N., AND TSCHARN, R. Multimodal Expression of Artificial Emotion in Social Robots Using Color, Motion and Sound. *Proceedings of the 2018 ACM/IEEE International Conference on Human-Robot Interaction*, 334–343 (2018). https://doi.org/10.1145/3171221.3171261

[41] MERAT, N., LOUW, T., MADIGAN, R., WILBRINK, M., AND SCHIEBEN, A. What externally presented information do vrus require when interacting with fully automated road transport systems in shared space? *Accident Analysis & Prevention 118* (2018), 244–252.

[42] MONDADA, L. Precision timing and timed embeddedness of imperatives in embodied courses of action: Examples from french. In *Imperative Turns at Talk: The Design of Directives in Action*, M.-L. Sorjonen, L. Raevaara, and E. Couper-Kuhlen, Eds., Studies in language and social interaction. John Benjamins Publishing Company, Amsterdam, Philadelphia, 2017, pp. 65–101.

[43] MONDADA, L. Contemporary issues in conversation analysis: Embodiment and materiality, multimodality and multisensoriality in social interaction. *Journal of Pragmatics 145* (May 2019), 47–62.

[44] MONDADA, L. Audible Sniffs: Smelling-in-Interaction. *Research on Language and Social Interaction 53*, 1 (Jan 2020), 140–163.

[45] MOORE, D., CURRANO, R., AND SIRKIN, D. Sound decisions: How synthetic motor sounds improve autonomous vehicle-pedestrian interactions. In *12th International Conference on Automotive User Interfaces and Interactive Vehicular Applications* (New York, NY, USA, 2020), AutomotiveUI '20, Association for Computing Machinery, pp. 94–103.

[46] MOORE, D., TENNENT, H., MARTELARO, N., AND JU, W. Making noise intentional: A study of servo sound perception. In *Proceedings of the 2017 ACM/IEEE International Conference on Human-Robot Interaction* (New York, NY, USA, 2017), HRI '17, Association for Computing Machinery, pp. 12–21.

[47] NORÉN, K. AND LINELL, P. "Meaning potentials and the interaction between lexis and contexts: an empirical substantiation", *Pragmatics*, vol. 17, no. 3 (2007), pp. 387–416.

[48] NORMAN, D. *The Design of Everyday Things: Revised and Expanded Edition.* Basic Books, 2013.

[49] OCHS, E., SCHEGLOFF, E. A., AND THOMPSON, S. A., Eds. *Interaction and Grammar*. Cambridge University Press, Cambridge, 1996.

[50] OGDEN, R. Audibly not saying something with clicks. *Research on Language and Social Interaction 53*, 1 (2020), 66–89.

[51] OKADA, M. Imperative actions in boxing sparring sessions. *Research on Language and Social Interaction 51*, 1 (Jan 2018), 67–84.

[52] OKADA, M. Lexical repetitions during time critical moments in boxing. *Language & Communication* (in press).

[53] PANARIELLO, C., SKÖLD, M., FRID, E., AND BRESIN, R. From vocal sketching to sound models by means of a sound-based musical transcription system. *Proceedings of the SMC Conference 2019. Sound and Music Computing.* (2019). https://www.smc2019.uma.es/articles/S2/S2_05_SMC2019_paper.pdf

[54] PARK, H. W., GELSOMINI, M., LEE, J. J., AND BREAZEAL, C. Telling stories to robots: The effect of backchanneling on a child's storytelling. In *2017 12th ACM/IEEE International Conference on Human-Robot Interaction (HRI)* (2017), IEEE, pp. 100–108.

[55] PEHKONEN, S. Response Cries Inviting an Alignment: Finnish huh huh. *Research on Language and Social Interaction 53*, 1 (Jan 2020), 19–41.

[56] PELIKAN, H. Transcribing human–robot interaction. In P. Haddington, T. Eilittä, A. Kamunen, L. Kohonen-Aho, T. Oittinen, I. Rautiainen, & A. Vatanen (Eds.), Ethnomethodological Conversation Analysis in Motion: Emerging Methods and New Technologies (1st ed.). Routledge. (2023). https://doi.org/10.4324/9781003424888

[57] PELIKAN, H., AND HOFSTETTER, E. Managing delays in human-robot interaction. *ACM Transactions on Computer-Human Interaction* (2022).

[58] PELIKAN, H. R., AND BROTH, M. Why that nao?: How humans adapt to a conventional humanoid robot in taking turns-at-talk. In *Proceedings of the 2016 CHI Conference on Human Factors in Computing Systems* (New York, NY, USA, May 2016), ACM, pp. 4921–4932.

[59] PELIKAN, H. R., AND JUNG, M. Designing robot sound-in-interaction: The case of autonomous public transport shuttle buses. In *ACM/IEEE International Conference on Human-Robot Interaction* (2023).

[60] PELIKAN, H. R. M., BROTH, M., AND KEEVALLIK, L. "Are you sad, cozmo?": How humans make sense of a home robot's emotion displays. In *Proceedings of the 2020 ACM/IEEE International Conference on Human-Robot Interaction* (2020), Association for Computing Machinery, pp. 461–470.

[61] PITSCH, K., KUZUOKA, H., SUZUKI, Y., SUSSENBACH, L., LUFF, P., AND HEATH, C. The first five seconds: Contingent stepwise entry into an interaction as a means to secure sustained engagement in hri. In *RO-MAN 2009 - The 18th IEEE International Symposium on Robot and Human Interactive Communication* (Sep 2009), IEEE, pp. 985–991.

[62] PORCHERON, M., FISCHER, J. E., REEVES, S., AND SHARPLES, S. Voice interfaces in everyday life. In *Proceedings of the 2018 CHI Conference on Human Factors in Computing Systems* (New York, NY, USA, 2018), CHI '18, ACM, pp. 640:1–640:12.

[63] READ, R., AND BELPAEME, T. How to use non-linguistic utterances to convey emotion in child-robot interaction. In *Proceedings of the Seventh annual ACM/IEEE International Conference on Human-Robot Interaction - HRI '12* (New York, New York, USA, 2012), HRI '12, ACM Press, p. 219.

[64] READ, R., AND BELPAEME, T. People interpret robotic non-linguistic utterances categorically. In *Proceedings of the 8th ACM/IEEE International Conference on Human-Robot Interaction* (Piscataway, NJ, USA, 2013), HRI '13, IEEE Press, pp. 209–210.

[65] READ, R., AND BELPAEME, T. Situational context directs how people affectively interpret robotic non-linguistic utterances. In *Proceedings of the 2014 ACM/IEEE International Conference on Human-Robot Interaction* (New York, NY, USA, 2014), HRI '14, Association for Computing Machinery, pp. 41–48.

[66] READ, R. G., AND BELPAEME, T. Interpreting non-linguistic utterances by robots. In *Proceedings of the 3rd International Workshop on Affective Interaction in Natural Environments - AFFINE '10* (New York, New York, USA, 2010), AFFINE '10, ACM Press, pp. 65.

[67] REYNOLDS, E. Emotional intensity as a resource for moral assessments: The action of "incitement" in sports settings. In *How Emotions Are Made in Talk*, J. S. Robles and A. Weatherall, Eds. John Benjamins Publishing Company, 2021, pp. 27–50.

[68] ROBINSON, F. A., BOWN, O., AND VELONAKI, M. Designing sound for social robots: Candidate design principles. *International Journal of Social Robotics 14*, 6 (2022), 1507–1525.

[69] ROBINSON, F. A., VELONAKI, M., AND BOWN, O. Smooth operator: Tuning robot perception through artificial movement sound. In *Proceedings of the 2021 ACM/IEEE International Conference on Human-Robot Interaction* (New York, NY, USA, 2021), HRI '21, Association for Computing Machinery, pp. 53–62.

[70] ROCCHESSO, D., LEMAITRE, G., SUSINI, P., TERNSTRÖM, S., AND BOUSSARD, P. Sketching sound with voice and gesture. Interactions, 22(1) (2015), 38–41. https://doi.org/10.1145/2685501

[71] SACKS, H. Notes on methodology. In *Structures of Social Action: Studies in Conversation Analysis*, J. Heritage and J. M. Atkinson, Eds. Cambridge University Press, Cambridge, UK, 1984, pp. 21–27.

[72] SACKS, H., SCHEGLOFF, E. A., AND JEFFERSON, G. A simplest systematics for the organization of turn-taking for conversation. *Language 50*, 4 (1974), 696.

[73] SAVERY, R., ROSE, R., AND WEINBERG, G. Establishing human-robot trust through music-driven robotic emotion prosody and gesture. In *2019 28th IEEE International Conference on Robot and Human Interactive Communication (RO-MAN)* (2019), IEEE Press, pp. 1–7.

[74] SIMONE, M., AND GALATOLO, R. Timing and prosody of lexical repetition: How repeated instructions assist visually impaired athletes' navigation in sport climbing. *Research on Language and Social Interaction 54*, 4 (Oct 2021), 397–419.

[75] SONG, S., AND YAMADA, S. Expressing Emotions Through Color, Sound, and Vibration with an Appearance-Constrained Social Robot. *Proceedings of the 2017 ACM/IEEE International Conference on Human-Robot Interaction*, 2–11 (2017). https://doi.org/10.1145/2909824.3020239

[76] SUCHMAN, L. A. *Plans and Situated Actions: The Problem of Human-Machine Communication*. Cambridge University Press, Cambridge, 1987.

[77] TEKIN, B. S. Cheering together: The interactional organization of choral vocalizations. *Language & Communication* (in press).

[78] TENNENT, H., MOORE, D., JUNG, M., AND JU, W. Good vibrations: How consequential sounds affect perception of robotic arms. In *2017 26th IEEE International Symposium on Robot and Human Interactive Communication (RO-MAN)* (Aug 2017), IEEE, pp. 928–935.

[79] TORRE, I., HOLK, S., CARRIGAN, E., LEITE, I., McDONNELL, R., AND HARTE, N. Dimensional perception of a 'smiling McGurk effect'. In *2021 9th International Conference on Affective Computing and Intelligent Interaction (ACII)* (Nara, Japan, Sep 2021), IEEE, pp. 1–8.

[80] TORRE, I., LATUPEIRISSA, A. B., AND McGINN, C. How context shapes the appropriateness of a robot's voice. In *2020 29th IEEE International Conference on Robot and Human Interactive Communication (RO-MAN)* (Naples, Italy, Aug 2020), IEEE, pp. 215–222.

[81] TUNCER, S., LICOPPE, C., LUFF, P., AND HEATH, C. Recipient design in human–robot interaction: The emergent assessment of a robot's competence. *AI & SOCIETY* (Jan 2023).

[82] WEATHERALL, A. "Oh my god that would hurt": Pain cries in feminist self-defence classes. *Language & Communication* (in press).

[83] WEATHERALL, A., KEEVALLIK, L., LA, J., STUBBE, M. AND DOWELL, T. The multimodality and temporality of pain displays. *Language and Communication*, 80: 2021, 56–70.

[84] WIGGINS, S. Talking with your mouth full: Gustatory mmms and the embodiment of pleasure. *Research on Language and Social Interaction 35*, 3 (Jul 2002), 311–336.

[85] WIGGINS, S., AND KEEVALLIK, L. Parental lip-smacks during infant mealtimes: Multimodal features and social functions. *Interactional Linguistics 1*, 2 (2021), 241–272.

[86] YILMAZYILDIZ, S., READ, R., BELPEAME, T., AND VERHELST, W. Review of semantic-free utterances in social human–robot interaction. *International Journal of Human-Computer Interaction 32*, 1 (Jan 2016), 63–85.

4

Talk to Me: Using Speech for Loss-of-Trust Mitigation in Social Robots

Amandus Krantz, Christian Balkenius and Birger Johansson

4.1 Introduction

For all their progress in recent years, the fields of artificial intelligence, robotics, and human–robot interaction (HRI) are still in their youths. While many implementations of robots and autonomous systems have a low rate of catastrophic failures that would cause damage to the system or its operator, they still have a somewhat high rate of smaller, temporary, errors of operation such as faulty navigation. Where to draw the line for what is an acceptable rate for these smaller errors in, for example, consumer robots is still a hotly debated topic,

DOI: 10.1201/9781003320470-4

with arguments often falling somewhere on the spectrum of "Human-level performance" to "Completely unacceptable" (see [15, 17, 19, 22] for discussions on advantages and disadvantages of erroneous behavior in robots).

Regardless of where the acceptable rate of error will fall, these systems are currently being actively sold and deployed in real-world situations throughout society (e.g. "self-driving" cars, virtual assistants, and robotic vacuum cleaners). Temporary errors, then, may have a real impact on how much a user trusts a system and how likely they are to use it again, which could cause frustration and waste resources as the system sits unused. It thus becomes important to understand how the trust relationship between a human user and a robot evolves and changes due to interaction, and how potential damage to this relationship might be averted.

In their influential analysis of factors that impact trust in HRI, Hancock et al. [10] showed that the primary driver of trust in HRI is the robot's performance. Traditionally, robots in HRI have had some clear function they are expected to perform (e.g. robotic arms on factory lines or military drones) and trust in these robots is often evaluated in terms of how likely the robot is believed to successfully perform its function, based on how it has performed in the past. However, as robots are becoming more common in society, they are treated more and more like social agents. For example, it has been pointed out that users have a tendency to draw on their experience of human-human interaction when interacting with robots and thus may not have completely rational expectations for how the robot is to behave [5, 6]. The advantage of designing robots that follow social norms is also argued for by Brinck et al. [4], who write that robots that follow social norms reduce the cognitive load of their users and operators who can fall back on a lifetime of instinct and familiar patterns of interaction, rather than having to learn new methods of interaction for their robot. Additionally, many theories of trust (E.g. [8, 12, 14]) point out that there is a social component to trust, often called affective trust, in addition to the more competency based component. The affective trust is based more on gut feeling and instinct rather than purely on rational reasoning about the system's past performance. If people tend to treat robots as social agents, it is reasonable to assume that this social component of trust plays a part in HRI as well, meaning it might be possible to use error-recovery strategies from human interaction.

In human-human interaction, one of the most effective strategies for disarming a tense situation in which one has made a mistake is to apologize and give a truthful explanation for why the mistake happened. If the apology is accepted and the explanation deemed reasonable, it is possible to avoid damage to the trust relationship or at least mitigating the effects of the damage. Cameron et al. [6] tested how such a strategy might work for a robotic guide which has navigated to the wrong floor. A number of human participants were shown videos of an HRI scenario where the robotic guide either says nothing, apologizes for the mistake, explains why the mistake happened, or both apologizes and explains why the mistake happened. They found that explaining why the mistake happened was beneficial for trust in the robot and its perceived capability, while just apologizing made the robot seem less capable but more likable.

Further highlighting the benefits of social behaviors in HRI, Rossi et al. [21] investigated how robotic social behaviors such as talking or gesturing affects trust and social acceptance of the robot. Participants were told to follow a robot as it guided them through a navigation task, where the robot could either exhibit social behaviors or non-social behaviors. They found that social behaviors had the best effect on trust and acceptance of the robot as a guide. When asked which social behavior they preferred robots had, the participants unanimously voted for speech.

On the other hand, Savery et al. [24] showed that non-linguistic musical prosody was viewed more favorably than speech in terms of trust. They asked participants to interact with a robot that was either using speech or non-linguistic musical notes to signal emotion. The participants were able to correctly identify the emotion that was being conveyed and additionally, on average, rated the non-linguistic robot higher in trustworthiness.

The results from Rossi et al. [21] point toward social behaviors being advantageous for social robots, with speech seemingly being particularly preferred among users, while the results from Savery et al. [24] and Cameron et al. [6] show that verbal communication is beneficial for trust in robots. However, unlike with most neurotypical humans, the ability to speak is not a given for robots or other artificial agents, which more often than not tend to be mute. How trust in robots is affected by possessing the ability to speak, without necessarily acknowledging an error, is to our knowledge still an unexplored area of research.

We designed two online independent measures experiments that together may help shed some light on this open question. Participants were asked to view a video of a robot exhibiting one of two different behaviors, and afterward evaluate their perceptions about the robot. Experiment 1 aimed to investigate how faulty and non-faulty gaze behaviors impact trust in HRI. The results of the experiment were ultimately inconclusive, showing no difference in trust between the two conditions. This was surprising, as faulty behavior has been shown to affect perceptions of robots [22] and negatively impact trust in HRI [23].

We suspected a potential cause of this surprising effect might be that a portion of the experiment involved the robot "speaking" since the participants were told that the robot's purpose was to explain facts. As mentioned, while the presence of speech may not affect trust in human-human interaction, speech may not always be expected in HRI. If that is the case, possessing the ability to speak could conceivably increase the perceived intelligence or capability of the robot, which has been shown to correlate with trust in robots [9]. Communicating in a way that relates to the robot's core functionality (Such as using speech when explaining facts) has also been shown to increase perceived intelligence [25].

To test our theory that speech impacts trust in HRI, we designed a follow-up experiment, Experiment 2, recreating Experiment 1 as closely as possible, but without the speech portion.

4.2 Methodology

4.2.1 Participants

The experiments were done with a total of 227 participants, 110 in Experiment 1 and 117 in Experiment 2. They were recruited from the online participant recruitment platform Prolific[1]. Participants were required to be fluent in English and naive to the purpose of the experiment (i.e. participants from Experiment 1 could not participate in Experiment 2), but otherwise no pre-screening of the participants was done. The mean age of the participants in Experiment 1 was 27 years (SD 7.73; range from 18 to 53), in Experiment 2 it was 39 years (SD 15.84; range from 18 to 75). In Experiment 1, the distribution of genders was 49.1% identifying as male, 50% identifying as female, and 0.9% preferring not to say. For Experiment 2, the distribution of genders was 53.3% identifying as male, 46.7% identifying as female, and 0% preferring not to say.

All participants were required to give their consent to participating in the experiment before beginning. None of the data collected could be used to identify the participants.

4.2.1.1 A Note on Online HRI Experiments

While it may be difficult to convey some subtler elements of HRI using online studies with video-displayed robots [1], it is still a commonly used approach and gives access to a much larger and more diverse group of potential experiment participants compared to live-HRI experiments. At the very least, the results from such studies can be used as guidance for experiments that may be worth replicating in live HRI studies [6].

4.2.2 Robot

The experiments were done using the humanoid robot platform Epi (see Figure 4.1), developed at Lund University [11]. The robot's head is capable of playing pre-recorded smooth and fluid movements with 2 degrees of freedom (yaw and pitch), and has a speaker built into its "mouth". The eyes of the robot also have 1 degree of freedom (yaw), adjustable pupil size, and adjustable intensity of its illuminated pupils. Using the control system Ikaros[2], also developed at Lund

---

[1]https://www.prolific.co
[2]https://github.com/ikaros-project/ikaros/

FIGURE 4.1
Epi, the humanoid robotics platform used in the experiment.

University, it is possible to use models of the human brain and cognitive systems for completely autonomous behavior, however only pre-recorded movements of the head and the speaker were used for the experiments.

4.2.3 Experiment Set-Up

Both experiments had a between-group design, where each participant was assigned to one of two conditions. Each condition had an associated video that the participants were told to base their evaluation of trust on. The videos showed the robot exhibiting either faulty or non-faulty gaze behaviors.[3]

In the non-faulty gaze behavior (see Figure 4.2), the robot starts looking into the camera. When an object is presented to the robot, the head moves until it appears to look at the object, holds the position for roughly 1 second, and moves back to its starting position, looking into the camera.

In the faulty gaze behavior (see Figure 4.3), the robot again starts looking into the camera. When the object is presented, the head moves in a random direction, rather than in the direction of the object. We chose this behavior over having the robot remain static, as it was important that the robot appeared to have the same capabilities in all conditions. All other behavior in the conditions with faulty gaze-behavior is identical to the non-faulty behaviors.

In Experiment 1, once the gaze behavior had been displayed, the robot would play a pre-recorded audio file of a computerized voice presenting a number of facts about the object that had been displayed. The speech makes no reference to whether or not the robot displayed a faulty or non-faulty behavior.

---

[3]Videos available at: http://www.soundandrobotic.com/ch4

FIGURE 4.2
Gaze positions of non-faulty gaze behavior.

FIGURE 4.3
Gaze positions of faulty gaze behaviors.

Care was taken to ensure that the robot's speech never overlapped with the movement of the head. All behaviors exhibited by the robot were pre-recorded and no autonomous behaviors were implemented.

4.2.4 Experiment Scenario

To avoid any observer effects, it was necessary to give the participants a scenario for which to judge the trustworthiness of the robot. As the purpose of Experiment 1 was to examine the effect of different gaze behaviors, the scenario was that the robot was being developed for a classroom setting, that its purpose was to answer children's questions, and that it was different voices we were comparing.

Experiment 2 had no speech component, so the participants were instead told that the robot was reporting which objects it was seeing to an unseen operator.

All participants were debriefed and told the real purpose of the experiment after completion.

4.2.5 Measures

How much we trust an agent changes over time as we progress through our interactions, both with the agent itself and other agents [3, 9]. Related to this notion, it has also been argued that knowing one's level of trust in an agent at any one time is not sufficient to draw any meaningful conclusions (e.g. [16]).

Rather than looking at trust at a single point in time, one should look at trends of trust, using measurements at multiple points in time, and draw conclusions based on how it has changed as a result of the interaction or stimulus. We thus measured the amount of trust the participants felt toward the robot twice; before and after the interaction. For the pre-interaction measurement, the participants evaluated the trust based on a static image of the robot (see Figure 4.1). For the post-interaction measurement, they based their evaluation on one of the previously described videos. The trust relation was measured using the 14-item sub-scale of the trust perception scale-HRI (TPS-HRI) questionnaire, developed by Schaefer [26]. In the questionnaire, the participants are asked to estimate how frequently they believe a robot will exhibit a certain behavior or characteristic, such as being dependable or requiring maintenance. The scale outputs a value between 0 and 100, where 0 is complete lack of trust and 100 is complete trust.

To measure the participants' impressions of the robot after the interaction we used the Godspeed questionnaire [2] which has the participants rate the robot on 5-point scales where the extremes of the scale have labels with semantically opposite meanings (E.g. "Unfriendly" and "Friendly"). Specifically, we used the Perceived Intelligence, Likeability, and Animacy sub-scales of the Godspeed questionnaire.

To control for any negative feelings the participants may have harbored toward robots before the experiment, the Negative Attitudes Toward Robots Scale (NARS) was used [28]. NARS gives an overview of both general negative feelings toward robots, and three sub-scales for negative feelings toward interaction with robots (S1), social influence of robots (S2), and emotions in robots (S3).

Since robot experience has been shown to affect feelings of trust toward robots [18], we also asked the participants how often they interact with robots and autonomous systems on a 5-point scale, where 1 was daily interaction and 5 was rare or no interaction.

4.3 Results

4.3.1 Trust

In Experiment 1, no significant difference was found between the faulty and non-faulty conditions (Mann-Whitney U, $p = 0.179$). However, once the speech of the robot was removed in Experiment 2, a significant difference was found (Mann-Whitney U, $p < 0.01$). Looking at the plot of differences in trust in Figure 4.4, this difference seems to be due to the faulty behavior reducing the trust, rather than the non-faulty behavior increasing the trust. No significant difference was found between the non-faulty conditions in Experiment 1 and Experiment 2 (Mann-Whitney U, $p = 0.230$).

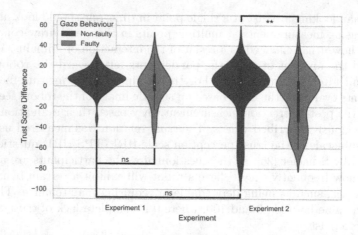

FIGURE 4.4
Comparison of differences in trust before and after interaction. A significant difference exists between the faulty and non-faulty conditions in Experiment 2.

4.3.2   Perceived Characteristics

Cronbach's Alpha with a confidence interval of 0.95 for all Godspeed questionnaires were in the 0.7–0.9 interval, indicating acceptable to good internal consistency. Both conditions from Experiment 1 rank higher than Experiment 2 in all measured characteristics.

4.3.3   Participant-Centric Metrics

Regarding participant-centric characteristics that may affect the trust in the robot, we controlled for mean age, gender distribution, pre-existing negative attitudes toward robots, and participant experience with robots and other autonomous systems.

4.3.3.1   Negative Attitudes Toward Robots

Figure 4.5 shows the Kernel density estimate of NARS and its three sub-scales. The full NARS scale and the two sub-scales S2 and S3 are roughly normally distributed, indicating that the participants had overall neutral feelings toward robots before starting the experiment. The sub-scale S1 skews slightly lower, indicating that the participants had slightly negative feelings toward social situations and interactions with robots.

No significant differences can be seen in negative attitudes between the two experiments.

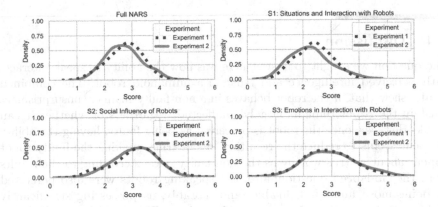

FIGURE 4.5
Kernel density estimate of NARS and its three sub-scales. A lower score indicates a more negative attitude.

4.3.3.2 Participant Age

Of the participant-centric metrics we controlled for, only age differed significantly between the two experiments, with the mean age being 12 years higher in Experiment 2. While age has been shown to have an impact on attitudes toward technology, with older people having a more negative attitude [7], the negligible difference in the distributions of the NARS scores (Figure 4.5) indicate that the difference in mean age between the experiments is likely not large enough to affect the results.

4.3.3.3 Participants' Experience with Robots

The participants in either experiment interact with robots, AI, and autonomous systems with roughly equal frequency (see Table 4.1), with the majority interacting with such systems daily.

TABLE 4.1
Proportions of how frequently the participants in either experiment interact with robots, AI, and other autonomous systems.

Frequency	Experiment 1	Experiment 2
Daily	40%	41%
Once a week	30%	22.2%
Once a month	13.6%	14.5%
Once a year	8.2%	11.1%
Never	8.2%	11.1%

4.4 Discussion

The question we sought to answer here was how the ability to speak interacts with perceived intelligence and trust in a humanoid robot. The combined results show that, if the robot behaves in a non-faulty manner, unsurprisingly, trust in the robot remains largely unaffected, regardless of whether it can speak. However, once the robot is perceived as being faulty, having the ability to speak seems to reduce the resulting loss of trust, making the faulty robot appear about as trustworthy as the non-faulty ones. According to the results from the Godspeed questionnaire, the speaking robots were also perceived as being more animated, likable, and, notably, as possessing significantly higher intelligence than the non-speaking robots. This could be an indication that, for humanoid robots, the ability to speak is perceived as a sign of high intelligence. Alternatively, the speaking robot may appear to be more sophisticated or be more capable than the non-speaking robot. Both high perceived intelligence and high capability are believed to have some correlation with a higher trust [9, 20, 27].

It is worth noting at this point that possessing the ability to speak is not necessarily indicative of "actual" intelligence. There are plenty of animals, such as corvids and primates, that are capable of tool-making and other intelligent behavior but are unable to speak. These findings therefore should only be applied to humanoid robots and even then should be interpreted carefully. While a speaking robot may appear more capable or intelligent than one that is mute, it is important to remember that the only aspect that is necessarily different between them is that the speaking robot is equipped with a speaker. They may still contain the same, potentially faulty, sensors, circuitry, and algorithms. Perceived intelligence in artificial agents can be deceptive and should thus not be confused with "actual" intelligence.

Nevertheless, these results highlight the benefits of implementing speech in a robot. Not only are robots with speech more likable according to the Godspeed questionnaire, perhaps because speech reduces the cognitive load by allowing users to take advantage of social norms as argued by Brinck et al. [4], but they are also trusted more in the event of an error in its operation. Why this effect exists is still unknown, but it is conceivable that speech makes the robot appear to be more human and thus more like a social actor, causing the user to be more lenient when evaluating its performance. The increase in the Animacy score of the Godspeed questionnaire for the speaking robots is in favor for this theory, as is the fact that people seemingly already have a tendency to treat robots as social actors (E.g. [6]).

In conclusion, we have presented results from two experiments in HRI that together suggest that a humanoid robot with the ability to speak may not suffer the same loss of trust when displaying faulty behavior as a robot without the ability to speak. We theorize that this effect is due to speech increasing a humanoid robot's perceived intelligence, which has been shown to correlate with trust in HRI [9]. Further research along these lines may help explain existing studies in HRI (see e.g. [6]) that indicate that a robot providing a verbal explanation for its errors is beneficial for user attitudes.

4.4.1 Limitations

There are some limitations that should be kept in mind when using these results. First, as mentioned, the experiments were done online using pre-recorded videos of the robot rather than direct human–robot interaction. The large amount of available participants should safeguard against false positives, however a live-HRI study may nevertheless yield different results.

Second, the experiment scenario was different between the two experiments, with participants in Experiment 1 being told that the voice was the focus of the study. This could potentially have caused participants to ignore the gaze behavior of the robot and focus solely on its voice, which was the same across the conditions.

Finally, the content of the robot's speech was not controlled for. It is conceivable that some part of the speech is signalling to some participants that the robot is highly capable or intelligent, causing the trust to increase.

4.4.2 Future Work

Several future research directions are available based on these results. The first step would be to address some of the identified limitations with the original study. Redoing the experiment with the same scenario for all conditions and/or controlling for the contents of the speech would be useful to further strengthen the hypothesis.

It would also be interesting to see if the same effect is present in a less controlled live-HRI scenario. It is conceivable that physically interacting with the robot would allow a participant to spot the "cracks" in the robot's behavior, which could negatively affect the perception of it. Alternatively, physical interaction could be more powerful than virtual interaction, since interacting with a humanoid robot in real life is a novel experience for many people. Related to this, it could be interesting to see whether the effect holds with more common, non-humanoid, robots as well.

Going along the line of impacting trust through perceived intelligence, it would be interesting to see what other characteristics of a robot can be used to increase or decrease perceived intelligence. For example, while the results of the initial gaze behavior study proved inconclusive due to the effects of the speech, investigating how different gaze behaviors affect perceptions of a humanoid social robot. An interesting question that could be answered is whether a robot that imitates human behavior appears more intelligent and capable (and thus more trustworthy), or if it is perceived as something "other" imitating humans and thus becomes alienating.

One could also investigate whether non-linguistic sounds such as grunts and sighs impact the perception and trust in a humanoid robot. Such sounds have been shown to be used for communication of information in animals and humans, and affect perceptions of characteristics such as size and aggressiveness (see e.g. [13]). Testing this in robots would be interesting since non-linguistic sounds can be used to signal the similar perceptive capabilities (e.g. object detection/identification) as speech, but without the linguistic capabilities necessary for speech. This could be contrasted with speech to see if it is truly speech that impacts trust, or if it is the capabilities that are implied by the speech.

Acknowledgments

This work was partially supported by the Wallenberg AI, Autonomous Systems and Software Program – Humanities and Society (WASP-HS) funded by the Marianne and Marcus Wallenberg Foundation and the Marcus and Amalia Wallenberg Foundation.

Bibliography

[1] BAINBRIDGE, W. A., HART, J. W., KIM, E. S., AND SCASSELLATI, B. The benefits of interactions with physically present robots over video-displayed agents. *International Journal of Social Robotics 3*, 1 (Jan. 2011), 41–52.

[2] BARTNECK, C., KULIĆ, D., CROFT, E., AND ZOGHBI, S. Measurement Instruments for the Anthropomorphism, Animacy, Likeability, Perceived Intelligence, and Perceived Safety of Robots. *International Journal of Social Robotics 1*, 1 (Jan. 2009), 71–81.

[3] BLOMQVIST, K. The many faces of trust. *Scandinavian Journal of Management 13*, 3 (Sept. 1997), 271–286.

[4] BRINCK, I., BALKENIUS, C., AND JOHANSSON, B. Making place for social norms in the design of human-robot interaction. In *What Social Robots Can and Should Do*, vol. 290 of *Frontiers in Artificial Intelligence and Applications*. IOS Press, Oct. 2016, pp. 303–312.

[5] BRINCK, I., HECO, L., SIKSTRÖM, K., WANDSLEB, V., JOHANSSON, B., AND BALKENIUS, C. Humans perform social movements in response to social robot movements: Motor intention in human-robot interaction.

In *2020 Joint IEEE 10th International Conference on Development and Learning and Epigenetic Robotics (ICDL-EpiRob)* (Oct. 2020), p. 6.

[6] CAMERON, D., DE SAILLE, S., COLLINS, E. C., AITKEN, J. M., CHEUNG, H., CHUA, A., LOH, E. J., AND LAW, J. The effect of social-cognitive recovery strategies on likability, capability and trust in social robots. *Computers in Human Behavior 114* (Jan. 2021).

[7] DE GRAAF, M. M. A., AND BEN ALLOUCH, S. Exploring influencing variables for the acceptance of social robots. *Robotics and Autonomous Systems 61*, 12 (Dec. 2013), 1476–1486.

[8] FISKE, S. T., CUDDY, A. J. C., AND GLICK, P. Universal dimensions of social cognition: Warmth and competence. *Trends in Cognitive Sciences 11*, 2 (Feb. 2007), 77–83.

[9] GLIKSON, E., AND WOOLLEY, A. W. Human trust in artificial intelligence: Review of empirical research. *Academy of Management Annals 14*, 2 (Mar. 2020), 627–660.

[10] HANCOCK, P. A., BILLINGS, D. R., SCHAEFER, K. E., CHEN, J. Y. C., DE VISSER, E. J., AND PARASURAMAN, R. A meta-analysis of factors affecting trust in human-robot interaction. *Human Factors 53*, 5 (Oct. 2011), 517–527.

[11] JOHANSSON, B., TJØSTHEIM, T. A., AND BALKENIUS, C. Epi: An open humanoid platform for developmental robotics. *International Journal of Advanced Robotic Systems 17*, 2 (Mar. 2020), 11.

[12] MARSH, S. P. *Formalising Trust as a Computational Concept.* PhD thesis, University of Sterling, Apr. 1994.

[13] MASSENET, M., ANIKIN, A., PISANSKI, K., REYNAUD, K., MATHEVON, N., AND REBY, D. Nonlinear vocal phenomena affect human perceptions of distress, size and dominance in puppy whines. *Proceedings of the Royal Society B: Biological Sciences 289*, 1973 (Apr. 2022), 20220429.

[14] MCALLISTER, D. J. Affect- and cognition-based trust as foundations for interpersonal cooperation in organizations. *Academy of Management Journal 38*, 1 (Feb. 1995), 24–59.

[15] MIRNIG, N., STOLLNBERGER, G., MIKSCH, M., STADLER, S., GIULIANI, M., AND TSCHELIGI, M. To Err Is Robot: How Humans Assess and Act toward an Erroneous Social Robot. *Frontiers in Robotics and AI 4* (2017).

[16] O'NEILL, O. Questioning Trust. In *The Routledge Handbook of Trust and Philosophy.* Routledge, 2020.

[17] RAGNI, M., RUDENKO, A., KUHNERT, B., AND ARRAS, K. O. Errare humanum est: Erroneous robots in human-robot interaction. In *2016 25th IEEE International Symposium on Robot and Human Interactive Communication (RO-MAN)* (Aug. 2016), pp. 501–506.

[18] ROGERS, K., BRYANT, D., AND HOWARD, A. Robot gendering: Influences on trust, occupational competency, and preference of robot over human. In *Extended Abstracts of the 2020 CHI Conference on Human Factors in Computing Systems* (New York, NY, USA, Apr. 2020), CHI EA '20, Association for Computing Machinery.

[19] ROSSI, A., DAUTENHAHN, K., KOAY, K. L., AND WALTERS, M. L. How the Timing and Magnitude of Robot Errors Influence Peoples' Trust of Robots in an Emergency Scenario. In *Social Robotics* (Cham, 2017), A. Kheddar, E. Yoshida, S. S. Ge, K. Suzuki, J.-J. Cabibihan, F. Eyssel, and H. He, Eds., Lecture Notes in Computer Science, Springer International Publishing, pp. 42–52.

[20] ROSSI, A., DAUTENHAHN, K., LEE KOAY, K., AND WALTERS, M. L. How social robots influence people's trust in critical situations. In *2020 29th IEEE International Conference on Robot and Human Interactive Communication (RO-MAN)* (Aug. 2020), pp. 1020–1025.

[21] ROSSI, A., GARCIA, F., MAYA, A. C., DAUTENHAHN, K., KOAY, K. L., WALTERS, M. L., AND PANDEY, A. K. Investigating the effects of social interactive behaviours of a robot on people's trust during a navigation task. In *Towards Autonomous Robotic Systems* (Cham, 2019), K. Althoefer, J. Konstantinova, and K. Zhang, Eds., Lecture Notes in Computer Science, Springer International Publishing, pp. 349–361.

[22] SALEM, M., EYSSEL, F., ROHLFING, K., KOPP, S., AND JOUBLIN, F. To Err is Human(-like): Effects of Robot Gesture on Perceived Anthropomorphism and Likability. *International Journal of Social Robotics 5*, 3 (Aug. 2013), 313–323.

[23] SALEM, M., LAKATOS, G., AMIRABDOLLAHIAN, F., AND DAUTENHAHN, K. Would you trust a (faulty) robot? Effects of error, task type and personality on human-robot cooperation and trust. In *10th ACM/IEEE International Conference on Human-Robot Interaction (HRI)* (Mar. 2015).

[24] SAVERY, R., ROSE, R., AND WEINBERG, G. Establishing Human-Robot Trust through Music-Driven Robotic Emotion Prosody and Gesture. In *2019 28th IEEE International Conference on Robot and Human Interactive Communication (RO-MAN)* (Oct. 2019), pp. 1–7.

[25] SAVERY, R., ZAHRAY, L., AND WEINBERG, G. Emotional musical prosody for the enhancement of trust: Audio design for robotic arm communication. *Paladyn, Journal of Behavioral Robotics 12*, 1 (Jan. 2021), 454–467.

[26] SCHAEFER, K. E. Measuring trust in human robot interactions: Development of the "trust perception scale-HRI". In *Robust Intelligence and Trust in Autonomous Systems*, R. Mittu, D. Sofge, A. Wagner, and W. Lawless, Eds. Springer US, Boston, MA, 2016, pp. 191–218.

[27] SCHAEFER, K. E., CHEN, J. Y. C., SZALMA, J. L., AND HANCOCK, P. A. A meta-analysis of factors influencing the development of trust in automation: Implications for understanding autonomy in future systems. *Human Factors 58*, 3 (May 2016), 377–400.

[28] SYRDAL, D. S., DAUTENHAHN, K., KOAY, K. L., AND WALTERS, M. L. The negative attitudes towards robots scale and reactions to robot behaviour in a live human-robot interaction study. *Adaptive and Emergent Behaviour and Complex Systems* (Apr. 2009).

5

Grounding Spoken Language

Cynthia Matuszek

5.1 Introduction

When we imagine interacting with robots in human environments, we imagine speech and language as a core modality. Human–robot interaction is a key area of study for using robots in human-centric environments such as schools, homes, and care facilities; in order to usefully engage with people in such settings, robots will need to be able to gracefully interact with the people around them, and natural language represents an intuitive, comfortable mechanism for such interactions. This chapter will discuss the role of natural language processing (NLP) in modern robotics and human–robot interaction, and specifically, how grounded language learning is a critical modality for robots to understand the world. While it is crucial to study language from an auditory perspective, understanding the underlying semantics – the linguistic meaning and intent of those speech acts – is necessary for smooth interaction with robots in human-centric environments.

Robots can use language as a mechanism of interaction but also as a tool for learning about the world around them. They can follow instructions [32, 79], respond to interactions using language (for example, by acknowledging commands or seeking clarification) [53], and use language to repair or reshape interactions [18, 28]. Frequently, these interactions are scripted: the human has a fixed set of possible commands, which they may be provided, or they

DOI: 10.1201/9781003320470-5

may need to learn from experimentation. However, for truly flexible agents, it is preferable to learn from interactions what words and instructions mean. Grounded language learning is specifically the process of learning language from interactions with the world, and learning about the world from language used to describe it [87]. The core idea is that treating language learning as a problem grounded in the physical world via robotic agents will improve the effectiveness of both robotic interaction and natural language understanding [66].

Speech is already a key mechanism for embodied devices such as home assistants, phones, and even game consoles. While automated speech recognition (ASR) has become a relatively mainstream technology and continues to improve, there are still substantial difficulties with using those technologies for robotics [62], including environmental noise, latency, and a lack of datasets and models specific to real-world environment interaction. Some of the complexities of managing speech in robotics have been discussed in other chapters. However, even when the difficulties of accessing speech are disregarded, there are substantial NLP-based challenges with using language as it pertains to robots.

In this chapter, language is considered as a mechanism of referencing concepts in the physical world, and how natural language processing dovetails with work on using speech and making sense of language in a robot's environment is explored. Grounded language acquisition as a research area is examined, and a case study is presented of learning grounded language from speech by examining the question from three distinct-but-related angles: the need for complex perceptual data (and a resulting corpus), the need to learn to interact directly from speech without using a textual intermediary, and the problem of learning grounded language from richly multimodal data.

5.2 Grounded Language Acquisition

Language is comprised of symbols, and understanding those symbols and their underlying meaning – the *symbol grounding problem* – is a core aspect of artificial intelligence as a field [14, 36]. The fundamental idea underlying grounded (or embodied) language is that language does not exist in a vacuum: it derives from, and refers to, objects and actions in the physical environment in which robots operate [87]. Accordingly, this language can be learned by connecting co-occurrences of language with physical percepts perceived by a robot. While a substantial body of this work is related to what is frequently referred to as "Vision-and-Language", in practice, this terminology often refers to tasks such as Visual Question Answering [3] and Visual Commonsense Reasoning [96], where no literal physical agent is necessarily involved. This is a richly studied area, with connections to language modeling, automatic speech recognition, human–robot interaction, learning in simulation, and vision-and-language navigation and manipulation, among others. This section provides a necessarily partial overview of recent work.

One way to examine the current state of the art in grounded language learning is to consider the related and overlapping sub-tasks which people use as testbeds. Grounded language can refer to language about a wide range of aspects of the environment: objects and their attributes [16,75], actions and tasks [20,41,57] – notably including vision-and-language navigation, a special case that has been studied since very early in the history of embodied language understanding (surveys: [35,69,93]) – or referring expression groundings [61,89], to name a few.

One significant body of physically situated language work revolves around the use of large pretrained vision-and-language models (VLMs). Contrastive language image pretraining (CLIP) encoders [72] have been successfully used for embodied navigation and vision-and-language navigation-related tasks [51,80] and tabletop manipulation-based instruction following [81]. Other grounded language learning-adjacent works depend on such large language models (LLMs) as BERT [30], including ViLBERT [58] and Embodied BERT [85], which focuses on object-centric navigation in the ALFRED benchmark [82]. In particular, LLMs are frequently used to help derive plans for following natural language instructions for completing tasks [45]. SayCan [1] uses a combination of LLMs to extract possible useful actions in the world given a goal, and uses affordances of a physical robot to determine, of those actions, which are feasible to perform. In [43], an LLM is used to generate steps toward a goal, incorporating perceptual feedback about the environment in order to improve long-horizon planning.

Other approaches focus on the use of smaller but more task-specific knowledge bases. In [16], few-shot learning of object groundings is accomplished by adding to a database of examples of simulated objects overlaid on real environments; [68] learn to interpret task instructions by probabilistically connecting instructions to background knowledge found in part in relational and taxonomic databases. In the class of interactive language grounding in robotics, a physical agent can follow instructions interactively [59] or can learn to improve its performance on a task based on communication with a person (e.g. [20]). Other work has focused on collections of instructions that pertain to a specific environment [4,63] and do not incorporate non-perceptual background knowledge.

Despite this preponderance of language-based approaches, the space of robots learning about the world via actual speech is comparatively nascent. Despite early work in learning the grounded semantics of spoken utterances [76, 95], most recent work on language grounding has focused on textual content, generally obtained via crowdsourcing [4,70] or from web-scale data such as image captions or tags [54]. In some cases, text is transcribed from spoken language, either using automatic speech recognition (ASR) [9] or manually [90]. However, when interacting with embodied agents such as robots, speech is a more appropriate modality than typed text. This leads to the core technical aspect of this chapter: a discussion of grounding language via speech.

5.3 Learning Grounded Language about Objects

This section describes a case study of attempting to understand unconstrained spoken language about objects. While object recognition based on vision is an extremely active area of research in both 2D and 3D contexts [7,71], using grounded language with robots in human environments introduces additional problems. First, although large pre-trained visual models contain extensive coverage of some object classes, they suffer from long tail problems when encountering rare objects in the environment or unusual exemplars of common objects [84]. Second, language about grounded concepts frequently evolves over the course of an interaction: people create new terminology and repurpose terminology on the fly during interaction [17,44], meaning that a robot may need to learn new and remapped terms in real time as interactions unfold. Finally, existing large vision-and-language models tend to be Western-centric [5,78], potentially limiting the usefulness of deployed systems in other cultural settings.

Learning to understand unconstrained language about objects may entail learning class names, but may also require learning to understand a variety of perceptually meaningful descriptors – for example, people may choose to describe objects based on color and shape, or on the material they are made from, e.g. ceramic or aluminum [75]. It is therefore necessary to learn the semantics of a variety of grounded terms above and beyond simple object names. This necessitates addressing a number of subproblems, including not only the collection of an appropriate dataset in order to benchmark the success of our efforts but also the development of mechanisms for learning from speech without a textual intermediary and learning from rich multimodal sensor data.

5.3.1 A Dataset for Multimodal Language Understanding

While there has been extensive work on understanding language using large-scale pretrained natural language models such as BERT [30], in practice, grounding requires that such language-focused models be augmented with perceptual data from robotic sensors [12], typically in the form of visual data (survey: [26]). In our use case of allowing people to use speech directly with robots, there are two significant difficulties with this approach: First, robots tend to have perceptual capabilities beyond vision, such as depth perception (potentially along with more exotic sensory capabilities such as thermal sensing); and second, HRI contexts provide a comparatively small amount of spoken training data, as distinct from the very large amounts of textual data that are available (for example, in the form of image captions).

Nonetheless, when interacting with robots and other embodied agents, it is natural for people to want to speak to them, and many deployed systems use speech [88]. Spoken language is critical for interactions in physical contexts, despite the inherent difficulties: spoken sentences tend to be less well framed than written text, with more disfluencies and grammatical flaws [74]. However, despite these differences, text is commonly used for grounded language learning, presumably due to its wide availability and comparative ease of computational processing (with a few exceptions, *inter alia* [10, 34, 48]). In this section, the development of the Grounded Language Dataset (GoLD)[1] is described. GoLD is a dataset of object descriptions for speech-based grounded language learning [49]. This dataset contains visual- and depth-based images of objects aligned with spoken and written descriptions of those objects collected from Amazon Mechanical Turk.

GoLD contains perceptual and linguistic data in five high-level categories: *food, household, medical, office*, and *tools*. In these groups, 47 object classes contain 207 individual object instances. These categories were selected to represent objects that might be found in common human-centric environments such as homes and offices, and contain multiple examples of objects that are typical for such contexts, such as apples, dishes, analgesics, staplers, and pliers. For each of these objects, approximately 450 vision+depth images were collected as the object was rotated on a turntable in front of a Kinect RGB+D camera. For each object, four images from distinct angles were chosen to represent object 'keyframes.' Using rotational visual data helps to avoid a known problem with many image datasets, namely their tendency to show pictures of objects from a limited set of angles [8, 89, 91], whereas a robot in a home might see an object from any angle.

Three distinct types of language data were collected/created for each object in this dataset. For each keyframe, approximately twenty spoken descriptions were collected, leading to a dataset of 16,500 spoken descriptions. For comparison purposes, transcriptions of these descriptions were generated using Google Speech to Text. Manual evaluation of a subset of these transcriptions suggests that approximately 80% were good enough for grounded language understanding. Another 16,500 textual descriptions were separately collected; these were not associated with the spoken descriptions or provided by the same set of Mechanical Turk workers (although some workers did work on both problems, they were not given aligned examples to label). The types of data present in GoLD are shown in Figure 5.1.

GoLD fits into a landscape of datasets used for grounded or embodied language learning, and extends that landscape by providing a very rich dataset, in which each object is associated with a large number of images, depth images, and spoken descriptions. While there are many spoken-language datasets in existence, they are frequently handcrafted for the specific task that the research seeks to accomplish, often leading to narrower applications, for example, question answering [47]. Meanwhile, recent large-scale datasets that include speech typically incorporate synthetically generated speech [27, 29, 38, 39], use generated spoken descriptions from the text captions [56, 94], or ask crowdsourced workers to read captions [37, 42]. This may remove agrammatical constructs,

---

[1]https://github.com/iral-lab/gold

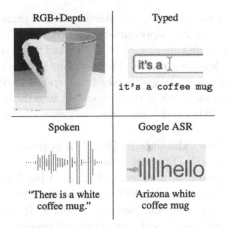

FIGURE 5.1
The GoLD dataset has RGB and depth point cloud images of 207 objects in 47 categories. It includes 8250 text and 4045 speech descriptions; all spoken descriptions include automatic transcriptions.

disfluencies, and speech repair, effectively gating the complexities of speech through written language. Other larger datasets exist that contain (real or virtual) scenarios in which embodied vision+depth sensor data can be extracted, such as the ALFRED benchmark [83]; however, aligned, unconstrained spoken language is rarely included.

5.3.2 Learning Multimodal Groundings from Spoken Language

Grounded language learning offers a way for robots to learn about dynamic environments directly from individual end users. However, as described above, much of the current work in this area uses text as the language input rather than speech. Grounded language learning that does incorporate speech frequently relies on automatic speech recognition (ASR). Off-the-shelf ASR systems often have substantial drawbacks when used in robotic settings [62]: they introduce latency to the system, work poorly in noisy environments (including the noise produced by the robot itself), and cannot use perceptual information about the environment to improve on recognition. In addition, current ASR systems work inconsistently across demographics such as gender, race, and native language [2, 13, 33, 86, 92], which can lead to failures of inclusive design.

Despite this gap, in many cases, the learning methods applied to acquiring grounded language are relatively agnostic to the type of input, relying on broad approaches such as manifold alignment [67]. As a result, text can be replaced with appropriately featurized speech as an input to the joint language learning model [50]. A number of existing pre-trained speech representation models are available to encode speech into appropriate featurizations, making it possible to encode the spoken descriptions in GoLD and treat those encodings as input to a combined language learning model. This section describes a learning method that performs object grounding directly from speech, without relying on a spoken intermediary; that work is then extended to a model that is capable of handling complex, multimodal input, even in cases where some sensory data becomes unavailable.

In order to learn from the data in GoLD, it is first necessary to featurize the disparate data types. Different featurizations are used for each of the vision, depth, and spoken language modalities. Visual features (image and depth) are extracted using ResNet152 pre-trained on ImageNet [40], and the last fully connected layer is removed to obtain 2048-dimensional features. RGB images are processed directly, while depth images are colorized before processing [75]. For the simple manifold alignment-based learning case, these vectors are concatenated to make a single visual vector. Speech is featurized using wav2vec 2.0 [6], in which audio is encoded via a convolutional neural network, then masked spans of the resulting speech representations in the latent space are input to a transformer network. Wav2vec 2.0 uses a two-stage training process: the first stage of training focuses on learning local patterns in the audio, such as phonemes, while the second stage focuses on learning patterns such as sentence structure. In our case, a pre-trained model was used, which was fine-tuned for speech recognition.

Manifold alignment: In the simple case, language groundings are learned from these encodings using a procedure known as manifold alignment, where featurized language and sensor data are treated as projections of some under-lying manifold in a shared, non-observable latent space, and the goal is to find the functions that approximate the manifold. In this approach, groundings are learned by attempting to capture a manifold between speech and perceptual inputs. The goal is to find functions that make projections from both domains 'closer,' in feature space, to other projections of the same class.

For textual language, the approach used is triplet loss: a geometric approach that has shown success in learning metric embeddings [31]. Triplet loss is a form of contrastive learning, which learns representations of data by comparing points to representationally similar or dissimilar points. The goal is to learn an embedding which 'pulls' similar points closer together in the feature space and 'pushes' dissimilar points further apart. Our manifold alignment learning uses triplets of the form (a, p, n), where a is an 'anchor' point, p is a positive instance of the same class as the anchor (for example, tomato), and n is a negative instance from a different class (for example, plate). The goal is to maximize for each triplet the distance between a and n while minimizing the distance between a and p. This is achieved through the loss function $\mathcal{L} = \max(0, d(f(a) - f(p)) - d(f(a) - f(n)) + \alpha)$, where f is the relevant embedding function for the input domain, d is a distance metric, and α is a margin imposed between positive and negative instances. Given our multimodal

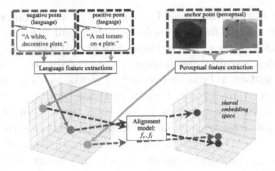

FIGURE 5.2

Triplet loss tries to minimize the distance between an anchor point, here the perceptual inputs of an object (right boxed object); an associated sample in another modality, here the language describing that object (middle boxed object); while maximizing the distance between the anchor point and a negative point in the other modality, such as the description of an unrelated thing (boxed). The learned embedding then should embed language "near" the things it describes and "far" from other things.

data, the embedding function f is the encoder that projects instances of a given modality into the shared manifold (see Figure 5.2 for an example).

This approach to learning language groundings directly from speech has proven to be quite successful [50] on a downstream object retrieval task, in which the system is presented with a query where a robot must choose an object to retrieve from a set of alternatives (where the queries are spoken natural language inputs such as "The black and gold University mug"). This learned model outperforms the system on the same task when it is performing retrieval based on transcriptions of spoken text, despite the fact that the model was initially designed for textual input [67].

Multimodal learning: As mentioned, although vision is a key way of perceiving the environment, it is not the only sensor available to robot platforms. The approach described above handles depth by concatenating the depth image to a scaled RGB image, and cannot incorporate additional modalities. Furthermore, it is still able to handle only a single communication modality (speech or text). However, people may wish to communicate in a multimodal fashion, in which case multiple interactive modalities should be taken into account simultaneously: while speech is an obvious mechanism for embodied interaction, there are cases when it is preferable to convey complex commands from a computer or via an interface on a phone. This section discusses an extension of the contrastive learning described above that begins to address these requirements.

In addition to handling multiple sensory and communication modalities, such a learning mechanism should be robust to missing modalities. A robot's percepts may be only partially available when handling a learning problem – sensors may fail, may be occluded at key moments, or may be missing from certain platforms; people may communicate via speech, gesture, text, or some subset of those. All of these desiderata taken together suggest the need for a broader learning mechanism that is capable of handling arbitrary numbers of modalities, robust in the face of modality dropouts, and able to learn from relatively small-scale, human-provided inputs.

Broadly speaking, our approach is to extend the idea of geometric loss by combining it with a cross-entropy based supervised contrastive loss function [52], in which labels are used to allow points belonging to the same class to be pulled to the same area in embedding space, while points belonging to other classes are pushed apart. It is a general version of multiple contrastive loss functions including triplet loss, as well as general contrastive loss [24]. A distance-based loss function is defined that can be used for an arbitrary number of modalities. Standard triplet loss, as described above, can be applied to only two modalities, and is not robust to sensor ablation.

To address this issue, pairwise distance optimization is used for all data points. During training, two different instances are sampled and their corresponding representations from all modalities into two sets – one positive set (referring to a specific object) and one negative set (referring to some randomly-selected different object). In our setting, every item in the positive set becomes an anchor once, and the distance is minimized between that item and other items in the positive set, while minimizing the distance between that item and all items in the negative set. This can be seen as an one-to-many relationship instead of the one-to-two relationship in the triplet loss formulation.

This approach is tested over the four main modalities of the GoLD dataset: RGB, depth, speech, and text. Encoding mechanisms appropriate to each modality are selected. BERT [30] embeddings are used to featurize textual input, and wav2vec2 [6] to extract audio embeddings from speech. To process images, ResNet152 [40] is used for both RGB and depth images, producing a 2048-dimensional embedding vector. The objective is then to first minimize the distance between each pair of positive points from heterogeneous modalities, and second, maximize the distance between each pair of positive and negative points from all modalities. This combined loss function results in a learning mechanism that outperforms supervised contrastive loss in both the speed of convergence during training, and number of data points required to build a model capable of performing a downstream object retrieval task.

5.4 Open Challenges

Although learning to understand and learn from grounded language is an active and successful field of research, a number of challenges remain to be addressed. Discussing open questions in a fast-moving field such as language grounding carries an element of risk. There has recently been a surge of rapid development in applications of NL technology and robotics, enabled by new technologies and very large-scale data sources, that would have been difficult to predict a small number of years ago. Nonetheless, and despite this promising uptick in progress, there remain significant barriers to deploying robots that understand, learn from, and interact using language in a physical context. In this section, some open challenges are briefly discussed, as well as some characteristics problems may have that make them difficult to address using currently popular approaches.

Some of the problem characteristics of note in this space are familiar from machine learning and robotics more generally, although grounded language offers its own unique difficulties in solving those problems. Some of these include *scalability*, or how learned models of grounded language can scale to a wide range of objects, tasks, and modalities; *generalization*, how such models can generalize to new examples of learned concepts and generalize across different robot platforms, including via few-shot and zero-shot learning; *multi-modality*, how robots using multiple complex sensors can interact with people using a variety of communication modalities; and *common sense reasoning*, in which systems can use an understanding of the broader world to solve otherwise under-specified problems.

First, despite the progress described above, there remain substantial problems involved in using actual speech with robots. [62] provides an overview of these difficulties, sorted into eight categories. These categories cover human–robot interaction focused questions (such as improving the modeling of social components of language), systems-level questions (such as the timing and latency difficulties of performing speech-based interaction in real time and developing improved learning models), and infrastructure-level suggestions for improving the context in which speech for robotics is studied. Despite the progress described above on using speech directly, challenges such as disambiguating speech in noisy environments remain.

A broad class of problems in this space includes developing models that can learn from a small amount of data or in unsupervised settings. While there is extensive work on learning from a small number of examples based on pre-training [55, 60, 77], the complexity of human spaces and robotic sensing make performing few-shot learning in idiosyncratic real-world settings a distinct challenge. There is a long tail of potentially out-of-distribution objects that may be encountered, sensors may give partial information, and people interacting with a robot will be understandably reluctant to provide a significant number of training examples. This ties into another difficulty, that of dealing with low-resource settings. For example, while there are a tremendous array of resources available for English and a few other major languages, the same is not always true of smaller languages or dialects of the sort that may be spoken in human-centric settings, or of idiosyncratic or ambiguous language.

Given their current popularity and effectiveness, it is particularly worth discussing the strengths and drawbacks of applying large language models and large vision-and-language models to grounded language. As described above, LLMs have demonstrated tremendous success on a wide variety of NL applications, including some language grounding problems. Nonetheless, while they are broadly good at producing output that seems correct at first glance, they do not necessarily fully grasp the semantics of complex grounded language [12] – such models have been trained on large amounts of (typically) textual data, but lack data to understand and reason about the physical world, making it difficult for them to understand contextual language about physical settings. Such models have shown some success in planning tasks where the goal is to follow high-level textual instructions (see section 5.2), but even these success stories may not generalize well to handling grounded language across domains or environments. LLMs and VLMs currently also have limited ability to handle multimodal sensor inputs of the sort that may occur in robotics settings, including auditory data.

Like many machine learning models, current approaches to grounded language learning tend to struggle with common sense reasoning [97], in which general, domain-agnostic background knowledge is key to understanding utterances. As an example, one description of an object might be "This is an apple, it's a kind of fruit." A robot with a good grounding system may learn the name and be able to identify apples subsequently, but will not be able to conclude that it is edible, or that it is similar to a banana. Another example has to do with a robot that has learned to hand someone a plate upon being instructed to do so, but would not from such a request conclude that the person is hungry or likely to engage in activities such as eating or setting the table. Efforts to combine common sense and language grounding exist [15, 23], but true common sense remains an elusive goal, as indeed it does in artificial intelligence generally.

There are also ethical questions and questions of bias and fairness associated with this problem area. There is a robust ongoing discussion in the machine learning community about discrimination and representation in machine learning technology, and the role of equitable development paradigms in addition to the deep-seated biases found in large data corpora (*inter alia*, [22, 25, 64]). These questions are highly relevant to the problem of making robots that learn from end users about their environment; a deployed system that works unevenly across different user demographics is inherently problematic, even if the system's average success is high. In considering this, it is necessary to bear in mind that discriminatory performance from machine learning models is not solely a product of unbalanced data [11, 21]. Model designs [65, 73], representational encoding choices [19], data collection methods [46], and learning paradigms all affect the inclusiveness of not only the results of machine learning, but the selection of the core questions being asked.

5.5 Conclusion

This chapter has discussed grounded language acquisition as a field where robotics and natural language understanding come together, and has discussed how learning to understand speech about the world plays a substantial role in human–robot interaction. A sampling of current work in this space has been described, with an emphasis on the challenges involved in going from understanding textual language to spoken language and in handling rich multi-modal perception and communication. The chapter closes with an overview of some of the many outstanding challenges in the general space of understanding grounded language, including dealing with speech, learning-derived problems such as generalization, and classical artificial intelligence problems such as incorporating common sense reasoning.

Language is not synonymous with sound: speech is a carrier for linguistic content, but only one of several mechanisms by which communicative content can be conveyed. Nevertheless, speech is an obvious, intuitive mechanism for human–robot interaction, tightly coupled with questions of language understanding and understanding the world from complex perceptual context. Grounded language understanding, particularly from speech, represents a rich, promising research space that is tightly interwoven with questions of sound in a robotic environment. There is extensive work in this area and in the related areas of spoken language processing and human–robot interaction, and this chapter attempts to provide an overview of some of the ways in which these elements come together.

Acknowledgments

We would like to extend heartfelt thanks to Francis Ferraro and Luke E. Richards for their assistance in preparing this manuscript. The case study described in this chapter is based on the work of a number of colleagues in the Interactive Robotics and Language (IRAL) lab, especially including Gaoussou Youssouf Kebe and Kasra Darvish, as well as Luke E. Richards, Padraig Higgins, Patrick Jenkins, Rishabh Sachdeva, Ryan Barron, John Winder, Don Engel, and Edward Raff. This material is based in part upon work supported by the National Science Foundation under Grant Nos. 1813223, 1920079, 1940931, 2024878, and 2145642.

Bibliography

[1] AHN, M., BROHAN, A., BROWN, N., CHEBOTAR, Y., CORTES, O., DAVID, B., FINN, C., FU, C., GOPALAKRISHNAN, K., HAUSMAN, K., HERZOG, A., HO, D., HSU, J., IBARZ, J., ICHTER, B., IRPAN, A., JANG, E., RUANO, R. J., JEFFREY, K., JESMONTH, S., JOSHI, N. J., JULIAN, R., KALASHNIKOV, D., KUANG, Y., LEE, K.-H., LEVINE, S., LU, Y., LUU, L., PARADA, C., PASTOR, P., QUIAMBAO, J., RAO, K., RETTINGHOUSE, J., REYES, D., SERMANET, P., SIEVERS, N., TAN, C., TOSHEV, A., VANHOUCKE, V., XIA, F., XIAO, T., XU, P., XU, S., YAN, M., AND ZENG, A. Do as i can, not as i say: Grounding language in robotic affordances, 2022.

[2] ALSHARHAN, E., AND RAMSAY, A. Investigating the effects of gender, dialect, and training size on the performance of arabic speech recognition. *Language Resources and Evaluation 54*, 4 (2020), 975–998.

[3] ANTOL, S., AGRAWAL, A., LU, J., MITCHELL, M., BATRA, D., ZITNICK, C. L., AND PARIKH, D. VQA: Visual question answering. In *Proceedings of the IEEE International Conference on Computer Vision* (2015), pp. 2425–2433.

[4] ARUMUGAM, D., KARAMCHETI, S., GOPALAN, N., WONG, L. L., AND TELLEX, S. Accurately and efficiently interpreting human-robot instructions of varying granularities. *arXiv preprint arXiv:1704.06616* (2017).

[5] ATWOOD, J., HALPERN, Y., BALJEKAR, P., BRECK, E., SCULLEY, D., OSTYAKOV, P., NIKOLENKO, S. I., IVANOV, I., SOLOVYEV, R., WANG, W., AND SKALIC, M. The inclusive images competition. In *The NeurIPS '18 Competition* (Cham, 2020), S. Escalera and R. Herbrich, Eds. Springer International Publishing, pp. 155–186.

[6] BAEVSKI, A., ZHOU, Y., MOHAMED, A., AND AULI, M. wav2vec 2.0: A framework for self-supervised learning of speech representations. In *Neural Information Processing Systems (NeurIPS)* (2020).

[7] BANSAL, M., KUMAR, M., AND KUMAR, M. 2d object recognition techniques: state-of-the-art work. *Archives of Computational Methods in Engineering 28* (2021), 1147–1161.

[8] BARBU, A., MAYO, D., ALVERIO, J., LUO, W., WANG, C., GUTFREUND, D., TENENBAUM, J., AND KATZ, B. Objectnet: A large-scale bias-controlled dataset for pushing the limits of object recognition models. In *Advances in Neural Information Processing Systems* (2019), pp. 9453–9463.

[9] BASTIANELLI, E., CASTELLUCCI, G., CROCE, D., IOCCHI, L., BASILI, R., AND NARDI, D. Huric: a human robot interaction corpus. In *LREC* (2014), pp. 4519–4526.

[10] BASTIANELLI, E., CROCE, D., VANZO, A., BASILI, R., NARDI, D., ET AL. A discriminative approach to grounded spoken language understanding in interactive robotics. In *IJCAI* (2016), pp. 2747–2753.

[11] BENDER, E. M., GEBRU, T., MCMILLAN-MAJOR, A., AND SHMITCHELL, S. On the dangers of stochastic parrots: Can language models be too big? In *Proceedings of the 2021 ACM Conference on Fairness, Accountability, and Transparency* (2021), pp. 610–623.

[12] BENDER, E. M., AND KOLLER, A. Climbing towards NLU: On meaning, form, and understanding in the age of data. In *Proceedings of the 58th Annual Meeting of the Association for Computational Linguistics* (Online, July 2020), Association for Computational Linguistics, pp. 5185–5198.

[13] BENZEGHIBA, M., DE MORI, R., DEROO, O., DUPONT, S., ERBES, T., JOUVET, D., FISSORE, L., LAFACE, P., MERTINS, A., RIS, C., ET AL. Automatic speech recognition and speech variability: A review. *Speech Communication 49*, 10-11 (2007), 763–786.

[14] BISK, Y., HOLTZMAN, A., THOMASON, J., ANDREAS, J., BENGIO, Y., CHAI, J., LAPATA, M., LAZARIDOU, A., MAY, J., NISNEVICH, A., PINTO, N., AND TURIAN, J. Experience grounds language. In *Proceedings of the 2020 Conference on Empirical Methods in Natural Language Processing (EMNLP)* (Nov. 2020), Association for Computational Linguistics.

[15] BISK, Y., ZELLERS, R., GAO, J., CHOI, Y., ET AL. Piqa: Reasoning about physical commonsense in natural language. In *Proceedings of the AAAI Conference on Artificial Intelligence* (2020), vol. 34, pp. 7432–7439.

[16] BLUKIS, V., KNEPPER, R., AND ARTZI, Y. Few-shot object grounding and mapping for natural language robot instruction following. In *Conference on Robot Learning* (2021), PMLR, pp. 1829–1854.

[17] BRENNAN, S. E., AND CLARK, H. H. Conceptual pacts and lexical choice in conversation. *Journal of Experimental Psychology: Learning, Memory, and Cognition 22*, 6 (1996), 1482.

[18] BUCKER, A., FIGUEREDO, L., HADDADINL, S., KAPOOR, A., MA, S., AND BONATTI, R. Reshaping robot trajectories using natural language commands: A study of multi-modal data alignment using transformers. In *2022 IEEE/RSJ International Conference on Intelligent Robots and Systems (IROS)* (2022), pp. 978–984.

[19] BUOLAMWINI, J., AND GEBRU, T. Gender shades: Intersectional accuracy disparities in commercial gender classification. In *Conference on Fairness, Accountability and Transparency* (2018), PMLR, pp. 77–91.

[20] CHAI, J. Y., GAO, Q., SHE, L., YANG, S., SABA-SADIYA, S., AND XU, G. Language to action: Towards interactive task learning with physical agents. In *IJCAI* (2018), pp. 2–9.

[21] CHAKRABORTY, J., MAJUMDER, S., AND MENZIES, T. Bias in machine learning software: Why? How? What to do? *arXiv preprint arXiv:2105.12195* (2021).

[22] CHANG, K.-W., PRABHAKARAN, V., AND ORDONEZ, V. Bias and fairness in natural language processing. In *Proceedings of the 2019 Conference on Empirical Methods in Natural Language Processing and the 9th International Joint Conference on Natural Language Processing (EMNLP-IJCNLP): Tutorial Abstracts* (2019).

[23] CHEN, H., TAN, H., KUNTZ, A., BANSAL, M., AND ALTEROVITZ, R. Enabling robots to understand incomplete natural language instructions using commonsense reasoning. In *2020 IEEE International Conference on Robotics and Automation (ICRA)* (2020), pp. 1963–1969.

[24] CHEN, T., KORNBLITH, S., NOROUZI, M., AND HINTON, G. A simple framework for contrastive learning of visual representations. *arXiv preprint arXiv:2002.05709* (2020).

[25] CHOULDECHOVA, A., AND ROTH, A. A snapshot of the frontiers of fairness in machine learning. *Communications of the ACM 63*, 5 (2020), 82–89.

[26] CHRUPAŁA, G. Visually grounded models of spoken language: A survey of datasets, architectures and evaluation techniques. *Journal of Artificial Intelligence Research 73* (2022), 673–707.

[27] CHRUPAŁA, G., GELDERLOOS, L., AND ALISHAHI, A. Representations of language in a model of visually grounded speech signal. In *Proceedings of the 55th Annual Meeting of the Association for Computational Linguistics (Volume 1: Long Papers)* (Vancouver, Canada, July 2017), Association for Computational Linguistics, pp. 613–622.

[28] CUI, Y., KARAMCHETI, S., PALLETI, R., SHIVAKUMAR, N., LIANG, P., AND SADIGH, D. "No, to the right" – online language corrections for robotic manipulation via shared autonomy. In *18th ACM/IEEE International Conference on Human Robot Interaction (HRI)* (2023).

[29] DE PONTE, F., AND RAUCHAS, S. Grounding words in visual perceptions: Experiments in spoken language acquisition. In *Proceedings of the Sixth Workshop on Natural Language for Artificial Intelligence (NL4AI 2022) co-located with 21th International Conference of the Italian Association for Artificial Intelligence (AI* IA 2022)* (2022).

[30] DEVLIN, J., CHANG, M.-W., LEE, K., AND TOUTANOVA, K. BERT: Pre-training of deep bidirectional transformers for language understanding. In *North American Chapter of the Association for Computational Linguistics (NAACL)* (Minneapolis, Minnesota, June 2019), Association for Computational Linguistics, pp. 4171–4186.

[31] DONG, X., AND SHEN, J. Triplet loss in siamese network for object tracking. In *European Conf. on Computer Vision* (September 2018).

[32] DUVALLET, F., WALTER, M. R., HOWARD, T., HEMACHANDRA, S., OH, J., TELLER, S., ROY, N., AND STENTZ, A. Inferring maps and behaviors from natural language instructions. In *Experimental Robotics: The 14th International Symposium on Experimental Robotics* (2016), Springer, pp. 373–388.

[33] FENG, S., KUDINA, O., HALPERN, B. M., AND SCHARENBORG, O. Quantifying bias in automatic speech recognition. *arXiv preprint arXiv:2103.15122* (2021).

[34] FLEISCHMAN, M., AND ROY, D. Grounded language modeling for automatic speech recognition of sports video. In *Proceedings of ACL-08: HLT* (2008), pp. 121–129.

[35] GU, J., STEFANI, E., WU, Q., THOMASON, J., AND WANG, X. E. Vision-and-language navigation: A survey of tasks, methods, and future directions. In *Association for Computational Linguistics (ACL)* (2022).

[36] HARNAD, S. The symbol grounding problem. *Physica D: Nonlinear Phenomena 42*, 1 (1990).

[37] HARWATH, D., AND GLASS, J. Deep multimodal semantic embeddings for speech and images. In *2015 IEEE Workshop on Automatic Speech Recognition and Understanding (ASRU)* (2015), pp. 237–244.

[38] HAVARD, W., BESACIER, L., AND ROSEC, O. Speech-coco: 600k visually grounded spoken captions aligned to mscoco data set. *GLU 2017 International Workshop on Grounding Language Understanding* (Aug 2017).

[39] HAVARD, W. N., CHEVROT, J.-P., AND BESACIER, L. Models of visually grounded speech signal pay attention to nouns: A bilingual experiment on english and japanese. In *ICASSP 2019 - 2019 IEEE International Conference on Acoustics, Speech and Signal Processing (ICASSP)* (2019), pp. 8618–8622.

[40] HE, K., ZHANG, X., REN, S., AND SUN, J. Deep residual learning for image recognition. In *2016 IEEE Conference on Computer Vision and Pattern Recognition (CVPR)* (Jun 2016), IEEE, p. 770–778.

[41] HERMANN, K. M., HILL, F., GREEN, S., WANG, F., FAULKNER, R., SOYER, H., SZEPESVARI, D., CZARNECKI, W. M., JADERBERG, M., TEPLYASHIN, D., ET AL. Grounded language learning in a simulated 3d world. *arXiv preprint arXiv:1706.06551* (2017).

[42] HSU, W.-N., HARWATH, D. F., SONG, C., AND GLASS, J. Text-free image-to-speech synthesis using learned segmental units. *arXiv abs/2012.15454* (2020).

[43] HUANG, W., XIA, F., XIAO, T., CHAN, H., LIANG, J., FLORENCE, P., ZENG, A., TOMPSON, J., MORDATCH, I., CHEBOTAR, Y., SERMANET, P., BROWN, N., JACKSON, T., LUU, L., LEVINE, S., HAUSMAN, K., AND ICHTER, B. Inner monologue: Embodied reasoning through planning with language models. *arXiv preprint arXiv:2207.05608* (2022).

[44] IBARRA, A., AND TANENHAUS, M. K. The flexibility of conceptual pacts: Referring expressions dynamically shift to accommodate new conceptualizations. *Frontiers in Psychology 7* (2016), 561.

[45] JANSEN, P. Visually-grounded planning without vision: Language models infer detailed plans from high-level instructions. In *Findings of the Association for Computational Linguistics: EMNLP 2020* (Online, Nov. 2020), Association for Computational Linguistics, pp. 4412–4417.

[46] JO, E. S., AND GEBRU, T. Lessons from archives: Strategies for collecting sociocultural data in machine learning. In *Proceedings of the 2020 Conference on Fairness, Accountability, and Transparency* (2020), pp. 306–316.

[47] JOHNSON, J. E., HARIHARAN, B., VAN DER MAATEN, L., FEI-FEI, L., ZITNICK, C. L., AND GIRSHICK, R. B. Clevr: A diagnostic dataset for compositional language and elementary visual reasoning. *CVPR* (2016), 1988–1997.

[48] KAMPER, H., SETTLE, S., SHAKHNAROVICH, G., AND LIVESCU, K. Visually grounded learning of keyword prediction from untranscribed speech. *arXiv preprint arXiv:1703.08136* (2017).

[49] KEBE, G. Y., HIGGINS, P., JENKINS, P., DARVISH, K., SACHDEVA, R., BARRON, R., WINDER, J., ENGEL, D., RAFF, E., FERRARO, F., AND MATUSZEK, C. A spoken language dataset of descriptions for speech-based grounded language learning. In *Proceedings of Neural Information Processing Systems (NeurIPS)* (2021).

[50] KEBE, G. Y., RICHARDS, L. E., RAFF, E., FERRARO, F., AND MATUSZEK, C. Bridging the gap: Using deep acoustic representations to learn grounded language from percepts and raw speech. In *Proceedings of the AAAI Conference on Artificial Intelligence* (2022), AAAI Press.

[51] KHANDELWAL, A., WEIHS, L., MOTTAGHI, R., AND KEMBHAVI, A. Simple but effective: Clip embeddings for embodied ai. In *Proceedings of the IEEE/CVF Conference on Computer Vision and Pattern Recognition (CVPR)* (June 2022), pp. 14829–14838.

[52] KHOSLA, P., TETERWAK, P., WANG, C., SARNA, A., TIAN, Y., ISOLA, P., MASCHINOT, A., LIU, C., AND KRISHNAN, D. Supervised contrastive learning. In *Advances in Neural Information Processing Systems* (2020), H. Larochelle, M. Ranzato, R. Hadsell, M. F. Balcan, and H. Lin, Eds., vol. 33, Curran Associates, Inc., pp. 18661–18673.

[53] KIRK, N. H., NYGA, D., AND BEETZ, M. Controlled natural languages for language generation in artificial cognition. In *2014 IEEE International Conference on Robotics and Automation (ICRA)* (2014), IEEE, pp. 6667–6672.

[54] KOLLAR, T., TELLEX, S., ROY, D., AND ROY, N. Toward understanding natural language directions. In *2010 5th ACM/IEEE International Conference on Human-Robot Interaction (HRI)* (2010), IEEE, pp. 259–266.

[55] LI, L. H., YATSKAR, M., YIN, D., HSIEH, C.-J., AND CHANG, K.-W. Visualbert: A simple and performant baseline for vision and language. *arXiv preprint arXiv:1908.03557* (2019).

[56] LIN, T.-Y., MAIRE, M., BELONGIE, S., BOURDEV, L., GIRSHICK, R., HAYS, J., PERONA, P., RAMANAN, D., ZITNICK, C. L., AND DOLLÁR, P. Microsoft coco: Common objects in context, 2015.

[57] LINDES, P., MININGER, A., KIRK, J. R., AND LAIRD, J. E. Grounding language for interactive task learning. In *Proceedings of the First Workshop on Language Grounding for Robotics* (2017), pp. 1–9.

[58] LU, J., BATRA, D., PARIKH, D., AND LEE, S. ViLBERT: Pretraining task-agnostic visiolinguistic representations for vision-and-language tasks. *Advances in Neural Information Processing Systems 32* (2019).

[59] LYNCH, C., WAHID, A., TOMPSON, J., DING, T., BETKER, J., BARUCH, R., ARMSTRONG, T., AND FLORENCE, P. Interactive language: Talking to robots in real time. *arXiv preprint arXiv:2210.06407* (2022).

[60] MAHMOUDIEH, P., PATHAK, D., AND DARRELL, T. Zero-shot reward specification via grounded natural language. In *International Conference on Machine Learning* (2022), PMLR, pp. 14743–14752.

[61] MAO, J., HUANG, J., TOSHEV, A., CAMBURU, O., YUILLE, A. L., AND MURPHY, K. Generation and comprehension of unambiguous object descriptions. In *Proceedings of the IEEE Conference on Computer Vision and Pattern Recognition* (2016), pp. 11–20.

[62] MARGE, M., ESPY-WILSON, C., WARD, N. G., ALWAN, A., ARTZI, Y., BANSAL, M., BLANKENSHIP, G., CHAI, J., DAUMÉ, H., DEY, D., HARPER, M., HOWARD, T., KENNINGTON, C., KRUIJFF-KORBAYOVÁ, I., MANOCHA, D., MATUSZEK, C., MEAD, R., MOONEY, R., MOORE, R. K., OSTENDORF, M., PON-BARRY, H., RUDNICKY, A. I., SCHEUTZ, M., AMANT, R. S., SUN, T., TELLEX, S., TRAUM, D., AND YU, Z. Spoken language interaction with robots: Recommendations for future research. *Computer Speech & Language 71* (2022), 101255.

[63] MATUSZEK, C., FITZGERALD, N., ZETTLEMOYER, L., BO, L., AND FOX, D. A joint model of language and perception for grounded attribute learning. *arXiv preprint arXiv:1206.6423* (2012).

[64] MEHRABI, N., MORSTATTER, F., SAXENA, N., LERMAN, K., AND GALSTYAN, A. A survey on bias and fairness in machine learning. *ACM Computing Surveys (CSUR) 54*, 6 (2021), 1–35.

[65] MITCHELL, M., WU, S., ZALDIVAR, A., BARNES, P., VASSERMAN, L., HUTCHINSON, B., SPITZER, E., RAJI, I. D., AND GEBRU, T. Model cards for model reporting. In *Proceedings of the Conference on Fairness, Accountability, and Transparency* (2019), pp. 220–229.

[66] MOONEY, R. J. Learning to connect language and perception. In *Proceedings of the 23rd National Conference on Artificial Intelligence - Volume 3* (2008), AAAI'08, AAAI Press, pp. 1598–1601.

[67] NGUYEN, A. T., RICHARDS, L. E., KEBE, G. Y., RAFF, E., DARVISH, K., FERRARO, F., AND MATUSZEK, C. Practical cross-modal manifold alignment for robotic grounded language learning. In *Proceedings of the IEEE/CVF Conference on Computer Vision and Pattern Recognition (CVPR) Workshops* (June 2021), pp. 1613–1622.

[68] NYGA, D., ROY, S., PAUL, R., PARK, D., POMARLAN, M., BEETZ, M., AND ROY, N. Grounding robot plans from natural language instructions with incomplete world knowledge. In *Conference on Robot Learning* (2018), pp. 714–723.

[69] PARK, S.-M., AND KIM, Y.-G. Visual language navigation: A survey and open challenges. *Artificial Intelligence Review* (2022), 1–63.

[70] PILLAI, N., AND MATUSZEK, C. Unsupervised selection of negative examples for grounded language learning. In *Proceedings of the AAAI Conference on Artificial Intelligence* (2022), AAAI Press.

[71] QI, S., NING, X., YANG, G., ZHANG, L., LONG, P., CAI, W., AND LI, W. Review of multi-view 3d object recognition methods based on deep learning. *Displays 69* (2021), 102053.

[72] RADFORD, A., KIM, J. W., HALLACY, C., RAMESH, A., GOH, G., AGARWAL, S., SASTRY, G., ASKELL, A., MISHKIN, P., CLARK, J., ET AL. Learning transferable visual models from natural language supervision. In *International Conference on Machine Learning (ICML)* (2021), PMLR, pp. 8748–8763.

[73] RAJI, I. D., SMART, A., WHITE, R. N., MITCHELL, M., GEBRU, T., HUTCHINSON, B., SMITH-LOUD, J., THERON, D., AND BARNES, P. Closing the ai accountability gap: Defining an end-to-end framework for internal algorithmic auditing. In *Proceedings of the 2020 Conference on Fairness, Accountability, and Transparency* (2020), pp. 33–44.

[74] REDEKER, G. On differences between spoken and written language. *Discourse Processes 7*, 1 (1984), 43–55.

[75] RICHARDS, L. E., DARVISH, K., AND MATUSZEK, C. Learning Object Attributes with Category-Free Grounded Language from Deep Featurization. In *2020 IEEE/RSJ International Conference on Intelligent Robots and Systems (IROS)* (Oct. 2020), pp. 8400–8407.

[76] ROY, D. Grounded spoken language acquisition: Experiments in word learning. *IEEE Transactions on Multimedia 5*, 2 (2003), 197–209.

[77] SADHU, A., CHEN, K., AND NEVATIA, R. Zero-shot grounding of objects from natural language queries. In *Proceedings of the IEEE/CVF International Conference on Computer Vision* (2019), pp. 4694–4703.

[78] SHANKAR, S., HALPERN, Y., BRECK, E., ATWOOD, J., WILSON, J., AND SCULLEY, D. No classification without representation: Assessing geodiversity issues in open data sets for the developing world. *arXiv preprint arXiv:1711.08536* (2017).

[79] SHE, L., CHENG, Y., CHAI, J. Y., JIA, Y., YANG, S., AND XI, N. Teaching robots new actions through natural language instructions. In *The 23rd IEEE International Symposium on Robot and Human Interactive Communication* (2014), IEEE, pp. 868–873.

[80] SHEN, S., LI, L. H., TAN, H., BANSAL, M., ROHRBACH, A., CHANG, K.-W., YAO, Z., AND KEUTZER, K. How much can clip benefit vision-and-language tasks? *arXiv preprint arXiv:2107.06383* (2021).

[81] SHRIDHAR, M., MANUELLI, L., AND FOX, D. Cliport: What and where pathways for robotic manipulation. In *Conference on Robot Learning* (2022), PMLR, pp. 894–906.

[82] SHRIDHAR, M., THOMASON, J., GORDON, D., BISK, Y., HAN, W., MOTTAGHI, R., ZETTLEMOYER, L., AND FOX, D. Alfred: A benchmark for interpreting grounded instructions for everyday tasks. In *Proceedings of the IEEE/CVF Conference on Computer Vision and Pattern Recognition* (2020), pp. 10740–10749.

[83] SHRIDHAR, M., THOMASON, J., GORDON, D., BISK, Y., HAN, W., MOTTAGHI, R., ZETTLEMOYER, L., AND FOX, D. ALFRED: A Benchmark for Interpreting Grounded Instructions for Everyday Tasks. In *The IEEE Conference on Computer Vision and Pattern Recognition (CVPR)* (2020).

[84] SKANTZE, G., AND WILLEMSEN, B. Collie: Continual learning of language grounding from language-image embeddings. *Journal of Artificial Intelligence Research 74* (2022), 1201–1223.

[85] SUGLIA, A., GAO, Q., THOMASON, J., THATTAI, G., AND SUKHATME, G. Embodied BERT: A transformer model for embodied, language-guided visual task completion. In *Novel Ideas in Learning-to-Learn through Interaction (NILLI) Workshop @ EMNLP* (2021).

[86] TATMAN, R. Gender and dialect bias in youtube's automatic captions. In *ACL Workshop on Ethics in Natural Language Processing* (2017).

[87] TELLEX, S., GOPALAN, N., KRESS-GAZIT, H., AND MATUSZEK, C. Robots that use language. *Annual Review of Control, Robotics, and Autonomous Systems 3* (2020), 25–55.

[88] THOMASON, J., PADMAKUMAR, A., SINAPOV, J., WALKER, N., JIANG, Y., YEDIDSION, H., HART, J., STONE, P., AND MOONEY, R. Jointly improving parsing and perception for natural language commands through human-robot dialog. *Journal of Artificial Intelligence Research 67* (2020), 327–374.

[89] THOMASON, J., SHRIDHAR, M., BISK, Y., PAXTON, C., AND ZETTLE-MOYER, L. Language grounding with 3d objects. In *Conference on Robot Learning* (2022), PMLR, pp. 1691–1701.

[90] THOMASON, J., SINAPOV, J., SVETLIK, M., STONE, P., AND MOONEY, R. J. Learning multi-modal grounded linguistic semantics by playing "i spy". In *IJCAI* (2016), pp. 3477–3483.

[91] TORRALBA, A., AND EFROS, A. A. Unbiased look at dataset bias. In *CVPR 2011* (2011), IEEE, pp. 1521–1528.

[92] VERGYRI, D., LAMEL, L., AND GAUVAIN, J.-L. Automatic speech recognition of multiple accented english data. In *Eleventh Annual Conference of the International Speech Communication Association* (2010).

[93] WU, W., CHANG, T., AND LI, X. Visual-and-language navigation: A survey and taxonomy. *arXiv preprint arXiv:2108.11544* (2021).

[94] YOSHIKAWA, Y., SHIGETO, Y., AND TAKEUCHI, A. STAIR captions: Constructing a large-scale Japanese image caption dataset. In *Proceedings of the 55th Annual Meeting of the Association for Computational*

Linguistics (Volume 2: Short Papers) (Vancouver, Canada, July 2017), Association for Computational Linguistics, pp. 417–421.

[95] YU, C., AND BALLARD, D. H. A multimodal learning interface for grounding spoken language in sensory perceptions. *ACM Transactions on Applied Perception (TAP) 1*, 1 (2004), 57–80.

[96] ZELLERS, R., BISK, Y., FARHADI, A., AND CHOI, Y. From recognition to cognition: Visual commonsense reasoning. In *Proceedings of the IEEE/CVF Conference on Computer Vision and Pattern Recognition* (2019), pp. 6720–6731.

[97] ZELLERS, R., BISK, Y., SCHWARTZ, R., AND CHOI, Y. SWAG: A large-scale adversarial dataset for grounded commonsense inference. In *Proceedings of the 2018 Conference on Empirical Methods in Natural Language Processing* (Brussels, Belgium, Oct.-Nov. 2018), Association for Computational Linguistics, pp. 93–104.

Part II

Non-Verbal Audio

6

Consequential Sounds and Their Effect on Human Robot Interaction

Aimee Allen, Richard Savery and Nicole Robinson

6.1 Introduction

Robots are more commonly being deployed into human-occupied environments such as workplaces, public spaces and homes, leading to humans and robots working together in close proximity. Effective human–robot interaction (HRI) is a key component in the successful use of robots in human spaces, and careful interaction design and deployment helps to create frequent or long-term interactions that are beneficial to people. Interaction design spans across both visual appearance and audible sounds, with most existing research focusing on design for visual components, such as the robot physical appearance, communicative gestures or facial expressions.

Sound is a highly influential component that underpins successful human–robot interaction, given that the human brain often prioritises sound inputs over visual inputs [16]. Sound in human–robot interaction can have strong benefits, such as helping people establish a sense of proxemic comfort and localization of other agents within our environment [5, 38]. Sound can also have negative consequences, such as if a sound is too loud for the situation, or the wrong sound is present during the interaction, leading to a sense of confusion or annoyance [24, 31, 41]. Despite the increased focus on sound research in robotics, the study of most non-natural-language robot sounds, including consequential sounds produced by the robot's actuators, is comparably rare within HRI research. This is despite the noted importance of non-language sound in human-human interactions [45] and the fact that consequential sounds are prevalent in almost every robot and thus every human–robot interaction.

Consequential sounds are of concern in human robot interaction because they are often perceived negatively by most people [10, 11, 24, 36]. As consequential sounds are currently extremely under-researched as noted in several publications [5, 9, 10, 23, 33], there is an important research gap around how effective sound design for consequential sounds will contribute to robot acceptance by people including their willingness to purchase, use, and work long-term with robots.

This chapter explains robot consequential sounds, and the resultant impact on human robot interaction. Different consequential sounds across a variety of robotic platforms are demonstrated, and the effect robot consequential sounds have on HRI are discussed, including techniques and implications when designing for consequential sounds in research or production soundscapes to improve human–robot interaction success.

6.2   What are Consequential Sounds?

The term 'consequential sounds' was first coined for the purposes of product design as "sounds that are generated by the operating of the product itself" [18]. In other words, consequential sounds are the **unintentional noises** that a **machine produces itself** as it **moves and operates**. Consequential sounds exist for any machine that has moving parts as they are the audible sounds that are generated when different mechanical parts or actuators operate together to make the machine work. Due to the machine's specific functional design, the machine can not perform normal operations without making these sounds.

Consequential sounds originate from **vibrations of components**, generating forces which create a sound compression wave. These vibrations can be caused by the friction of turning parts, including friction within fast-spinning components known as ego-motion sounds [42], interactions with air/fluids in pneumatic/hydraulic actuators, resonance in other materials (e.g. chassis or joints) or other vibrations of actuators. Some examples are the whirring of a refrigerator condenser, soft static ticking of an idle television or computer screen, the spinning thumps of a washing machine's large motor or the buzz of a vibrating electric shaver or toothbrush [26]. Consequential sounds are generated internally by the machine, rather than by what the machine interacts with, and most often will be audible to nearby people. These sounds exist before any additional extra sounds are implemented, such as the use of audible tones to signal to the user that the machine has completed a task.

Many consequential sounds consist of two main components; broadband noise with some specific narrow-band harmonic noise. Repeatedly spinning components (such as motors) tend to generate broad spectrum sounds, with narrow tones/harmonics often coming from fast bursts and sudden stops of actuator motions. The precise spectra of these have been shown to vary not just by actuator type but also quite dynamically with changes in the rotation speed of the motor [43].

There are two key criteria to help distinguish consequential sounds from other machine generated sounds:

Criteria 1: Consequential sounds must be **unintentional**. Therefore, this criteria will exclude all types of intentional sounds that a machine may produce. For example, sounds that are intentionally programmed into the machine to interact with a human, such as sounds that produce verbal speech, non-speech sound, music, or beeps/tones, including sounds used to communicate a state, function or affect of a robot. Any sound that an engineer or designer has chosen to add to a machine and use through a speaker are not consequential sounds.

Criteria 2: Consequential sounds must be **produced by the machine itself**, rather than something the machine is interacting with at the time, so this excludes noises such as vibrotactile sounds. Vibrotactile sounds are the noises generated from the vibrations which occur when two objects touch. For example, if the machine makes contact with the environment, or interacts with another object or person [21]. In the case of robots, this could include the sound produced when a robot walks or rolls across a surface, grasps items, or collides with a basket when picking an object. Vibrotactile sounds are not consequential sounds, however these sounds are likely to be perceived similarly to consequential sounds. Additionally, the same interactions which cause vibrotactile sounds often alter consequential sounds through changes in strain and friction inside the actuators.

6.3   Psychoacoustics: Human Perception of Sound

A human's perception of a robot's sound will be critical to the user experience and success of robots that interact with people. Thus, a basic understanding of how people perceive sound is crucial for anyone designing or researching

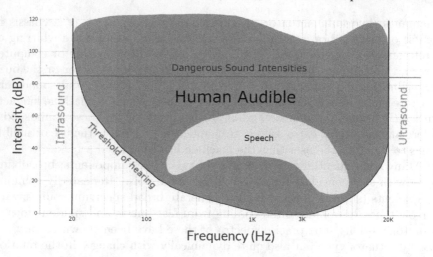

FIGURE 6.1
The intensity(dB) of sound which can be heard by humans (shaded region) across the typical human audible range (20Hz–20kHz). Human speech is typically centered around the middle of this range (banana-shaped region).

robots intended for human robot interaction. Psychoacoustics is the study of human sound perception and audiology, specifically the human psychological response to different sounds. The normal human hearing range is between 20Hz and 20kHz [5, 7], with this frequency range decreasing as a person ages. Humans find it easier to hear sounds in the middle of this spectrum (around 2kHz), which is the frequency range of human speech. Toward the outside of the human audible range, a higher volume (intensity) of sound is required to be above the threshold of hearing and thus be heard [5] (see Figure 6.1). Hertz (Hz) have an exponential relationship with the human perception of pitch, with each doubling of hertz (such as 440 to 880) representing an increase of one octave. Sounds of any frequency above a certain intensity (roughly 85dB) are dangerous as they can very quickly damage hearing.

Sound contains two main **objective components**: frequency (how fast the sound is oscillating in Hz) and intensity (strength of power in dB). Sound contains three main **perceptual components**; pitch, loudness and timbre. Pitch is how high or low the frequency is perceived, loudness is how strongly the intensity is perceived and timbre is a mixture of most of the subjective elements of sound such as how abrupt (sharp) or how pleasant the sound is to listen to [8]. Sharpness of a sound tends to decrease with a wider bandwidth of mixed frequencies. In sound design, it is critical to measure and understand the objective physical sound elements (frequency and intensity), and to also gauge the subjective sound elements as to how sounds are heard and perceived by humans (pitch, loudness, pleasantness) [12].

In general, humans find broadband noises (less sharp sounds) more pleasant than pure tones [8] and particularly favor those sounds where the frequencies (harmonics) are evenly spaced across the human-audible range [5]. Consequential sounds made by continuously spinning actuators (such as motors driving wheels of a mobile robot) tend to be mostly broad spectrum. People often respond more positively to rhythmic predictable sounds than sudden acute sounds such as alarms, car horns, or the sudden startup noises of a machine [16]. These sounds tend to disrupt focus by eliciting an instinctive danger or warning response. Sounds of this type when produced by robot actuators have the potential to both distract co-inhabitants, or ruin co-ordination between human–robot teams [16] by risking interrupting the natural human psychology to form spontaneous synchronised movements and emotional connections with a coworker.

People often encounter unpleasant sounds, with noise pollution being very common in many human environments [41]. Often the noise pollution is at least partially due to consequential sounds such as noise from cars, aeroplanes and trains; and machine consequential sounds such as air-conditioners. These persistent, annoying sounds can have strong mental and physiological effects on people including; annoyance and irritability (with social consequences), high stress (increased cortisol levels), anxiety, cognitive impairment, distraction and reduced productivity, and sleep disruption [4, 14, 41]. If robots become as well utilised as these other technologies, then their prevalence, and thus regularity of their consequential sounds, may lead to a noise pollution classification and negative perceptions. An example is the increasing noise complaints related to drones being used for delivery in residential areas [15].

6.3.1 Human Perception of Consequential Sounds

How consequential sounds are perceived, and thus how they could be used or altered, will vary across a number of individual and environmental factors. Common negative perceptions of consequential sounds are often not entirely due to the objective intensity levels or spectra of the sound, but rather the perception of the noise varies depending on the context of environment and tasks being undertaken [24, 31, 41]. Trying to work or rest at home typically leads to people perceiving sounds as louder and more annoying than when they are doing leisure activities [41]. For example, abrupt changes in sound (sharp sounds) can contribute to causing distraction or confusion, so abrupt consequential sounds can lead to a negative response in these contexts [38]. Type of environment, time of day, use of space, and the interactions between agents (both human and robot) within that space are all known to change sound perception [24, 41]. Context may include variables such as current task or mindset of a human (concentration versus socialising), the environment they are within (work, public or at home), as well as the expectations of the robot and its purpose (what is the robot assumed to be doing) within the context. The same high intensity machine sound can have a negative effect (noise pollution) whilst concentrating or sleeping, but may be seen as positive whilst interacting with the machine, such as correlating with power in a car or motorised tool [24].

Another factor is personality driven preferences of an individual listener [12]. Some people are consistently more or less annoyed by sounds than others based on their global 'noise sensitivity' [31], meaning that some people are more sensitive to sounds and being interrupted by them. Personality attributes such as level of introversion/extraversion are known to contribute to an individual's global noise sensitivity. This means consequential sounds made by robots may be viewed differently by different individuals.

There is also a spatial component to consider, given that sound propagates through space and can be experienced differently by different occupants [24] (a further study on spatial sound can be read in Chapter 4). People tend to prefer sounds from embodied (particularly moving) agents to convey a sense of proxemics, i.e. a sound that denotes their presence and relative positioning [38]. Humans are familiar with sharing environments with biological agents, and are accustomed to regular, rhythmic sounds such as breathing or rustling sounds that other people or nearby animals make [16]. In addition, it is often best when auditory and visual systems reinforce each other, and as such it may be unnerving to see but not hear a robot's presence, or to hear its presence more strongly than it is visually apparent [16]. Altering consequential sounds to create a consistent proxemic sound may increase comfort levels and therefore be less likely to create violations of 'personal space'.

6.4   Product Design for Consequential Sounds

Consequential sounds have been studied as a part of product sound design for decades due to their influence on a person's opinion of a product, and thus a person's likelihood of purchasing or using the product [6,18,19]. To achieve the desired product sound dimensions for the intended user perception, researchers, designers and engineers often focus on individual components and design principles to improve the sound experience. Product design theory describes several dimensions of product sound that are relevant to consider: strength, annoyance, amenity, and information content [19]. Strength or magnitude includes both objective dB intensity ratings as well as a loudness perception. Annoyance is a perceptual element consisting of factors such as sharpness, roughness and noisiness. Amenity or how pleasing the sound is to a person is very subjective and thus a challenge to measure, but includes elements such as rhythmic/regular sounds, harmonious qualities, and contextual appropriateness of the sound. Information content refers to the properties of the sound which communicate what or where the product is and its current task, performance and condition, and often consists of many intentional sounds alongside the consequential sounds.

To reduce the impact of consequential sounds during sound design, researchers, designers and engineers often focus on sound dampening or cancelling methods within specific sub-components. Example sound reduction and cancelling methods include: passive noise reduction by adding sound adsorption layers to enclose and dampen actuators, and active noise control, which makes use of added small sensors and speakers to help counter-act machine-like sounds. For example, cars often make use of active noise control techniques to reduce a narrow band of undesirable low frequency noises such as ground sounds and engine hum [30]. Microphones and vibration sensors are placed in multiple locations around the car to measure noise signals, and the car radio speakers each play sound of the opposite phase to cancel the measured signals. Many very similar technologies exist which are capable of generating opposite phase signals 2ms after the undesirable road noises are detected. Another common noise reduction technique is carefully designing the materials or shape of components to reduce and alter sound to a more pleasant spectrum. For example, in order to combat increasing complaints of noise pollution of drones in residential areas, researchers investigated using odd numbers of blades on propellers to generate a more even broad spectrum of noise [17]. Additional techniques used for managing consequential sounds are expanded on in Section 6.6.

6.5 Consequential Sound Spectrums of Robots

Robots produce a large spectrum of consequential sounds. These sounds must be considered not only in the design of the robot but also when planning implementation of a robot into physical spaces, as well as subsequent perception of the robot by users during human–robot interactions. The type of robot and its specific design are large factors which influence the spectrum of consequential sounds generated by a robot. There are clear differences in the type and frequency of consequential sounds generated by different robot form factors and motions. For example, humanoid robots moving their limbs to communicate with people, slow spinning motors to locomote a wheeled robot, walking actuations of legged robot, fast spinning props on a drone, or industrial robots conducting pick and place activities. Other significant contributors to the types of consequential sounds generated include: the number and type of actuator (motors etc), actuator positioning within the robot, and the shape and material composition of components surrounding or touching the actuators.

A large challenge for improving consequential sounds produced by robots is simply the shear volume of different actuators and robots that need to be addressed. Furthermore, an individual robot's consequential soundscape will also change over time, depending on its current operation and composition. For instance, a robot moving at rapid speeds at infrequent times compared to slow constant motion. Some of the common variables that impact and

TABLE 6.1

Variables that impact consequential sound generation.

Variable	Description of Effects
Component choice (actuator type and specs)	Different actuators create different base sounds. In general, higher power creates louder sounds. Electric motors are often the quietest actuators for their size/power
Product layout (how components are arranged relative to each other)	Components can generate consequential sounds by passing compression waves through each other, thus component relationship matters
Turning speed of actuators	Typically faster speeds generate more even sounds. At high speeds, these sounds can be negatively perceived as a frequent buzz
Full revolutions (continuous spinning) versus smaller angle motions (e.g. precise positioning of stepper motors)	Continuous rotations generate smoother sounds. Stop-start motions of small angles generate acute sounds
Material choices for inactive components (especially housings/chassis)	Certain shapes and surfaces can cause additional vibrations, resonate or amplify the actuator noises
Current power levels (important for DC powered devices)	As the battery drains, the voltage supplied to other components changes, and this alters the consequential sounds produced. Variable power supplies create variable sounds. Straining actuators can become louder and less pleasant if power is insufficient
Temperature of environment and components	Temperature changes how much strain the actuators are under, as well as expanding or contracting certain elements causing a change in friction generated sounds
Surfaces standing on/or objects interacting with (excludes vibrotactile sounds created by contact with the objects themselves)	Properties of contact object changes the exertion of the actuators and thus the sounds they produce
Imperfections in the robot as it ages e.g. dents in chassis	Minor shape differences change the acoustic properties of any resonance and sound amplification of the other consequential sounds

alter consequential sounds can be found in Table 6.1. Some variables or components have a larger effect than others on the spectra of consequential sounds produced. However, even minor changes can have an impact on the interaction experience, given that humans are able to perceive these sound differences [32], which has been shown in existing research to influence their opinions of a robot [5, 9, 10, 23, 33].

6.5.1 Existing HRI Research Involving Consequential Sounds

One of the first robotics papers to focus on "consequential sound" [24], used an online Amazon MTurk study to compare non-contextualised DC servo motor sounds on their own without robots. Participants were asked to compare videos of pairs of DC motors with dubbed sound i.e. with the 3D room ambient sound qualities removed. Consistency was found within participant ratings for preferred sounds, but sound preferences across participants were not consistent, suggesting a need for identifying more globally accepted pleasant or neutral noises. A second experiment overlaid consequential sounds from low versus high quality robotic arms onto videos of a high quality KUKA desktop robotic arm to investigate differences in perception of the robot when consequential sounds changed [36]. All consequential sounds showed a reduction in aesthetic ratings compared to silent videos, and the consequential sounds from the higher quality robot (which matched the video footage) correlated with higher ratings for competence of the robot. Another video-based study attempted to break down how specific variance in sound attributes (intensity and frequency) of consequential sounds affects perception of robots [46]. Videos of a UR5 robot arm had their natural consequential sounds manipulated up and down in terms of sound intensity (volume) and frequency (pitch). Results suggest that quieter robots are less discomforting, and higher frequency sounds correlate with positive perceptions such as warmth.

Several in-person HRI experiments have shown that consequential sounds can interact with other sounds or robot gestures to confuse the interpretation of affect [9, 11]. In one study, the low-frequency consequential sounds of a NAO humanoid robot created a strong arousal, negative valence affect [11]. This led to the robot's other sounds or gestures being perceived as frustrated, regardless of the intended affect. Another study examined whether augmenting consequential sounds of a soft robotic manipulator could change how happy, angry or curious the robot seemed [9]. Participants wore headphones which both deadened existing DC motor and pneumatic consequential sounds, and played additional affective sounds to supplement the movements of the robot. These changes to the consequential sounds altered the perceived valence of the robot to be happier, less angry and more curious.

Other research has investigated improving consequential sounds of a micro-drone within a naturalistic indoor home-like setting [44]. Three sound conditions: addition of birdsong, addition of rain sounds and no additional sounds, were tested at three distances from participants: near, mid and far. The masked consequential sounds were preferred at all distances as they were perceived as more pleasant than the unaltered consequential sounds. Which masking sound was preferred varied depending on the participant's distance from the drone, as well as their prior experience with related sounds i.e. existing associations with birds and rain.

Some non-sound focused HRI studies are also beginning to note the interaction effects of consequential sounds on other elements. In one study [37], participants deliberately limited robot motions specifically to avoid generating disliked consequential sounds, citing reasons such as "It even comes down to just how loud the motors were the first time it moved, that's very abrupt in a sonic way". In another context, autonomous vehicle researchers found that people use familiar consequential sounds, such as car engine noises, to locate and predict vehicle movements [22]. Fake consequential sounds were generated for a 'too quiet' hybrid autonomous car using chords of pure tones with frequency modulation based off the current car's speed, and attaching the speaker directly against the chassis to intentionally generate realistic resonance. People reported preferring interactions with the car that had these added sounds, as they found the intent of the car to change speed or yield was easier to predict.

The above studies help to illustrate the effects that consequential sounds of robots can have within HRI, and thus the importance for researchers and engineers to have an understanding of the variety of consequential sounds produced by different robots, and how to work with these consequential sounds. To further this understanding, consequential sounds generated by a selection of five example robots are herein presented and compared.

6.5.2   Consequential Sounds of Different Robots

A brief description of five different form-factor robots and their consequential sounds is presented below. Footage of these robots demonstrating a large range of different consequential sounds across a 30-45 second window has been supplied as supplementary material www.soundandrobotics.com/ch6, with a visual of each robot included in Figure 6.2.

Go1 Quadruped is a medium sized robot manufactured by Unitree [40] for both research and consumer use. The robot weighs 12 kg unloaded and has similar standing dimensions to a small-medium sized dog at approximately 65cm(L) x 28 cm(W) x 40-46 cm(H). Go1's consequential sounds are predominantly generated by its 12 brushless DC motors with three in each of its 4 legs. These sounds can vary considerably between gaits (e.g. walking, running or stair climbing), and more plyometric movements (jumps, twists and thrusts to stand on its hind legs). The robot has notable consequential sounds from the cooling fans for the multiple on-board computers. In addition to consequential sounds, the robot also produces significant vibrotactile sounds generated from the footpads interacting with ground surfaces.

Pepper is a social humanoid robot created by SoftBank Robotics [29] with a wheeled mobile base, an anthropomorphic upper torso, arms and head as

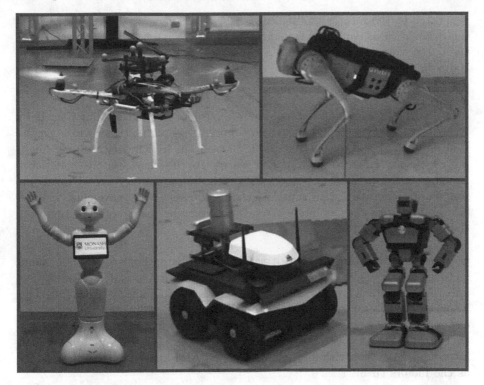

FIGURE 6.2
Robots as seen in supplementary material videos [1]. From top left to bottom right: Custom multi-rotor drone; Go1 EDU PLUS with 2D Lidar quadruped (Unitree); Pepper Social Robot (SoftBank Robotics); Jackal mobile UGV robot (ClearPath Robotics); Yanshee Humanoid Robot (UBTECH).

well as a tablet-like screen on its chest. Pepper stands 120 cm tall, and weighs approximately 28 kg. Most of Pepper's consequential sounds are of a fairly low intensity as most of the actuators are relatively low power. The Pepper is designed to make expressive motions with the head and arms. Consequential sounds are often produced by the friction between sections of chassis as the robot moves around, with the motors generating a soft electric whir as each motor switches on and off. There are a total of 20 actuators across the arms, head, and wheeled base, with the majority in the arms (6 per arm).

A custom multi-rotor drone (quadcopter) was designed and built for HRI research on developing semi-automated piloting drone software to assist inexperienced pilots [2]. The 'very small' quadcopter weighs 1.8kg and measures 38cm(L) x 38cm(W) x 30cm(H). The four actuators are off-the-shelf DC brushless motors which were designed for continuous spin uncrewed aerial vehicle (UAV) applications. The intensity of the consequential sounds generated by these rotor motors mask any sounds generated by other parts of the drone such as compute, resonance of frame, or mild vibrotactile sounds of the battery against the chassis. The general high intensity of these consequential sounds makes any changes in sound profile from other variables hard to detect.

Jackal UGV is an autonomous mobile outdoor robot created by ClearPath Robotics [28]. It is an entry-level field robotics research platform which weighs 17kg and is approximately 50cm long. Jackal has four identical, large, high-powered motors used to drive its four wheels. As the wheels tend to spin many times in the same direction to locomote the robot, mostly broad spectrum sounds are generated, with acute stop/start sounds as the movements change. Sound properties such as intensity and pitch vary by speed and direction that the robot is moving in.

Yanshee Humanoid Robot is a small, table-top humanoid robot manufactured by UBTECH [39]. It is an open-source platform intended for education and research which weighs 2.05kg and stands approximately 37cm tall. The Yanshee produces high-intensity sound for such a small robot due to its inexpensive DC motors creating ego-motion sounds from the many points of friction. This robot has a large number of actuators (17 servo motors), which are coupled together with aluminum alloy and ABS links that tend to resonate as the motors turn.

6.5.3 Case Study: Comparison of Consequential Sounds across Different Robots

A common method for the analysis of sound is through the use of a spectrogram, which produces a visual representation of the sound frequency content and intensity over time. Figure 6.3 shows a spectrogram for each of the previously described robots. The y axis of a spectrogram represents the frequency, usually between 0 and 20,000 Hertz, to cover the range of human hearing. The x axis of a spectrogram displays the time, and is typically considered in milliseconds or seconds, dependant on the analysis purpose. The color intensity or brightness of the spectrogram represents the intensity or amplitude of the frequency content of the audio signal. Brighter regions indicate higher sound intensity or amplitude, while darker regions indicate lower sound intensity or amplitude.

There are many standard features that can be extracted from a spectrogram. One common feature is the spectral centroid, which measures the center of gravity of the frequency distribution in a signal, providing a measure of the "brightness" or "darkness" of the sound. Two other features include spectral bandwidth and spectral flatness, which provide additional information about the frequency content of a signal. Spectral bandwidth measures the range of frequencies present in a signal, and can provide insight into the "sharpness"

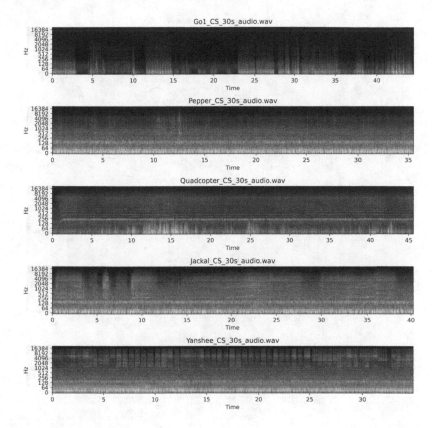

FIGURE 6.3
Spectrograms of the five case study robots.

or "dullness" of the sound. Spectral flatness measures the degree to which a signal's energy is spread evenly across its frequency spectrum, and can provide insight into the "tonality" or "noisiness" of the sound. Figure 6.4 displays the spectrograms for the case study robots, with overlaid spectral centroid and bandwidth. Whilst there are other methods to visualise and analyze audio, spectrograms are a good start as they can be easily generated, and provide many easily analysable features within an easy to understand visualization. The properties from spectrograms can additionally be used for machine learning analysis and generation.

From a visual analysis comparing the robots' spectrograms, multiple features of note become clear. The Go1 quadruped, shows clear fluctuations in the spectral centroid with each step the robot takes. This spectrogram also indicates that there are very low noise levels between steps, and that the sound from each step primarily takes place in the lower frequencies. Much of Pepper's spectrogram comes from underlying low frequencies of ambient room noise,

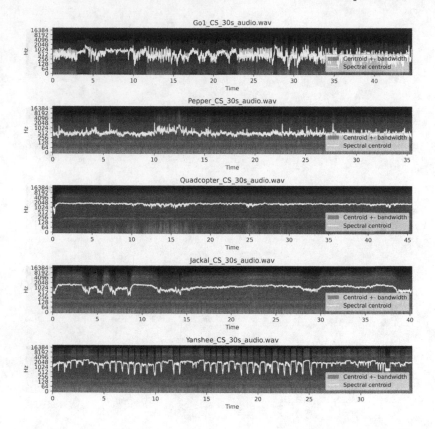

FIGURE 6.4
Spectrograms of the five case study robots: with spectral centroid and spectral bandwidth overlaid.

rather than the consequential sounds themselves. From an audio engineering perspective, it would be common practice to first apply a low-pass filter to remove the lower frequency noise before analysing the robot's consequential sounds. Within boosts of intensity in the higher frequency bands of the spectrogram, very uniform intensity regions can be seen, and with no clear pitches. The quadcopter has a very consistent centroid, with only occasional fluctuations. Of note is the reoccurring lines in the spectrogram, which are heard audibly as the pitch of the hum from the drone. Addressing this reoccurring harmonic series would be important when designing sounds for a drone. The Jackal has the widest, most noise-like signal compared to the other robots, with a more consistent intensity across all frequencies. Audibly, this is perceived as the sound being relatively un-pitched. The Yanshee has a very broad sound across the spectrum, with emphasis on each movement. Importantly for the Yanshee, the higher more piercing sounds can be seen on the spectrogram between 4096 and 8192 Hz, which would be a key consideration for any sound alterations.

Each robot has a unique audio signature in terms of the frequency distribution and intensity of their sound, which becomes more readily apparent through visualization. The analysis of these robots' audio spectrograms provides insight into the characteristics of their consequential sounds, and thus any potential features of these sounds to target for control or alteration.

6.6   Capturing and Altering Consequential Sounds

Given the large range of consequential sounds robots are capable of producing, there are many difficulties and challenges to consider when working with consequential sounds. Whilst the objective existence of consequential sounds can not be changed, there are both hardware and software options available to control or alter the consequential sounds produced. Recommendations for handling the many variables which can impact and change consequential sounds produced by a robot can be found in Table 6.2. Being mindful of these recommendations will help generate reduced and consistent sounds, allowing for accurate capture of consequential sounds, which is necessary prior to applying any further augmentation techniques. Many current industry practices for working with the unintentional sounds generated by a product or machine, may prove useful as a starting point or part-solution for altering consequential sounds in robotics.

To capture consequential sounds, at least one quality microphone must be used to record the sounds. If possible, a condenser microphone with full human audible spectrum range of 20Hz–20kHz should be used, although 60Hz–18KHz should suffice. Many consumer-grade microphones come with built-in filtering software to reduce the recording to only include frequencies of sound common in human speech. As such, these microphones should be avoided as it is more effective to capture the raw sound across the entire human audible spectrum. In most circumstances, it is ideal to place the microphone as close as possible to the sound source (i.e. the robot or actuator) and in such cases an omnidirectional (flat recording spectrum in all directions) microphone is ideal. In cases where the microphone is placed near a camera, a directional or cardioid microphone facing toward the robot may be better, to avoid picking up room ambient sounds. When intending to mount the microphone onto the robot, other criteria should be considered such as minimising the payload. Microphone placement is extremely important and should be selected based off how the user will perceive the sounds, how discreet the microphones should be during the interaction, the types of changes that will be made to the consequential sounds, and how many microphones are available. In general, the microphones should be placed at any point where consequential sounds could be audibly perceived

TABLE 6.2

Recommendations for generating reduced and consistent consequential sounds.

Variable	Recommendations
Component choice (actuator type and specs)	Be mindful of the sounds generated by chosen actuators, choose quieter actuators if possible, but don't sacrifice functional requirements for less sound
Product layout (how components are arranged relative to each other)	Avoid shapes and arrangements which resonate at frequencies which the actuators typically move at
Turning speed of actuators	Speed requirements are mostly fixed by functional movement requirements. Use sound alteration techniques when useful
Full revolutions (continuous spinning) versus smaller angle motions (e.g. precise positioning of stepper motors)	Try to reduce actuator accelerations when starting/stopping actuators such that sound changes are less abrupt
Material choices for inactive components (especially housings/chassis)	Avoid using materials which conduct sound well or that naturally resonate at frequencies which the actuators typically move at
Current power levels	For critical human-perception use-cases (such as experiments and product trials), attempt to maintain close to full charge when possible. For many robots, above 80% power should be ideal, or above 60% for high-drain applications
Temperature of environment and components	Minimise strong environmental temperature changes by running experiments in controlled indoor environments or on days where temperatures are mid-range and fairly consistent
Surfaces standing on/or objects interacting with	This may vary substantially with robot use cases. Sound deadening materials can be attached to contact points such as footpads and grippers
Imperfections in the robot as it ages e.g. dents in chassis	If a robot component begins creating undesirable noises from age, that component should likely be replaced

from, whilst avoiding anything that adds noise to the signal e.g. clipping from microphone being too close to fans or moving parts, or nearby cables causing electrical interference. Common scenarios and suggested microphone locations can be found in Table 6.3. There are also software requirements to record

TABLE 6.3
Recommended microphone placements to record consequential sounds in HRI scenarios.

Scenario	Microphone Placement
Video or online study	At location of person i.e. near camera
Multi-person and general in-person studies	As close as possible to actuators (omni-directional microphone recommended)
Individual customised sound perception (single participant)	Close to the single person
Recording consequential sounds with intent to alter them	As close as possible to actuators (omni-directional microphone recommended)

consequential sounds. Fortunately, most recording software covers the full audible human spectrum range 20Hz–20kHz, and there exists a variety of professional recording software, hobbyist phone apps or Python libraries which should be suitable depending on individual requirements such as budget and onboard versus offline recording requirements. The following techniques can help to improve the effect that consequential sounds can have.

Technique 1. Actuator Choice: Many current solutions focus on hardware design for sound, i.e controlling for consequential sounds by pre-designing actuators and other components to minimise any potentially negative sounds. Whilst useful, this is not feasible to do for every different actuator and robot, may necessitate non-desirable compromises on other functional requirements, and doesn't allow for adaptability for variables which change consequential sounds and their perception. Currently, most off-the-shelf actuators have not considered sound during their design. Therefore, choosing robot actuators that have in-built noise control might necessitate non-desirable compromises on other functional requirements. Whilst it would be possible to control for consequential sounds in the manufacturing process for future or custom actuators, it would still be challenging to predict the full effects of sonic interactions across every robot using a specific actuator across a multitude of contexts. Some existing industry sound reduction techniques (such as dampening) may also add unnecessary weight to the robot, increasing power requirements and potentially putting further strain on the actuators increasing generation of consequential sounds.

Technique 2. Audio Recording and Analysis: A good option to record consequential sounds is to use a Digital Audio Workstation (DAW), which is a piece of software used for recording, editing and producing complex recorded audio. DAWs are typically used by professional sound engineers for music production and sound effect generation, with Ableton, Avid Pro Tools, Logic Pro and Cubase being some of the most popular. A DAW can be used to look at the full frequency spectrum of sounds, and allow identification of good broad spectrum consequential sounds versus any acute/abrupt sounds which might be worth altering. Unless access to a pro-level DAW is readily available, these may not be ideal for consequential sound analysis as they can be expensive, have high learning curves, and may not perform well in real-time on the onboard compute within a robot. For HRI applications, the Audacity DAW is recommended for recording of consequential sounds, as it runs well on typical embedded architectures including Linux/Ubuntu, Intel/ARM processors, Raspberry Pi's and Nvidia Jetsons. In addition to Audacity, recommended tools to analyse consequential sounds for HRI include Sonic Visualiser https://www.sonicvisualiser.org and Librosa for python-based analysis. Producing a spectrogram (as shown in section 6.5.2) can be a good start for analysing consequential sounds.

Technique 3. Masking: A widely accepted psychoacoustic method for hiding any negatively perceived sound is masking [7, 27, 34], A more pleasant sound (masker) is used to reduce the detection of an unpleasant sound (maskee), thus improving the overall perception. It is worth noting that the masker sound does not necessarily have to be loud and add noise, as it can have an effect by containing similar frequencies as the sound to mask, even if it is the same (or slightly lower) intensity. Whilst masking is not often used in current robotics practice, this does have the potential to be applicable to a large range of different robotic platforms, adaptable to specific contexts and customizable to individual preferences. It is known that broad-spectrum sounds tend to work best for masking [12, 27] i.e. non-pure tones covering a wide frequency band of sounds across the entire human audible spectrum, however more research is required on what other properties make good consequential sounds masks. Two particular attributes which require further investigation are specific sounds that most people enjoy and the timing with which to produce these sounds. Regarding noise types to use for masking, standard broad spectrum noises have shown good promise, especially pink and brown (Brownian) noise bands with sound intensities matched to average human loudness perceptions [5, 27]. In terms of timing, it is currently unknown if temporal (slightly before or after the sound), simultaneous (during the sound) or using the masker continuously to feign proxemics is most effective.

Technique 4. Other Software Adjustments: In addition to masking, there are several other promising software techniques to alter consequential sounds in an adaptive way that could be retrofitted to existing robots. Most of these are yet to be applied to consequential sounds in practice. Adaptive techniques are particularly useful when working with consequential sounds as it is hard to tell what the full sound profile of the robot will be during design (i.e. prior to construction) [24]. Another benefit of using many of these techniques is the real-time adaptability to allow for a variety of contexts, and the ability to personalise sound alterations to preferences of individual users or co-inhabitants. To make use of real-time sound augmentation techniques, additional software (to generate sound alterations) and hardware (speaker to emit the generated sounds) may be required. Selection criteria for speakers are similar to requirements for microphones, with speaker being placed as close as possible to sound sources whilst avoiding anything that adds noise to signal e.g. cabling too close to each other, or the microphone feeding the speaker.

Ego-motion sound detection: Existing software based noise mitigation techniques used to improve speech recognition [13, 43] or for contact detection [21] may be useful for handling consequential sounds. In the first case, the real-time changes in noises a robot produces from ego-motion (the friction sounds within motors) can be captured and separated into groups via intensity and localization on the robot using techniques such as Blind Source Separation(BSS) [43]. Noise cancelling techniques to reduce environmental noises within each group can be determined to improve the recorded sound to be parsed for speech recognition. Many newer robots (especially robotic arms and teleoperated robots) are being built with internal accelerometers for measuring ego-vibrations when contacting objects in the environment [21]. Audio processing algorithms have been successfully used to isolate the noises generated by vibrations of the robot's own actuators(ego-vibrations) from contact with the environment sounds(vibrotactile sounds) to increase accuracy in detecting these contacts. Both these techniques work well for removing noise from output sound files, and have potential as the first step of a larger consequential sound solution. There is already significant research on recording and identifying ego-noise from robots [13, 21, 42, 43], however research is scarce regarding the next step, which is using the identified consequential sounds as data to inform other techniques capable of augmenting the audible consequential sounds, which are produced by a robot. For example, this data could be used to inform masking or noise cancellation techniques.

Reducing variance in consequential sounds: Many robots have consequential sounds which alternate between complete silence when stationary, to very acute and high intensity sounds when the robot suddenly begins moving. Both of these are often perceived negatively, leading to people feeling uncomfortable, distracted from a task, or even scared of the robot. One way to alter these consequential sounds could be to add constant sounds to create a sense of proxemics for the robot, giving the robot a consistent passive noise, and making the active motor noises less obtrusive. Other promising methods to minimise noise in less controlled 3D spaces such as outdoors and public spaces are also being researched such as estimating and cancelling time-variant sound in a sphere traveling out from the primary source [20].

Lastly, it is important to review local and federal laws and regulations related to product sound in the jurisdictions the robot will be deployed in. Many places have noise pollution and consumer sound protection laws for maximum sound intensities, and some machines such as autonomous and electric cars also have stipulations on minimum sound volumes.

6.7 Design Implications for Consequential Sounds in HRI

Given the prevalence of consequential sounds in every human–robot interaction, there are notable design decisions regarding how to create successful experiences with robots that either address or negate the impact of consequential sounds. Human perception of consequential sounds can have a notable effect on robot interactions and subsequent acceptance, particularly if these perceptions are not appropriately addressed when robots are deployed into human spaces.

1. Choosing to leave the consequential sounds alone or augment the sounds: An important initial decision for sound designers and researchers is between choosing to cover up consequential sounds, to leave them unaltered, or perhaps alter some sounds and leave others unchanged. Researchers and engineers should aim to identify the relevant sound attributes which may be appropriate or inappropriate for long-term robot use (see section 6.3.1). How consequential sounds might impact human perception, or affect interaction outcomes should be carefully considered. Altering consequential sounds can make the sound more pleasing to people, so any consequential sound alterations should be focused on less desirable sounds without removing perception of the desirable consequential sounds. Leaving some consequential sounds may help to amplify the experience of working with a non-human agent, by allowing people to hear machine-like sounds and thus associate the robot as a mechanical device. This could reduce some false expectations of human-like capabilities in a robot [3]. Whatever the decision, it is important to accurately capture consequential sounds to both test their prevalence, as well as to make effective improvements. Sound alterations that could be considered include softening contrasts between silence and abrupt sounds, and masking of undesirable sounds including sharp sounds and those centered consisting of limited frequencies such as pure tones without harmonics.

2. Early exposure to accurate consequential sounds to aid in long-term adoption: Given the consistent presence of consequential sounds, sound designers and engineers deploying robots into human spaces should aim to maintain the robot's eventual consequential sounds (including decided upon sound alterations) in any research, case studies, marketing or promotional material. While some of these sounds may be reduced, augmented or altered within specific use cases, its important to note that users should at least be well-aware of the expected sounds they are likely to encounter during long-term interactions with a robot. Most sound research is still conducted with the participant wearing headphones or via pre-recorded videos where a de-contextualised fully controllable sound is used. These scenarios do not capture or display consequential sounds faithfully to participants. Assessing user experience of a robot that does not have consequential sounds may unintentionally be producing a biased response, given that the robot may be perceived within a video as ideal for a scenario, meeting all the functional and aesthetic requirements, but once deployed, is abandoned due to the sound profile not being contextually appropriate.

3. Multiple groups of people to accommodate in one setting: There are two groups of people who are affected by robot consequential sounds: people directly using a robot (i.e. those intentionally interacting with the robot) and people who are sharing an environment with a robot (i.e. are colocated within the same space as a robot). Different user groups will experience these sounds in different manners, whether the sound causes notable distraction for people sharing a space with a robot, or the sound becomes part of the interaction experience with the user. Robots in shared spaces may cause disruption to people, or otherwise negatively effect a well-designed robot interaction [38]. A compromise needs to be made to establish a soundscape which minimises negative perception across all concurrent robot stakeholders.

4. Expectations of real robot sounds: When deciding how to alter or present consequential sounds, engineers and researchers should consider what expectations their robot users may already have of the sounds prior to first impression of the robot. Often consequential sounds are stripped from video promotion, or a musical soundtrack is played over the top, leading to users being unaware that robots even make consequential sounds. This could lead to a high probability that expectations will be mis-aligned, thus that the robot does not sound correct on the first encounter. This could contribute to someone becoming uncomfortable, and forming a negative initial association with the robot, as "expectation confirmation theory" [35] has been violated. People may habituate over time [25] and become comfortable with or enjoy these sounds, but this does require that the person persists with interacting with the robot long enough for the sounds to be familiar, which is typically not the case in experiments. When possible, research on sound should be done in person in a real 3D space. This will allow for correctly gauging the effects of the full sound spectrum by considering 3D sound effects, and proxemic effects. If initial tests must be conducted using video clips, then sound tests should be included at the start of any experiment to verify what frequencies and intensities of sound people can hear within the videos. This allows for control of variables such as sound equipment settings, personal sound sensitivity, and hearing capabilities.

6.8 Research Potential for Improving Consequential Sounds

Given the extensive prevalence of consequential sounds in human–robot interaction, this is clearly an important area for future research. Below are several clear opportunities for potential research avenues to explore to further understand and improve how consequential sounds impact robot engagement and perception.

1. Real world or naturalistic lab settings: Experimental verification of results are not often conducted using in-person naturalistic environments. Most existing research on robot sounds uses only pre-recorded footage where the consequential sounds have been completely stripped [24], which is clearly not comparable to a real-world situation. Other research involves masking consequential sounds by playing higher intensity sound directly into participants ears through headphones [9]. However, it is not feasible to have humans augment their own hearing by wearing headphones whenever interacting with a robot in a real setting. Thus research is required on pragmatic solutions to improving consequential sounds directly on the robot itself, in real world or naturalistic lab settings designed to imitate the home, workplace and public spaces.

2. Verification of results on different robots: Due to financial costs in acquiring or accessing robots, as well time costs in familiarization of setting up new robotic platforms, most robotics research is done on a single robot. As robots are known to each produce different consequential sounds, it is likely that findings of useful techniques and preferred sounds could vary between different robots, with what works well on a quadruped, either not transferring to a humanoid or UGV or needing alterations to work successfully on different platforms. Thus further research using multiple robots to verify results would be useful. An ideal circumstance here would be more standardization in research to allow for collaborations where researchers could exchange their setups so other researchers could verify the results using a different robot with which they are already familiar.

3. Development of full solutions for real-time adaptations of consequential sounds: As noted in Section 6.6, there are currently no end-to-end solutions for capturing and augmenting variable consequential sounds in real-time. Research on real-time sound alterations that can change between context and individual personal preferences will be immensely useful once robots are deployed heavily in offices, homes and public spaces. Additionally, if robots are able to control their sounds produced contextually, there is potential to further enhance environments beyond just improving consequential sounds. For example, developed algorithms to alter consequential sounds could be extended to include sounds to improve mood or concentration, target an individual's health conditions or positively augment the consequential sounds of other machines within the shared environment.

4. Larger and multiple-participant studies: In contrast to individual adaptations useful for cases where the robot has a primary interaction target, it would be beneficial to the field to conduct multiple-participant studies to uncover which techniques are effective for groups of people, such as is common in offices or public spaces. In addition, studies with larger numbers of participants would help to identify sound alterations, which are more globally accepted by people.

6.9 Conclusion

Consequential sounds produced by robots is clearly a persistent phenomenon which will continue to occur over time for all robots and all human robot interactions. Whilst there are several part solutions available to address consequential sounds, there is no known full solution for even a single robot, let alone the millions of robots that are or will be collocated with people in the near future. Consequential sounds vary between different robot platforms but also over time with the same robot, so it is important to accurately collect and analyse the sounds specific to each robot and contextual use. There are a variety of techniques which researchers and engineers can use to capture and alter robot consequential sounds in order to improve their perception, and HRI experiences. More research is required to further streamline techniques for consequential sound alterations in order to produce refined techniques that can work in the real world for a variety of different contexts, with different robots, and be personalizable to different people.

Attributions

This book chapter (and associated research) was partly supported by an Australian Government Research Training Program (RTP) Scholarship.

Bibliography

[1] ALLEN, A., SAVERY, R., AND ROBINSON, N. Sound and Robotics: Chapter 6 supplementary material, www.soundandrobotics.com/ch6, (2023).

[2] BACKMAN, K., KULIC, D., AND CHUNG, H. Learning to assist drone landings. *IEEE Robotics and Automation Letters 6* (2021), 3192–3199.

[3] BARTNECK, C., KULIĆ, D., CROFT, E., AND ZOGHBI, S. Measurement instruments for the anthropomorphism, animacy, likeability, perceived intelligence, and perceived safety of robots. *International Journal of Social Robotics 1* (2009), 71–81.

[4] BASNER, M., BABISCH, W., DAVIS, A., BRINK, M., CLARK, C., JANSSEN, S., AND STANSFELD, S. Auditory and non-auditory effects of noise on health, The Lancet (2014), 1325–1332.

[5] CHA, E., FITTER, N. T., KIM, Y., FONG, T., AND MATARIC, M. J. Effects of robot sound on auditory localization in human-robot collaboration. *ACM/IEEE International Conference on Human-Robot Interaction* (2018), 434–442.

[6] VAN EGMOND, R. The experience of product sounds, Product Experience (2007), 69–89.

[7] FASTL, H., AND ZWICKER, E. Masking, Psychoacoustics: Facts and Models, (2007), 61–110.

[8] FASTL, H., AND ZWICKER, E. Sharpness and sensory pleasantness, Psychoacoustics: Facts and Models, (2007), 239–246.

[9] FREDERIKSEN, M. R., AND STOEY, K. Augmenting the audio-based expression modality of a non-affective robot. In *2019 8th International Conference on Affective Computing and Intelligent Interaction, ACII 2019* (2019).

[10] FREDERIKSEN, M. R., AND STOY, K. Robots can defuse high-intensity conflict situations, In *2020 IEEE/RSJ International Conference on Intelligent Robots and Systems*, IROS2020. pp. 11376–11382.

[11] FRID, E., BRESIN, R., AND ALEXANDERSON, S. Perception of mechanical sounds inherent to expressive gestures of a NAO robot - implications for movement sonification of humanoids. In *Proceedings of the 15th Sound and Music Computing Conference: Sonic Crossings, SMC 2018* (2018), pp. 43–51.

[12] HALL, J. L. Auditory psychophysics for coding applications, Digital Signal Processing Handbook (1999), Chapter 39.

[13] INCE, G., NAKADAI, K., RODEMANN, T., TSUJINO, H., AND IMURA, J. I. Whole body motion noise cancellation of a robot for improved automatic speech recognition. *Advanced Robotics 25* (2011), 1405–1426.

[14] JARIWALA, H. J., SYED, H. S., PANDYA, M. J., AND GAJERA, Y. M. Noise pollution & human health: A review Proceedings of Noise and Air Pollution: Challenges and Opportunities (2017).

[15] JOKISCH, O., AND FISCHER, D. Drone sounds and environmental signals – a first review. Proceedings of ESSV2019_84 Studientexte zur Sprachkommunikation: Elektronische Sprachsignalverarbeitung (2019), pp. 212–220.

[16] JOUAITI, M., AND HENAFF, P. The sound of actuators: Disturbance in human - robot interactions? In *2019 Joint IEEE 9th International Conference on Development and Learning and Epigenetic Robotics, ICDL-EpiRob 2019* (2019), pp. 75–80.

[17] KLOET, N., WATKINS, S., WANG, X., PRUDDEN, S., CLOTHIER, R., AND PALMER, J. L. Drone on: A preliminary investigation of the acoustic impact of unmanned aircraft systems (UAS). In *24th International Congress on Sound and Vibration, ICSV 2017* (2017).

[18] LANGEVELD, L., VAN EGMOND, R., JANSEN, R., AND ÖZCAN, E. Product sound design: Intentional and consequential sounds, Advances in Industrial Design Engineering, (2013), 47–73.

[19] LYON, R. H. Product sound quality - from perception to design. *Sound and Vibration 37* (2003), 18–22.

[20] MA, F., ZHANG, W., AND ABHAYAPALA, T. D. Active control of outgoing noise fields in rooms. *The Journal of the Acoustical Society of America 144* (2018), 1589–1599.

[21] MCMAHAN, W., AND KUCHENBECKER, K. J. Spectral subtraction of robot motion noise for improved event detection in tactile acceleration signals. , In *Proceedings of EuroHaptics*, (2012), 326–337.

[22] MOORE, D., CURRANO, R., AND SIRKIN, D. Sound decisions: How synthetic motor sounds improve autonomous vehicle-pedestrian interactions. , In *Proceedings - 12th International ACM Conference on Automotive User Interfaces and Interactive Vehicular Applications, AutomotiveUI 2020*, 94–103.

[23] MOORE, D., AND JU, W. Sound as implicit influence on human-robot interactions. In *ACM/IEEE International Conference on Human-Robot Interaction* (2018), pp. 311–312.

[24] MOORE, D., TENNENT, H., MARTELARO, N., AND JU, W. Making noise intentional: A study of servo sound perception. In *ACM/IEEE International Conference on Human-Robot Interaction Part F1271* (2017), pp. 12–21.

[25] MUTSCHLER, I., WIECKHORST, B., SPECK, O., SCHULZE-BONHAGE, A., HENNIG, J., SEIFRITZ, E., AND BALL, T. Time scales of auditory habituation in the amygdala and cerebral cortex. *Cerebral Cortex 20* (11), (2010), 2531–2539.

[26] DEVICE ORCHESTRA. We will rock you on 7 electric devices, https://www.youtube.com/watch?v=Hh9pm9yjmLs, 2019.

[27] RICHARDS, V. M., AND NEFF, D. L. Cuing effects for informational masking. *The Journal of the Acoustical Society of America 115* (2004), 289–300.

[28] CLEARPATH ROBOTICS. Jackal UGV, https://clearpathrobotics.com/jackal-small-unmanned-ground-vehicle/.

[29] SOFTBANK ROBOTICS. Pepper the humanoid and programmable robot, https://www.aldebaran.com/en/pepper

[30] SAMARASINGHE, P. N., ZHANG, W., AND ABHAYAPALA, T. D. Recent advances in active noise control inside automobile cabins: Toward quieter cars. *IEEE Signal Processing Magazine 33* (2016), 61–73.

[31] SCHÜTTE, M., MARKS, A., WENNING, E., AND GRIEFAHN, B. The development of the noise sensitivity questionnaire. *Noise and Health 9* (2007), 15–24.

[32] SNEDDON, M., PEARSONS, K., AND FIDELL, S. Laboratory study of the noticeability and annoyance of low signal-to-noise ratio sounds. *Noise Control Engineering Journal 51* (2003), 300–305.

[33] SONG, S., AND YAMADA, S. Expressing emotions through color, sound, and vibration with an appearance-constrained social robot. In *ACM/IEEE International Conference on Human-Robot Interaction Part F1271* (2017), pp. 2–11.

[34] TANNER, W. P. What is masking? *Journal of the Acoustical Society of America 30* (1958), 919–921.

[35] TAYLOR, J. L., AND DURAND, R. M. Effect of expectation and disconfirmation on postexposure product evaluations: A casual path analysis. *Psychological Reports 45* (1979), 803–810.

[36] TENNENT, H., MOORE, D., JUNG, M., AND JU, W. Good vibrations: How consequential sounds affect perception of robotic arms. In *RO-MAN 2017 - 26th IEEE International Symposium on Robot and Human Interactive Communication*, (2017), 928–935.

[37] TIAN, L., CARRENO-MEDRANO, P., ALLEN, A., SUMARTOJO, S., MINTROM, M., CORONADO, E., VENTURE, G., CROFT, E., AND KULIC, D. Redesigning human-robot interaction in response to robot failures : a participatory design methodology, In *CHI '21: ACM CHI Conference on Human Factors in Computing Systems*, May 08–13, (2021).

[38] TROVATO, G., PAREDES, R., BALVIN, J., CUELLAR, F., THOMSEN, N. B., BECH, S., AND TAN, Z. H. The sound or silence: Investigating the influence of robot noise on proxemics. In *RO-MAN 2018 - 27th IEEE International Symposium on Robot and Human Interactive Communication* (2018), pp. 713–718.

[39] UBTECH. Yanshee website, http://global.ubtechedu.com/global/pro_view-4.html.

[40] UNITREE. Go1 - unitree, https://www.unitree.com/en/go1.

[41] DE PAIVA VIANNA, K., ALVES CARDOSO, M., AND RODRIGUES, R. M. C. Noise pollution and annoyance: An urban soundscapes study. *Noise and Health 17* (2015), 125–133.

[42] PICO VILLALPANDO, A., SCHILLACI, G., HAFNER, V. V., AND GUZMÁN, B. L. Ego-noise predictions for echolocation in wheeled robots, In *Artificial Life Conference (ALIFE)*, (2019), pp. 567–573.

[43] WANG, L., AND CAVALLARO, A. Ear in the sky: Ego-noise reduction for auditory micro aerial vehicles. In *2016 13th IEEE International Conference on Advanced Video and Signal Based Surveillance*, AVSS (2016), pp. 152–158.

[44] WANG, Z., HU, Z., ROHLES, B., LJUNGBLAD, S., KOENIG, V., AND FJELD, M. The effects of natural sounds and proxemic distances on the perception of a noisy domestic flying robot. *ACM Transactions on Human-Robot Interaction* (2023).

[45] YILMAZYILDIZ, S., READ, R., BELPEAME, T., AND VERHELST, W. Review of semantic-free utterances in social human-robot interaction. *International Journal of Human-Computer Interaction 32* (2016), 63–85.

[46] ZHANG, B. J., PETERSON, K., SANCHEZ, C. A., AND FITTER, N. T. Exploring consequential robot sound: Should we make robots quiet and kawaii-et? In *IEEE International Conference on Intelligent Robots and Systems (IROS)* (2021), pp. 3056–3062.

7

Robot Sound in Distributed Audio Environments

Frederic Anthony Robinson, Mari Velonaki and Oliver Bown

7.1   Introduction

While a robot's ability to listen spatially has been an active field of research for a while, robots emitting spatial sound have received significantly less attention, as social robots are typically treated as a single sound source with access to a single loudspeaker. However, with the growing availability of audio-enabled networked devices in the home, one may imagine a future where (i) these devices become coordinated networks of distributed speakers emitting spatial sound throughout the home, and where (ii) social robots will be able to emit sound across these systems. This raises interesting questions around the role and benefit of spatial cues in a domestic robot's communication.

DOI: 10.1201/9781003320470-7

From a sound designer's perspective, a robot that has access to more than one speaker can begin to emit spatial sound and thereby draw from an extended pool of perceptual effects when communicating with humans. Spatial sound – positioning different sounds at different locations around the listener – creates a more immersive experience, enabling the creation of unreal, yet convincing sound environments [35]. Experiencing ambient soundscapes in this manner, for example, can make people feel calm or vibrant [12], more or less safe [37], or even affect how they recover in health care facilities [29]. More generally, reproducing sound in higher spatial fidelity has shown to substantially contribute to perceived audio quality [36].

From a human–robot interaction (HRI) perspective, robot sound being able to *leave* a robot's body has an additional implication. Using an anthropomorphism, the robot may be considered a *body* (hardware) and a *mind* (software) that can be disembodied. While many of the robot mind's capabilities are shared with other entities like voice assistants, the robot's physical body is a distinctive characteristic. The unique implications of a robot's physical presence have been thoroughly explored by human–robot interaction (HRI) researchers. See, for example, the work by Kiesler and colleagues, who compared how humans interact with a physical robot and a robot on a screen [24]. Among other things, they found that participants were simply more engaged with the physical robot. More recently, Li surveyed a large number of experimental works investigating this phenomenon and concluded that robots are perceived more positively and are more persuasive when physically present [28]. Sound, and particularly sound distribution, plays an interesting role in this mind body dichotomy, as robot sound can be attributed to a physical robot when it is emitted by the same, but it may well be attributed to some other source when it is perceived from somewhere else in the environment. As a result, distributed sound may turn this *physical presence* from the inherent characteristic of a social robot into a temporary state that can be changed at will. This poses interesting questions around a robot's identity, such as "How is a robot's perceived identity affected when it is no longer associated with its physical structure, and might this be desirable in certain interaction scenarios?"

Re-embodiment, the transfer of one single social intelligence between several physical robots, is a design paradigm that has recently received attention among HRI researchers. Luria and colleagues, for example, had researchers and designers explore the "vast design space of social presence flexibility" in a User Enactments study [30, p. 635]. Reig and colleagues explored user acceptance of a robot personality moving between multiple robot bodies in a service context [32]. Both studies reported that participants showed general acceptance of the practice. While these works consider the transfer of a specific type of sound, the voice, between multiple specific locations, robot bodies, it is not far-fetched to consider the distribution of robot sound across arbitrary locations. The latter has, in fact, been explored by Iravantchi and colleagues, who used an ultrasonic array, a highly directional loudspeaker, to project speech onto arbitrary objects in a living space, giving a "voice to everyday objects" [23, p. 376].

In the case where no such specialised equipment is available, one might ask where these speakers in environment should come from. Can we expect to have loudspeakers present in a robot's environment, and can we expect a robot to have access to these loudspeakers? Strong indicators that both of these prerequisites will eventually be met can be found in current developments in the internet of things (IoT) ecosystem. A growing number of hardware devices with networking capabilities permeates public space, work places, and the home [3]. Many of these devices have loudspeakers built into them, and efforts have begun to use these capabilities to create distributed audio playback systems. Electronics manufacturer LG, for example, patented an implementation of smart connected distributed audio systems in the home using miscellaneous networked devices such as TVs, fridges, and, interestingly, also robotic vacuum cleaners. Their implementation utilised approximate position tracking and streaming via the 5G network [25]. It is reasonable to expect that robots will eventually be able to make use of these systems, and the various challenges and opportunities surrounding this new design space are, in our view, worthy of exploration. In light of these considerations, this chapter aims to map out a design space for spatial sound in human–robot interaction by asking the following question:

"How might a robot's auditory communication be enhanced by being distributed across loudspeakers in the environment?"

To explore this question, we first conduct interviews with researchers and practitioners with distributed audio and sound installation experience, identifying design themes for applying interactive sound installation techniques in the context of human–robot interaction. These insights, combined with the authors' own expertise in interactive immersive audio environments, then inform the creation of a virtual distributed robot sound prototype. This process includes the ideation, and realization of spatial robot sound. After presenting the prototype and its design process, we reflect on lessons learnt and propose a generalised design framework for spatial robot sound. Rather than attempting to capture all of the experimentation presented in this chapter, the framework provides a pragmatic formalization of what we argue to be key aspects of spatial robot sound.

7.2   Research Context

In order to provide additional context for this research – how human–robot interactions might benefit from distributed robot sound – two questions need to be answered, "What is the current research on distributed sound in the home?" and "What is the state of current research around distributed sound in human–robot interaction?"

7.2.1 Distributed Sound in the Home

A discussion of the technologies around networked audio systems in the home is beyond the scope of this chapter. It is, however, worthwhile to consider current research around the design challenges faced in this context. A notable concept in this space is what Francombe et al. coined *media device orchestration* (MDO) to describe the "concept of utilizing any available devices for optimum reproduction of a media experience" [19, p. 3]. Their motivation is partly based on user studies suggesting that more general spatial audio experiences such as envelopment and image depth are more relevant to the user than precise localization [31]. This potentially makes MDO a viable alternative to precisely calibrated loudspeaker systems that can reproduce spatial audio in high resolutions and may enable novel and improved multimedia experiences. Listening tests utilising MDO showed, that low-channel count sound material could be successfully augmented by being distributed across various devices, such as phones and tablets, traditional stereo systems, and TVs [43]. An early audio drama by the BBC which utilised the technology was well received among listeners [18]. More experiences have been created since then. Discussing the concrete implementation of this type of audio content into a broad range of loudspeaker setups in homes, Francombe et al. state that "it is important to understand both the required metadata and rendering methods to best select devices for different object types and audio signal features" [20, p. 424]. Defining a media format to enable mass distribution remains an ongoing challenge, raising questions concerning both content creation as well as the role of the end-user, who in the current implementation has to report the device positions manually [21]. The researchers have created a toolkit that allows designers to create their own interactive spatial sound experiences [4].

7.2.2 Distributed Sound in HRI

When considering the question how robots might fit into such an environment, it should be noted that robots have long been applied within smart environments in health and aged care contexts [10]. Bordignon et al. state, that the areas of robotics and ambient intelligence are "converging toward the vision of smart robotic environments" [8, p. 3101], and so far, various research in this domain has focused on establishing the required technological and conceptual frameworks [11,16,41,42]. Sound has, to our knowledge, seen little consideration in this context, and research on robot sound with a spatial component is generally sparse. The context it has appeared in thus far is mainly that of localization – using the spatial cues embedded in a robot's sound to make it easier to localise by a human sharing the same space. The benefit of this is improved spatial coordination and lower chances of collisions [17]. In a study investigating the effect of sound on human–robot collaboration, Cha and colleagues continuously sonified robot movement and found that specific sound profiles made the robot easier to localise for the human [14]. In this work the robot sound source and the physical location of the robot were treated as one and the same. A disconnect between the two, however, was discussed by industry sound designers interviewed in a previous studies of ours [33]. Toy robot Cozmo, for example, has a complete parallel music system with orchestral recordings emitted by a companion app on a mobile device. Robot companion Vector had the opportunity to emit sound via surrounding smart devices, but its designers decided against it in order to not distract from its pet-like character. Due to its speaker characteristics, the voice of social robot Jibo sometimes appears to come from the walls around the robot, which was described by its sound designer as a design flaw.

7.3   Expert Interviews

Due to the exploratory nature of this research, we chose to collect qualitative interviews on the participants' design practice in order to gather data that is both rich and detailed [2,39]. The expert interviews were motivated by two key questions, "How is the medium of sound applied in immersive environments?", and "How might practices from that domain be translated into the HRI context?" Themes emerging from the analysis of these interviews should then provide a series of design considerations around the use of interactive sound in immersive environments, which could be used to inform the design of a spatial robot sound prototype. The semi-structured interviews were loosely based on a shared set of questions to ensure all relevant areas are covered [6]. However, the questions were adjusted to target relevant projects from the interviewees respective portfolios. Participants were also encouraged to add any information they felt was not covered by the questions to reduce the questions' role in shaping the responses [22]. This approach allowed us to (i) ask detailed questions on the participants' design process for existing works, and (ii) encourage them to envision how these processes might be applied to the context envisioned in this paper: spatial robot sound for the home. It should be noted that interviewees were not familiar with this new context. It was therefore the interviewer's task to encourage participants to apply experiences from prior work to this new context without preempting their own opinions on the subject [27]. The interviews took between 45 and 75 minutes and were conducted remotely via video call. The resulting six hours of audio recordings were then transcribed.

7.3.1   Thematic Analysis

To analyse the transcripts, we used thematic analysis, an inductive technique to identify and cluster themes from interview data [9]. This process involves the careful examination of qualitative data to identify patterns and extract meaning [26]. This was done in several stages. In a first step, we became familiar with the data by reading the transcripts to understand the overall message behind participants' answers. In a second step, the transcripts were codified, meaning important features of the data were marked with succinct tags. The codes were then separated into categories to identify broader patterns in the data. Finally, we generated themes to describe the categories. To provide additional context with these themes, we additionally quote participants, as recommended by Bechhofer and Paterson [5].

7.3.2 Participants

The participants include designers and researchers working with interactive immersive sound environments. This can range from technical work such as speaker planning and system design, through conceptual design, to sound design practice. We chose experts with this background because of the common ground between their disciplines and the focus of this paper: using interactive spatial sound to create rich and engaging experiences for the listener.

 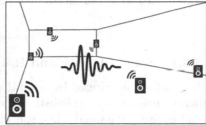

FIGURE 7.1
Left: An exhibition space. Right: Distributed sound sources create an immersive environment.

An example of this kind of work can be found in site-specific media installations, where designers use distributed and interactive audio throughout museum spaces to create themed, coherent sound and music environments that reinforce core messages of the exhibition [15]. An environment with distributed sound sources is illustrated in Figure 7.1. Several of the interview participants had previously worked on projects like these, which additionally touched on themes relevant to the HRI context. Among others, the projects included (i) a large, cloud-like audiovisual installation that used interactive sound to convey notions of intelligence and animacy, (ii) an exhibition about the history of automata and robots, featuring sound for the specific machines exhibited, as well as enveloping soundscapes, and (iii) a museum that had a bodiless artificial intelligence follow visitors through the exhibition area.

7.3.3   Themes

Five main themes emerged during analysis of the expert interviews. They are *sonic identity and robot fiction, functions of sound, roles and affordances of interactive sound, roles and affordances of distributed sound,* and *technical considerations,* and are shown in Figure 7.2.

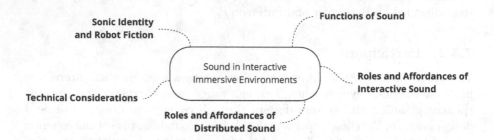

FIGURE 7.2
Five themes emerging from expert interviews with researchers and designers working with sound in interactive immersive environments.

7.3.3.1   Sonic Identity and Robot Fiction

A key notion mentioned by several of the experts was the idea of creating a coherent, consistent, and holistic sonic identity, both when discussing immersive sound environments and when speculating on the embedding of robotic agents into them. Motivations for this are not only a more conceptually refined design but also potential branding considerations. This idea of sonic identity matches well with the notion of robot fiction, mentioned by robot sound designers working toward creating a believable robotic character [33]. Sonic identities are created by putting all emitted sound in relation to each other, or, in the words of one expert, by "making them harmonize".

Often, references to existing sound sources are used to achieve this. Creating sound for an exhibition on robotics, one expert drew from recordings of historical industrial equipment. Creating a soundscape whose particular timbre and internal rhythm was defined by this idiosyncratic but real sound source was "immediately enhancing the visitors' projection". In their words: "You know this is not real. This is quite clearly artificial, but you want to believe in it. You want it to be real". This referencing was also said to potentially constrain the designer's creative freedom. One expert noted how building sonic identities around rigid references can provide a "narrow" design space. As an example, they mentioned the sound of an electric vehicle, which can be a digital recreation of a combustion engine, or it can be "invented from the ground up". A middle ground was discussed by a third expert, who was tasked with creating affective sound for an installation containing a responsive structure meant to be perceived as animate. They recorded the human voice non-verbally expressing various emotional states, and then processed them to turn them into utterances of an artificial character. In this instance, loosely referencing the human voice provided an "organic" source material, while the processing of the sound material made it part of a non-human sonic identity. As a result, the responsive structure was perceived as an "independent thing that is [...] not just trying to imitate".

Opinions differed on how much interactivity or personalization should be allowed to work against or even break sonic identities. One expert noted that a user's influence on the soundscape should be limited in a way that does not allow for the fiction to be broken. The core message of the sound is unchangeable, and interactivity is a thin layer above it. Another expert felt the sonic identity of a robot should be able to be easily changed and personalised. According to them, people may want different things out of the same robot model – or even the exact same robot – and sound is a key differentiator here, as it is the only thing that can be significantly transformed at any point during the robot's deployment.

Space is a relevant parameter for playing with the notion of sonic identity, as sound distribution has previously been flagged as a sound characteristic that can impact a robot's fiction. One expert described a robot embedded in a sounding IoT environment as both a self-contained character and an interface through which one could interact with their smart home. Both of these cases imply certain sound distributions. Robots Cozmo and Vector's sound designer Ben Gabaldon stressed how his robots could have emitted sound through smart devices in their environment, but this was avoided specifically to not break the robots' fiction [33].

7.3.3.2 Functions of Sound

Experts highlighted three general aims in their sound designs: (i) conveying and causing various emotional states and moods, (ii) providing orientation and guiding a listener's attention, and (iii) creating associations to external concepts, environments or things. Those who worked on robotics-related projects specifically highlighted the use of associations to convey animacy – "how can you deceive someone into thinking that this is more lifelike than it is?" – and to differentiate the various elements that come together to form a robotic agent. One expert suggested using recordings of real sound sources to convey information about the robot body, and purely synthetic sound and musical patterns to convey information about the robotic mind, such as data flow, networks and connections between several robots.

Another expert proposed five categories of sound to be used in a robot's communication: speech, sounds of the body, alert sounds, sounds used for orientation, and moods and atmospheres. For the latter, they suggested the use of music for conveying mood, and more realistic ambient soundscapes for illustration, noting that both of these are a promising new way of communicating that could be unique to the machines, as neither are part of a human's communication. Several experts mentioned the potential of combining these two by embedding musical material into ambient soundscapes to create abstract but emotional environments.

7.3.3.3   Roles and Affordances of Interactive Sound

In the context of this chapter, *interactive sound* describes sound that is in some way mapped to states and actions of a human, robot, or the environment. The experts discussed various ways to make sound responsive, both in simple and complex ways. Data that could be used to affect sound includes the number of people present, weather data, time of day, distance to humans, and temperature. Sound parameters to be affected could include volume, timbre, choice of sound material, and audio effect processing. One expert described how these mappings could be combined to build sound environments that are perceived as a "basic intelligence". They also mentioned, however, that one can "take a very simple piece of sensor information like distance [and] use it to communicate a lot". In an application of this, they used distance information to make a responsive structure convey increasingly negative and agitated emotional states as it was being approached, communicating not only that it is aware of its surroundings but also hinting at underlying personality traits. In another interactive installation, they distinguished between three different types of interactivity, (i) *sharing*, where the installation communicates but does not respond to the listener, (ii) *subtle interactivity*, where the listener feels like they affect the soundscape but do not know how specifically, and (iii) *play*, where there is an explicit interaction between machine and human.

When asked about specific generative techniques, one expert described the use of several mapping layers to first translate arbitrary robot sensor data into parameters that are relevant to the interaction, before then using those parameters to control sound output. For example, we may not be interested in a robot wheel's rotations per minute, but observing the motion of all wheels over time, and combining them with information about the surface it is moving on, may then be translated to a parameter called "exertion". This information can then be mapped to various sound characteristics to convey the notion of strain. As a result, a simple model of a biological muscle is inserted between sensor data and sound mappings, essentially allowing the designer to define interactive behavior in two stages, first translating raw robot data into meaningful descriptors of robot behavior or interaction scenario, and then translating these descriptors into sound. It also means that the first half of this responsive sound system is independent of the actual sound generation methods being used. The expert notes that this allows the designer to freely choose between different approaches like real-time synthesis, or sample-based playback.

7.3.3.4 Roles and Affordances of Distributed Sound

The potency of immersive sound to create a range of powerful listening experiences and was a recurring theme among experts, who highlighted how individual sound sources from locations around the listener can together form rich and believable environments. The ability to place sound at different locations around the listener allows for a "huge level of complexity" to emerge. One expert mentioned how the use of "tangential sounds" played through loudspeakers across a space can "give this sense of detail that [people] are used to in a real environment", allowing the listener to "project a lot further". Another expert mentioned how distributing sound across a space not only allows for the creation of detailed environments but also for the creation of clear distinction and separation. Referencing a prior project, they described how individual voices of a recorded string quartet were separated spatially, thereby making it easier for the listener to focus on individual instruments.

When asked about the potential of distributed or spatial sound for robot communication, many experts discussed the boundaries between the robot sound and sound in the environment. One described a project in which they placed microphones inside of an animatronic and projected the internal sounds of the machine into the surrounding space. When the animatronic would perform gestures, the sounds associated to these gestures would then be amplified across the room, at times in a dramatic fashion. As a result, the gestures of the machine were staged in a way that made them highly noticeable and relevant to the listener. Another expert described a project that, instead of using the space to accentuate actions by the physical robot, used space to blur the boundaries between robot and its environment. A media exhibit that was meant to appear animate had built-in speakers, as well as speakers across the exhibition space. When the exhibit communicated, the entire space appeared to resonate.

A third expert described the possibility of a robotic sound source temporarily leaving its physical body to travel to locations in the home that the body might not be able to reach in time, or at all. As an example, one might imagine a robot planning to enter the kitchen, and before doing so, sending only its voice onto a loudspeaker in the kitchen to announce its arrival. The expert described this disembodied state as the robot being *schizophonic*, a term coined by composer R. Murray Schaefer, which refers to sound being separated from its original source, a misalignment between the auditory and the visual [38].

7.3.3.5 Technical Considerations

Experts were asked to elaborate on technical considerations when working with sound across an immersive sound environment, and what they would consider noteworthy when imagining robots in such a context. A common concern was variety in speaker distribution to create depth and give designers access to a high number of diverse spatial locations. One expert illustrated this with an example where loudspeakers distributed across the ceiling may evenly cover an area, but still provide sound from only one general direction, above. In their words, "we don't want the room to be in the ceiling". Another expert noted that with envelopment being a key perceptual attribute, distributions do not necessarily need to resemble standard speaker layouts such as 5.1 to be effective. This differs from the well-established audio research paradigm of localization accuracy, where the quality of a spatial sound experience is often quantified by measuring the smallest noticeable offset of degrees between two sound sources. The expert instead suggested to spatialize sound across broad and clearly distinguished categories, such as close vs distant, point source vs enveloping, and moving vs static.

Focusing on the audio capabilities of the robot itself, one expert highlighted the dispersion characteristics of a robot's speaker – meaning whether a loudspeaker emits sound in a targeted or omnidirectional fashion. A robot emitting sound in all directions will have better speech intelligibility for people not faced by the speaker, but this might not be desirable when they are not part of the interaction. A robot with a highly directional speaker, in turn, will have to adjust its orientation to face a specific listener that is to be addressed. In their eyes, distributing robot sound to other speakers could then be a way to apply these effects in a targeted and refined way. However, this reliance on using external speakers for communication would also limit functionality in deployment locations that do not have additional speakers in the environment.

7.4   Distributed Robot Sound Prototype

After presenting the various themes emerging from the expert interviews, we now take key notions from these findings and work them into a design prototype: a distributed robot sound profile. Based on the themes, A) *sonic identity and robot fiction*, B) *functions of sound*, C) *roles and affordances of interactive sound*, D) *roles and affordances of distributed sound*, and E) *technical considerations*, we chose the following design goals:

- Create a holistic robot sound set based on a core fiction (A), which incorporates utterances, music, movement sound, and ambient sound (B).

- Design the robot sound set to take into account sound emitted by both the robot and the surrounding space (D).

- Implement the sound set in a virtual environment, so that spatial distribution and responsive behavior can be prototyped without being restricted by technical constraints (E).

- Make the sound set responsive to user presence and activity (C).

FIGURE 7.3
An overview of the prototype design process.

An excerpt of the resulting prototype is shown in https://www.robinson.
audio/distributedrobotsound Video Excerpt 1. It showcases the sound of
various behaviors of robot mind and body in a virtual environment. It also
shows the spatial distribution of sound elements, by placing them around a
centered listener. The video has a binaural spatial audio soundtrack which,
when listened to with headphones, positions sounds around a virtual listener
standing in the center of the room. As a result, sound taking place in the top
half of the room is positioned in front of the listener, while any sound in the
bottom half is positioned in the back.

The following section details the prototyping process and discusses the
various design decisions made. Based on a robot sound design process presented
in prior work [33], we followed a five-step design process (see Figure 7.3). It
involves (i) the creation of what we call an *Audio Design Board*, (ii) the
collection of sound sources making up the *sound identity*, (iii) the creation of
a flow diagram we call *Audio Interaction Flow*, (iv) the *production of audio
assets*, and (v) the final *implementation* of said assets into a virtual robot. The
Audio Design Board collects a number of high-level metaphors that informed
the robot characteristics, personality, and behaviors we were to design for, and
the associations we wanted to evoke. To create the robot's sound identity, we
collected sound materials (recordings) and sound sources (synthesis methods),
which together would make up how the robot sounds. The identity should
sound coherent and consistent and be appropriate for and reflective of the
robot's application context and capabilities. The next step was the creation of
what we call an Audio Interaction Flow (see Figure 7.4), a modified user flow
diagram that represents the entire robot sound set, and how it responds in
different interaction scenarios. This flow diagram then formed the foundation
for the asset production, the creation of all individual sound loops and events to
be emitted by the robot and its environment. Finally, during implementation,
the sounds were reassembled in an interactive audio engine. In the following
sections, we would like to specifically address two steps in particular: the Audio
Design Board, and the Implementation.

Audio Flow

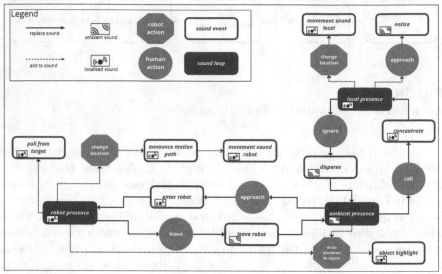

FIGURE 7.4
The prototype's Audio Interaction Flow, a flow diagram giving an overview of all sound assets and their interactive relationships.

7.4.1   Audio Design Board

Figure 7.5 shows the various ideas, metaphors, and associations that formed the foundation of the prototype design. Their high-level categorization is based on the candidate design principles we identified in previous work [33]: fiction, source, scope, interactivity, and production. The principles provided a framework to assist in creating a complete and internally consistent robot sound set which, in this case, should be emitted both by the robot itself and across its environment.

We chose the **fiction** to be that of a playful and theatrical abstract digital entity. The robot fiction should be technological in nature, and abstract enough to take different forms, such as freely moving across the room, or being embodied as a robotic agent. We also wanted the entity to be expressive, as it would allow us to more freely explore different sound design options instead of working with a more muted sound pallet.

The robot sound **source** should comprise both sound emitted by the physical robot body, as well as any position in the environment. The "true form" of the entity should be apparent when it was freely moving around. Sonically, this should be reflected by sounds with a smeared, floating, and pulsing quality, as if hard to grasp or clearly make out. Entering the robot body would then be reflected in sound through a more grounded, low-frequency range,

Audio Design Board

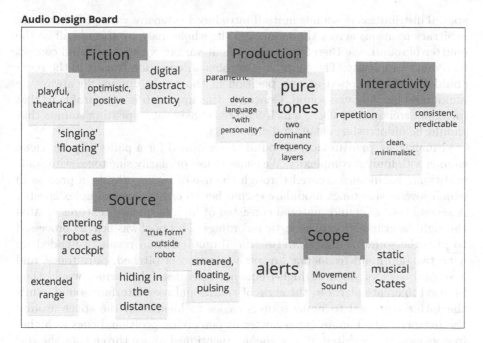

FIGURE 7.5
An Audio Design Board bringing together a collection of metaphors and associations, derived from five top-level design themes identified in interviews with robot sound designers [33].

and more tactile sound with clearer transients and more natural sounding material. However, this range of sounds – from smearing to tactile – should be derived from the same core sound material through editing and processing, giving a coherent core identity to the overall sound pallet.

The **scope** of the robot sound should comprise movement sound, static musical states, and alerts. The latter could also be considered semantic-free utterances, but we chose to stay closer to more traditional UX sound conventions, because the entity was not exclusively embodied, but could also emit sound through smart speakers.

Interactivity refers to the way robot sound is modified through parameters in the environment or through randomness to create a more varied and rich listening experience. For this prototype, we decided to favor the simple playback of single sound files, and not introduce any additional variation. Any robot action would therefore always result in the same respective sound event being played back. We chose this for two reasons. The first was the prior mentioned link to smart devices, which encouraged the use of a UX sound-based language which rarely features this kind of variety. The second reason was that the spatial distribution of sounds in itself introduced extensive variety. Robot body, arbitrary positions across the room, and the whole space itself could all be the emitter of sound. We therefore chose to limit variety within the sound content to reduce complexity. The sound of the robot mind concentrating in the room could therefore happen at any position in the room, but would always be announced by the same sound. Generally, this approach is comparable to that for commercial robot Kuri, which features consistent, repeating sounds that should be unintrusive, clean, and simple [33].

Finally, the **production** of sound assets aimed for a pallet of pure, clean sounds with limited complexity. We chose to use primarily sine tones with some additional harmonics created through the use of FM synthesis, a process in which several sine tones modulate each other to create more complex spectra. A second pool of sound material consisted of high frequency textures created through the granulation of acoustic recordings. This pool was heavily processed to give the sound a digital and artificial quality. As a result, we ended up with two dominant frequency layers, a selection of pitched, pure tones, and a second layer of smeared high-frequency textures. The former would then be used to create alerts in the style of traditional user interface sound, while the latter was used to make sounds easier to localise in the space around the listener – the human ear is better at perceiving positional cues in higher frequencies. We revisited all five themes mentioned above throughout the rest of the design process, both to get inspiration for sound choice and processing, and to ensure that the sounds we created were consisted with the overall design goals established in the beginning.

7.4.2   Implementation

The interactive behavior of the prototype can be summarised as follows: The robot sound has three major states, which we describe as *presences*. A *robot presence* is emitted when the robot mind is present in the robot body. In this state, the robot can experience a *pull from a target* location – for example, when being asked to move somewhere – and it can *announce a motion path*, which is then subsequently accompanied by a *movement sound*. The second state is the *ambient presence*, in which the robot mind is present and available across the entire space. Leaving and entering the physical robot body are accompanied by a *leave robot* and *enter robot* sound event respectively. When embodied or present across the space, the robot can *highlight an object* in the environment with a sound event. The third state is a *local presence*, during which the robot mind is not embodied, but localised at a specific point in the environment. This state is accessed through a *concentrate* sound event, and left through a *disperse* sound event. When in the local state, the robot mind has its own disembodied movement sound, and can acknowledge events happening around it.

After producing audio assets for all of these states and events, we implemented them into a robot to experience them in context. For this chapter we chose to implement a virtual prototype, rather than deploying a physical robot. This was done for two reasons:

(i) Earlier work co-authored with Albastaki and colleagues demonstrated the value of Virtual Experience Prototypes (VEPs) as tools for prototyping and evaluating robot behaviors, including robot sound [1]. We created a virtual version of a physical robot and embedded it in a virtual environment which resembled a prior deployment location of the physical robot. We found VEPs to be "lightweight in development and deployment" enabling evaluations that are "location-independent with a broad pool of potential participants" [1, p. 84]. We also found that impressions of the virtual robot closely resembled those of the physical robot in an earlier real-world deployment. While remote data collection was not part of our goals for this chapter – a physical deployment of the robot will be presented in future work – we were interested in using the VEP's inherent speed and ease of use for our prototyping workflow, and wanted to gather early impressions of the final outcome ourselves.

(ii) The second reason for virtual deployment was the focus of the research: distributed sound in the robot's environment. With the goal being the exploration of arbitrary sound distributions, we wanted to work in a prototyping environment that supports this. Different from the very site-specific loudspeaker distributions in prior media-installation work – and loudspeaker distributions across a robot's body described in previous work [34] – we wanted to work with abstract sound trajectories that might apply to the arbitrary loudspeaker distributions promised by the current IoT developments in the home (see Section 7.2.1). A virtual sound environment provided this abstraction.

We therefore embedded a 3D model of robotic artwork Diamandini, discussed in prior work [34], within a custom-made virtual environment created in the Unity game engine. The virtual environment, which is modeled after the University of New South Wales' National Facility for Human–Robot Interaction Research, is shown in Figure 7.6. Mouse-based interactions with the virtual prototype are shown in https://www.robinson.audio/distributedrobotsound Video Excerpt 1. The remainder of this section will describe the various components making up the virtual environment. An overview of the system is shown in Figure 7.7.

- The virtual environment contains a 3D model of the robot Diamandini and various visual elements that showcase the location of the robot mind at any given time and visualise the distribution of key sound sources. When the robot mind is embodied in the robot body, the model lights up. When the robot mind is present across the space (ambient presence), the entire room is filled with small glowing particles. When the robot mind, and by extension its sound, concentrates at a particular location, these particles concentrate as well.

- We programmed the model and the visualizations to respond to mouse and keyboard interactions. Clicking on any location in the space moves the robot to that location (when it is embodied or localised) or makes it

FIGURE 7.6
Prototyping spatial robot sound in a virtual environment.

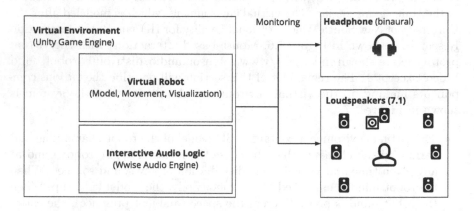

FIGURE 7.7
Overview of the virtual environment. Spatial sound can be monitored either
through headphones or loudspeakers.

concentrate at that location (when it is an ambient presence across the space). Double clicking any location in the space makes the robot mind leave the robot body (when it is embodied) or disperses it (when it is a local presence). Clicking on the robot model makes the robot mind enter the body, and various click and key combinations announce motion paths or highlight objects.

- The soundscape of the virtual environment can be heard via the standard audio output of the game engine. This means playback can be heard via headphones using binaural rendering, which renders a 3D audio environment to two headphone channels. This can be heard in https://www.robinson.audio/distributedrobotsound Video Excerpt 1. Audio can also be played back via a 7.1 loudspeaker system in our own, or any other studio with a surround sound system. This is how we worked with the prototype. It also means that running the virtual environment in a room with a surround sound system projects the virtual space into the listening space and thereby provides a way to listen in on how any designed sound set might sound in a home environment.

- We used the Wwise audio engine, embedded within Unity, to specify interactive audio behavior. One particular benefit of this modular approach is that the robot sound is not hard coded into the virtual environment and can therefore be taken out of the game engine and used in any other context. A design tested and finalised inside the virtual environment can therefore easily be exported and used within an actual physical robot. This is the approach taken for a physical deployment discussed in future work.

7.5 Findings and Discussion

After presenting the prototype, this final section contains a critical reflection of the design work done in this chapter, and subsequently generalises the findings in form of a design framework for distributed robot sound. Together, these two sections address the question "How might a robot's auditory communication be enhanced by being distributed across loudspeakers in the environment?" Finally, the section discusses limitations and future work related to the research in this chapter.

7.5.1 Lessons Learnt

7.5.1.1 Distributed Robot Sound

Needless to say, extending robot sound to any location in the environment results in large number of design possibilities. If we, for example, assume a rough spatial resolution of 1 meter – meaning we can move a sound across a space in 1 meter increments – this would give us around 30 possible locations for any sound emitted in a standard-size living room. Some of these locations are functional, such as highlighting relevant objects in the environment. Many others could serve a more aesthetic function, such as enveloping the listener or adding an additional creative dimension to sound events. A key parameter in this context is the level of spatial complexity in the sound design. Not all 30 positions in the above-mentioned example are relevant, and we aimed to identify a level of abstraction that would maintain a clarity of expression while still making use of this additional design dimension. Throughout this process, we experimented with various spatial effects that did not make it into the final design. The fundamental effect we noticed was that a sound's spatial movement path was influenced by the presence of a robot. Placing a robot in front of the listener led to sound in that general area be attributed to the robot. As a result, absolute sound positions, such as *left, middle, right* were more likely be perceived in relation to the robot (*outside the robot, inside the robot*). This manifested, among others, in the following ways:

- Making a sound circle around the listener is a popular technique in spatial sound design, because it highlights the spatial nature of the audio playback. When using this effect while a robot was in the field of view, sound positioned behind the robot was attributed to the robot. Any gradual circular movement was therefore broken up.

- More generally, the difference between sound located near the robot and sound located at its exact position was negligible. To tease out a difference in location between robot sound and environment sound, the distances had to be an estimated 3 meters or more.

- Rapid jumps between locations, such as the robot and an object in the environment, or simultaneous playback from multiple sources would perceptually fuse into a single global sound event and again be attributed to the robot, instead of being perceived as several distinct sources. This could be mitigated by either adding pauses between sound events at different locations, like a *call and response* pattern, or by giving the two sound sources significantly different timbres.

After excluding these and similar effects, we ended up working with a smaller number of key perceptually relevant locations and events. Those were a sound's *source*, *target*, and *transitions* between a source and target, or *across the entire space*. In other words, a sound was either *at* an object of interest, *moving away or toward* it, or it was *everywhere*. In the latter case, this could mean it was a global event fully in focus, or a subtle background ambient. These categories are reflected in a design framework in Section 7.5.2. Even with this rather reduced design space, the number of combinations is still substantial, allowing us to use spatial location to clearly differentiate sound events with similar timbres.

7.5.1.2 Applying Sound Installation Practice in Human–Robot Interaction

To conclude the lessons learnt, we will revisit the themes that emerged from the interviews in Section 7.3 to investigate how well they translated into actionable guidelines for the prototyping process and, by extension, how applicable and useful a sound installation perspective is to the HRI context.

When discussing **sonic identity and robot fiction**, experts emphasized how sound should be built around a core sound identity, noting how this helps create more coherent designs, while at the same time narrowing the design space. In the case of this prototype, we created a sound identity that took UX sound common among smart devices as a key reference. By mostly adhering to the conventions in that space, the sound set was relatively consistent. The key difference to more traditional UX sound was that we wanted to make a clear distinction between sound located in the robot, and sound emitted across the environment. We did so by making the sound inside the robot more mechanical and tactile, while making the sound in the environment more digital and smeared. An issue we faced in that context was that sound that was fitting for the sound identity was not necessarily fitting for the functions that sound needed to fulfill. An example of this was that many UX sound sets feature pure electronic tones with little high-frequency content. High-frequency content, however, was needed to make sounds easier to localise in space (see, for example, [13]). This presented a friction between functional and conceptual requirements for the design, and the audio processing and additional sound material we added to meet the functional requirements was one way to make this compromise. Generally, this theme's notions around identity, core fiction, and coherence were all reflected in the expert interviews in this chapter, prior robot sound expert interviews [33], our own prior design practice and the design work showcased in this chapter. We consider it highly applicable to the HRI context.

When discussing **functions of sound**, experts mentioned conveying emotion, providing orientation, and causing associations as key functions. They also suggested to choose sound materials based on the associations to be caused, and highlighted music and soundscapes as two unique ways to communicate beyond human speech. In the case of this prototype, we used musical elements throughout most of the design, taking traditional UX sound conventions and combining them with harmonic soundscapes. Providing orientation and drawing the listener's attention to specific locations was a key concern, and this was done through alert sounds with specific spatial properties. We also used sound material to create specific associations to a physical robot body (through tactile mechanical recorded sound) and to a disembodied digital entity (through heavily processed, smeared sound events). We consider these notions readily applicable to the HRI context. The possibility of using spatial sound to (i) subtly communicate through background ambient sound and (ii) guide the listener's attention to specific locations in the space, where both key motivations for the work in this chapter. While both of these have clear creative rationales – immersive musical soundscapes are pleasant to listen to, and space adds an exciting creative dimension to sound design – we also argue that those in particular have applications within more functional HRI contexts; that being able to subtly convey certain moods or draw attention to certain objects relevant to HRI scenarios is a valuable tool in an HRI designer's pallet.

When asked about **roles and affordances of interactive sound** – meaning sound that is in some way mapped to states and actions of human, robot, or the environment – the experts discussed various types of interactivity with varying degrees of complexity, illustrating their ideas with examples from their prior work. With this chapter's focus being on spatial sound, we implemented only basic interactivity. Key events were mapped to human presence. For example, approaching the robot body woke up the ambient presence and caused it to enter the physical robot. There were also several instances of mapping continuous robot movement data to the intensity of dedicated movement sounds. While we consider the various notions highlighted in the interviews to be generally applicable, the more complex ideas of interactivity were not explored in this prototype. Using distance as a parameter to accentuate human presence and robot awareness was, however, explored in previous work [34].

Discussing the **roles and affordances of distributed sound**, experts noted a significant increase in detail and complexity that comes with adding a spatial dimension to the sound design. They also highlighted its role in separating two sound sources to create noticeable differences between them even if they have similar timbres. One expert also suggested the use of spatial sound to separate an agent from its physical structure. In the case of this prototype we explored this complexity and found that when using it in the context of an agent, whether embodied or not, the space of potential sound locations had to be reduced, as not all design possibilities were equally relevant or effective. The use of spatial sound to create detailed immersive soundscapes that convincingly transport the listener to a different place is common in the installation context, but less relevant in this HRI context. The notion of separating sound sources was much more applicable. Clear distinction and disembodiment provided a wealth of design possibilities, and link well with HRI research into re-embodiment.

When asked about **technical considerations**, experts highlighted two key points, which partially contradicted each other: (i) To create quality designs, sound practitioners need access to a large number of loudspeakers to create a high-resolution spatial image. (ii) Rather than a complete high-resolution spatial image, designers only need the technological environment to create a few key experiences: enveloping sound or sound emitted from a point source, close or distant sound, and moving or static sound. In our prototyping work we chose to avoid potential technical constraints around spatial resolution by working within a virtual environment. Regarding the question of a high-resolution image versus a few key locations, the robot context provided interesting insights here. Due to the fact that sound was designed in the context of a robot, spatial locations were perceived in relation to it, which came with an inherent loss in spatial resolution. Some locations were inherently more relevant than others. Even though we had high-resolution spatial sound at our disposal, we ended up reducing the spatial resolution of our design, which leads us to believe that potential constraints in spatial resolution are less of an issue than some of the experts believe.

7.5.2 Design Framework

The core focus of this chapter – experimentation with a wide range of spatial behaviors – will now be generalised and simplified in the form of a design framework, a set of spatial sound distributions and trajectories we consider relevant to human–robot interactions. Its purpose is not to list or summarise the spatial sound work done for this specific prototype, but rather to map out a broader and simplified design space that explores the question "What are the ways distributed sound could be used in a robot's auditory communication?" These distributions and behaviors may then hopefully prove useful to HRI designers considering the use of spatial audio. The framework considers three *key locations* of robot sound (see Figure 7.8) which we consider as the most suitable simplification of the design space, and then applies this thinking to *four examples of spatial sound events* (see Figure 7.9). We argue that spatial robot sound should be broken down into these three key locations – the robot, objects of interest in the environment, and the space itself – and that transitions between these locations create spatial relationships which are relevant to HRI scenarios.

7.5.2.1 Key Locations

Based on the lessons learnt in Section 7.5.1.1 - not all sound locations are equally relevant to HRI and noticeable to the listener – we can render the broad space of possibilities down to three key sound-emitting locations. Figure 7.8 shows these locations: the robot, any position in the environment, and the space itself. If we imagine a social intelligence – let us again call it the robot mind – moving between these locations, we can also see them as states. The robot mind is either present within the robot body, at an arbitrary position in the environment, or everywhere. As a next step, we define transitions between these three states. While these are not strictly necessary, they allow the designer to convey continuity and direction between states. The language describing the transitions between these three states is heavily inspired by Smalley's spectromorphology [40]. While it ultimately references sound events, it uses descriptions of non-sounding phenomena.

Having the robot mind located within the **robot** body results in the standard scenario of an embodied agent. From there, the robot mind can

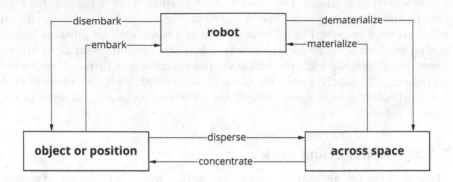

FIGURE 7.8

Three sound locations relevant to Human–Robot Interaction. Sound is emitted either by the robot itself, by an object or at any position in the environment, or across the space, without a discernible source.

disembark and travel to an arbitrary position or object in the environment. An example case for this would be the robot being asked to follow the user and continue a conversation in a room upstairs which the physical robot cannot access. To do so, the robot mind would disembark the robot body and move to a smart speaker in another room. The robot mind could also *dematerialize*, leaving the robot body to be present across the entire space. An example case of this would be the robot mind leaving the robot body to go into an inactive state, from which it can be awoken via voice commands.

Placing the robot mind at an arbitrary **object or position** brings the listening situation closer to interactions with current voice assistants in the home. The obvious object is, of course, a speaker, but voice interactions with other objects have been previously explored (see, for example, Iravantchi and colleagues' Digital Ventriloquism [23]). From this position, the robot mind can leave in two ways. It can leave its position and *embark* to become an embodied agent. An example case is a voice assistant taking on a physical form to do housework. The robot mind can also *disperse* to be present across the space, when it is not needed.

Lastly, the robot mind can be present **across the space**. In this case, there is no specific position the listener could discern as the sound source. Rather, sound is present in the entire space. From this state, the robot mind can *materialize* in the physical robot. An example case would be a robot mind on standby with a sonic presence across the space, which, after being activated by a voice command, confirms the command and subsequently moves into the robot body to carry out the request.

7.5.2.2   Four Examples of Spatial Sound Events

Keeping these key locations in mind, we now look at four examples for how a robot sound event could be designed spatially. To do so, we consider a sound moving between a source and a target. A source could either be a physical robot or a robot mind at any object or position in the environment. A target could be any object or location relevant the human–robot interaction scenario. Next to these two positions, we also consider the intensity of the sound used, as it allows us to emphasize and de-emphasize different locations along the sound's spatial trajectory. With these basic building blocks we can then attempt to convey relationships between the source and the target. Figure 7.9 shows a graphical representation of this. The simple visual language is based on the work of Blackburn [7], who created a rich and detailed visual representation of Smalley's spectromorphology, which we discussed in previous work [34]. It should be noted that the use cases in these examples are meant to illustrate the possible role of key sound locations, and not be an exhaustive list of possible applications.

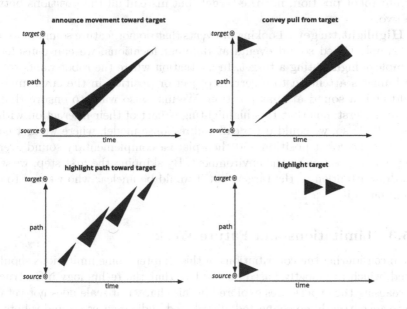

FIGURE 7.9
Four spatial sound events. Sound moves between source and target (y-axis) over a certain amount of time (x-axis) and changes intensity over time (thickness of shape).

Announce movement toward target – In this example, a robot plans a motion trajectory and announces it to any nearby listeners. The sound event begins with an alert sound located at the robot to draw attention to it, and then moves a second sound along the path the robot intends to take. Along this path, the sound loses intensity, placing emphasis on the initial robot location, while also indicating the general direction to robot plans to move in.

Convey pull from target – This example is an inversion of the above, but instead of placing emphasis on the initial robot location, emphasis is put on the goal and the message to be conveyed is not that the robot intends to move away from the source position, but instead that the robot is being pulled from a target position. To convey this, the listener's attention is first drawn to the target position, after which a second sound moves from the robot's initial location to the target while gradually increasing in intensity.

Highlight path toward target – In this example, we again have the situation of a robot announcing a movement path, but this time the actual path is the key information we want to bring across. Like an auditory equivalent of led strips in a plane that lead to the exit doors, we want to communicate the path a robot is going to take before it sets out on its journey. To do so, we move a sound from the source location to the target location and increase the sound's intensity along the path traveled. This then emphasizes neither the current robot position, nor the target, but instead all the positions between the two.

Highlight target – Looking at events that do not feature sound movement, but simply timed sound events at different locations, we can consider the example of highlighting a target. In a situation where the robot wants to draw the human's attention to a certain object or position in the environment it could emit a sound at that position. We may also want to ensure that the human understands that this highlighting is part of their interaction with the robot. To do so, we could use a call and response model, where we first emit a sound at the robot position and then place a complementary sound event at the relevant position in the environment. By skipping the first step, we would still draw attention to the target, but it would be unclear who wanted to draw attention to it.

7.5.3   Limitations and Future Work

When considering the contributions of this chapter, some limitations should be noted, which are mostly based on the fact that the technology infrastructure for realising the experiences explored in this chapter at scale does not yet exist. There are currently no connected, distributed audio devices around robots, and therefore, a key underlying question is, how many of the affordances assumed by this work can be expected to manifest.

One possible limitation regarding the spatial resolution was previously mentioned in Section 7.5.1.2. While it is reasonable to be sceptical about the spatial resolution which might eventually be available to creators who design for this context, the core set of spatial experiences proposed in this chapter are still possible to achieve even when emitted across a loudspeaker system with reduced spatial resolution, and the design framework presented takes this into account. Another, potentially more severe limitation is the difference of timbre and balance across the devices making up the distributed audio system. While this challenge of unbalanced devices is primarily technical in nature, there are some design aspects to consider. In current work around Media Device Orchestration, the design of entertainment experiences takes the nature of their playback devices into account. This means that creators are encouraged to assign sound that requires full-range playback, like music, is sent to a device with the appropriate frequency range like a soundbar, while sound with a more limited spectrum, like a recording of a person whispering, can be sent across smaller devices like mobile phones. This practice is, in fact, reflected in this chapter's prototype, which assumes a full-range speaker with adequate low-end for the robot body itself, while all sound in the environment is designed to be thinner and can therefore be emitted by smaller speakers. However, this was evaluated in a virtual environment without any frequency range constraints and it therefore remains to be seen, how much this potential inconsistency across devices needs to be considered during the sound design process. In a worst case scenario one might imagine that individual audio devices in the home have such different timbral characteristics, that a sound moving from one corner of the room to another sounds so different throughout its journey, that the listener hears it as several devices each emitting their own, individual audio alert. This focus on individual devices as opposed to individual sounds moving through the space would make much of the effects explored in this chapter impossible to achieve.

While general impressions of the spatial work for this prototype will be gathered upcoming studies, there are a number of open questions to be addressed in future studies that could isolate and investigate specific effects. Relevant questions include (i) how different a robot-internal and robot-external sound set can be, before they are assigned to two separate agents, (ii) how close a sound can be to a physical robot, before it is attributed to it, and (iii) how much sound design complexity people consider appropriate in a real-world context with other people and sound sources present.

7.6 Conclusion

In this chapter, we explored the role of spatially distributed sound in human–robot interaction by creating a spatial sound set to be emitted by a robot and its environment. We first interviewed researchers and practitioners who work with distributed audio and sound installations and identified five themes relevant to distributed sound in the context of HRI: *sonic identity and fiction, functions of sound, roles and affordances of interactive sound, roles and affordances of distributed sound,* and *technical considerations.* We then combined these themes with our own experience in interactive immersive audio environments to create a distributed robot sound prototype. The prototyping process was comprised of five stages: defining *fiction*, developing a *sound identity*, creating an *audio interaction flow*, *asset production*, and *implementation* into a virtual simulation environment. We then reflected on lessons learnt during this design process, evaluating the various tools and approaches we used. Finally, we proposed a generalised design framework for spatial robot sound, arguing that spatial robot sound can be broken down into three key locations – the robot, objects of interest in the environment, and the space itself – and that transitions between these locations create spatial relationships, which are relevant to HRI scenarios.

Bibliography

[1] ALBASTAKI, A., HOGGENMUELLER, M., ROBINSON, F. A., AND HES-
PANHOL, L. Augmenting Remote Interviews through Virtual Experience
Prototypes. In *32nd Australian Conference on Human-Computer Interaction (OzCHI '20), Sydney, NSW, Australia* (New York, NY, USA, Dec.
2020), ACM, p. 9.

[2] ALSAAWI, A. A critical review of qualitative interviews. *European Journal
of Business and Social Sciences 3*, 4 (2014).

[3] ATZORI, L., IERA, A., AND MORABITO, G. The Internet of Things: A
survey. *Computer Networks 54*, 15 (Oct. 2010), 2787–2805.

[4] BBC. Audio Orchestrator, Sept. 2021.

[5] BECHHOFER, F., AND PATERSON, L. *Principles of research design in the
social sciences.* Routledge, 2012.

[6] BERG, B. L., LUNE, H., AND LUNE, H. Qualitative research methods
for the social sciences. Publisher: Pearson Boston, MA.

[7] BLACKBURN, M. The Visual Sound-Shapes of Spectromorphology: an
illustrative guide to composition. *Organised Sound 16*, 1 (2011), 5–13.
Publisher: Cambridge University Press.

[8] BORDIGNON, M., RASHID, J., BROXVALL, M., AND SAFFIOTTI, A.
Seamless integration of robots and tiny embedded devices in a PEIS-
Ecology. In *2007 IEEE/RSJ International Conference on Intelligent
Robots and Systems* (San Diego, CA, USA, Oct. 2007), IEEE, pp. 3101–
3106.

[9] BRAUN, V., AND CLARKE, V. Using thematic analysis in psychology.
Qualitative Research in Psychology 3, 2 (Jan. 2006), 77–101.

[10] BROXVALL, M., GRITTI, M., SAFFIOTTI, A., BEOM-SU SEO, AND YOUNG-JO CHO. PEIS Ecology: integrating robots into smart environments. In *Proceedings 2006 IEEE International Conference on Robotics and Automation, 2006. ICRA 2006.* (Orlando, FL, USA, 2006), IEEE, pp. 212–218.

[11] BUI, H.-D., AND CHONG, N. Y. An Integrated Approach to Human-Robot-Smart Environment Interaction Interface for Ambient Assisted Living. In *2018 IEEE Workshop on Advanced Robotics and its Social Impacts (ARSO)* (Genova, Italy, Sept. 2018), IEEE, pp. 32–37.

[12] CAIN, R., JENNINGS, P., AND POXON, J. The development and application of the emotional dimensions of a soundscape. *Applied Acoustics 74*, 2 (Feb. 2013), 232–239.

[13] CARLILE, S., DELANEY, S., AND CORDEROY, A. The localisation of spectrally restricted sounds by human listeners. *Hearing Research 128*, 1-2 (1999), 175–189. Publisher: Elsevier.

[14] CHA, E., FITTER, N. T., KIM, Y., FONG, T., AND MATARIĆ, M. J. Effects of Robot Sound on Auditory Localization in Human-Robot Collaboration. In *Proceedings of the 2018 ACM/IEEE International Conference on Human-Robot Interaction - HRI '18* (Chicago, IL, USA, 2018), ACM Press, pp. 434–442.

[15] DE MARCO, R. *Sound Scenography. The Art of Designing Sound for Spaces*. Avedition Gmbh, Stuttgart, Oct. 2021.

[16] DO, H. M., PHAM, M., SHENG, W., YANG, D., AND LIU, M. RiSH: A robot-integrated smart home for elderly care. *Robotics and Autonomous Systems 101* (Mar. 2018), 74–92.

[17] DRURY, J., SCHOLTZ, J., AND YANCO, H. Awareness in human-robot interactions. In *SMC'03 Conference Proceedings. 2003 IEEE International Conference on Systems, Man and Cybernetics. Conference Theme - System Security and Assurance (Cat. No.03CH37483)* (Washington, DC, USA, 2003), vol. 1, IEEE, pp. 912–918.

[18] FRANCOMBE, J., AND HENTSCHEL, K. Evaluation of an immersive audio experience using questionnaire and interaction data. In *Proceedings of the 23rd International Congress on Acoustics* (Aachen, Germany, 2019), p. 8.

[19] FRANCOMBE, J., MASON, R., JACKSON, P. J., BROOKES, T., HUGHES, R., WOODCOCK, J., FRANCK, A., MELCHIOR, F., AND PIKE, C. Media Device Orchestration for Immersive Spatial Audio Reproduction. In *Proceedings of the 12th International Audio Mostly Conference on Augmented and Participatory Sound and Music Experiences - AM '17* (London, United Kingdom, 2017), ACM Press, pp. 1–5.

[20] FRANCOMBE, J., WOODCOCK, J., HUGHES, R., MASON, R., FRANCK, A., PIKE, C., BROOKES, T., DAVIES, W., JACKSON, P., COX, T., FAZI, F., AND HILTON, A. Qualitative Evaluation of Media Device Orchestration for Immersive Spatial Audio Reproduction. *Journal of the Audio Engineering Society 66*, 6 (June 2018), 414–429.

[21] FRANCOMBE, J., WOODCOCK, J., HUGHES, R. J., HENTSCHEL, K., WHITMORE, E., AND CHURNSIDE, T. Producing Audio Drama Content for an Array of Orchestrated Personal Devices. In *Audio Engineering Society Convention 145* (2018), Audio Engineering Society.

[22] GOMM, R. *Social Research Methodology: A Critical Introduction.* Macmillan International Higher Education, 2008.

[23] IRAVANTCHI, Y., GOEL, M., AND HARRISON, C. Digital Ventriloquism: Giving Voice to Everyday Objects. In *Proceedings of the 2020 CHI Conference on Human Factors in Computing Systems* (2020), pp. 1–10.

[24] KIESLER, S., POWERS, A., FUSSELL, S. R., AND TORREY, C. Anthropomorphic interactions with a robot and robot–like agent. *Social Cognition 26*, 2 (2008), 169–181. Publisher: Guilford Press.

[25] KIM, Y. J., KIM, B. H., AND KIM, Y. K. Surround audio device and method of providing multi-channel surround audio signal to a plurality of electronic devices including a speaker, Dec. 2019.

[26] LAWLESS, B., AND CHEN, Y.-W. Developing a method of critical thematic analysis for qualitative communication inquiry. *Howard Journal of Communications 30*, 1 (2019), 92–106. Publisher: Taylor & Francis.

[27] LEAVY, P. *The Oxford Handbook of Qualitative Research.* Oxford University Press, USA, 2014.

[28] LI, J. The benefit of being physically present: A survey of experimental works comparing copresent robots, telepresent robots and virtual agents. *International Journal of Human-Computer Studies 77* (2015), 23–37. Publisher: Elsevier.

[29] LOUPA, G. Influence of Noise on Patient Recovery. *Current Pollution Reports* (2020), 1–7.

[30] LURIA, M., REIG, S., TAN, X. Z., STEINFELD, A., FORLIZZI, J., AND ZIMMERMAN, J. Re-Embodiment and Co-Embodiment: Exploration of social presence for robots and conversational agents. In *Proceedings of the 2019 on Designing Interactive Systems Conference - DIS '19* (San Diego, CA, USA, 2019), ACM Press, pp. 633–644.

[31] MASON, R. How Important Is Accurate Localization in Reproduced Sound? In *Audio Engineering Society Convention 142* (2017), Audio Engineering Society.

[32] REIG, S., LURIA, M., FORBERGER, E., WON, I., STEINFELD, A., FOR-
LIZZI, J., AND ZIMMERMAN, J. Social Robots in Service Contexts:
Exploring the Rewards and Risks of Personalization and Re-embodiment.
In *Designing Interactive Systems Conference 2021* (2021), pp. 1390–1402.

[33] ROBINSON, F. A., BOWN, O., AND VELONAKI, M. Designing Sound
for Social Robots: Candidate Design Principles. *International Journal of
Social Robotics* (June 2022).

[34] ROBINSON, F. A., VELONAKI, M., AND BOWN, O. Crafting the Language
of Robotic Agents: A vision for electroacoustic music in human–robot interaction. *Organised Sound* (2022), 1–13. Publisher: Cambridge University
Press.

[35] ROGINSKA, A., AND GELUSO, P., Eds. *Immersive Sound: The Art and
Science of Binaural and Multi-channel Audio*. Routledge, Taylor & Francis
Group, New York; London, 2018.

[36] RUMSEY, F., ZIELIŃSKI, S., KASSIER, R., AND BECH, S. On the relative
importance of spatial and timbral fidelities in judgments of degraded
multichannel audio quality. *The Journal of the Acoustical Society of
America 118*, 2 (Aug. 2005), 968–976.

[37] SAYIN, E., KRISHNA, A., ARDELET, C., BRIAND DECRÉ, G., AND
GOUDEY, A. "Sound and safe": The effect of ambient sound on the
perceived safety of public spaces. *International Journal of Research in
Marketing 32*, 4 (Dec. 2015), 343–353.

[38] SCHAFER, R. M. *The New Soundscape*. BMI Canada Limited Don Mills,
1969.

[39] SEIDMAN, I. *Interviewing as Qualitative Research: A Guide for Researchers in Education and the Social Sciences*. Teachers College Press,
2006.

[40] SMALLEY, D. Spectromorphology: explaining sound-shapes. *Organised
Sound 2*, 2 (1997), 107–126. ZSCC: 0000945.

[41] TURCU, C., TURCU, C., AND GAITAN, V. Integrating robots into the
Internet of Things. *International Journal of Circuits, Systems and Signal
Processing 6*, 6 (2012), 430–437.

[42] WONPIL YU, JAE-YEONG LEE, YOUNG-GUK HA, MINSU JANG, JOO-
CHAN SOHN, YONG-MOO KWON, AND HYO-SUNG AHN. Design and
Implementation of a Ubiquitous Robotic Space. *IEEE Transactions on
Automation Science and Engineering 6*, 4 (Oct. 2009), 633–640.

[43] WOODCOCK, J., FRANCOMBE, J., HUGHES, R., MASON, R., DAVIES,
W. J., AND COX, T. J. A quantitative evaluation of media device

orchestration for immersive spatial audio reproduction. In *Audio Engineering Society Conference: 2018 AES International Conference on Spatial Reproduction-Aesthetics and Science* (2018), Audio Engineering Society. ZSCC: 0000010.

8

Navigating Robot Sonification: Exploring Four Approaches to Sonification in Autonomous Vehicles

Richard Savery, Anna Savery, Martim S. Galvão and Lisa Zahray

DOI: 10.1201/9781003320470-8

8.1   Intro

Autonomous cars are one of the most rapidly growing areas of robotics, and may be the first robots to be fully integrated into society [5]. Autonomous cars are also one of the few robotic platforms that have clear applications in personal, industrial and commercial settings. For these reasons, deep consideration of how we interact with autonomous cars is crucial, with extensive research already existing, focusing on areas such as ethics [14], communication methods [20], environmental impact [19] and many others.

Sound has played a significant role for enhancing the driving experience and improving safety. This can be through roles such as safety alerts, including seat belt warnings, or notifications for crossing lanes. More recently, sonification of car data has emerged as a growing area of research, due to rich layers of available real-time data and growth in production of autonomous vehicles [15].

In this chapter, we will explore the concept of sonification and its application in the context of autonomous cars, examining the benefits, challenges, and future directions of this emerging field. We present four different approaches, three using simulations of cars and one creating a sonification from a real world dataset. Our goal was to describe divergent approaches and cover a range of future possibilities when sonifying car data.

The first project, Unity Algorithmic Music, utilizes the Unity game engine to create a three-dimensional driving simulator on a dirt road with surrounding grass regions and mountains. The sound design is based on the cycle of fifths chord progression in C major, with two instruments assigned to the chords and the subdivision speed determined by the car's speed. The second project, Unity Deep Learning, uses the same Unity engine for car simulation, but instead uses a bidirectional Long short-term memory (LSTM) network, to map movements from the car to melodies, all performed through a virtual string orchestra. The third project, Unreal MetaSounds, is built in the Unreal Game Engine and employs Unreal's MetaSounds framework to create a data-driven synthesizer whose parameters are determined by the car's relationship to its environment. The final project, Sonification with Beads, uses the Berkley Deep Drive dataset to process violin samples and create compositions based on user interactions with the dataset.

Autonomous car sonification provides a unique opportunity to convey data and driving information to passengers. While the four projects presented in this chapter showcase different approaches to sonification, they collectively demonstrate the potential for sonification in the context of autonomous cars. More broadly, this chapter describes the potential for sonification in robotic systems, and how data taken from robot systems can be used to create compelling auditory feedback.

8.2   Related Work

8.2.1   Sonification

Sonification focuses on the use of sound to represent data or information. It is often described as the us of "nonspeech audio to convey information... for purposes of facilitating communication or interpretation" [11]. In Walker's theory of sonification, they describe four key purposes for sonification: alarm, status, art and entertainment, and data exploration [23]. Alarm sonifications can be used in many settings, such as to report patient heart and oxygen rates [10], or monitoring business processes in factories [9]. Sonification for status updates can focus on supplying background information, such as network traffic flow [3]. For art and entertainment, sonification can be used for musical installations [21], or to augment sports, such as backing tracks for soccer [17]. Data exploration sonification can aid both novice analysis of data [22] or provide new insights to those already familiar with data [8].

8.2.2   Robot Sonification

Robot sonification involves translating robot movements, sensor data or other robot data sources into audio feedback. This sonification is used for a range of purposes, including all categories from Walker's theory of sonification. For example, robot sonification has been used to improve social communication for children with Autism Spectrum Disorder (ASD), incorporating sonification to foster emotional and social communication [26]. Frid et al. conducted two experiments to investigate the perception of mechanical sounds produced by expressive robot movement and blended sonifications thereof. They found that blended sonification can successfully improve communication of emotions through robot sounds in auditory-only conditions [7]. Zahray et al. designed and evaluated six different sonifications of movements for a robot with four degrees of freedom to improve the quality and safety of human–robot interactions, recommending that conveying information in a pleasing and social way is important to enhance the human–robot relationship [25]. Other research has investigated the perception of synthesized robot sounds and the materiality of sound made by a robot in motion. This research found that participants preferred more complex sound models for the sonification of robot movements and that sound preferences varied depending on the context in which participants experienced the robot-generated sounds [13].

8.2.3   Autonomous Vehicle Sonification

A recent paper listed multiple key challenges and opportunities for the expanding use of autonomous car sonification. These challenges focus on the user experience design of highly automated cars, including avoiding motion sickness, ensuring trust, supporting accurate mental models, and providing an enjoyable experience. The authors propose that auditory displays could help address these issues, and suggest that continuous sonic interaction may be more effective than traditional discrete cues [12]. Specific studies have focused on a range of issues, one study aimed to investigate whether auditory displays can enhance users' trust in self-driving vehicles. Results suggest that the designed auditory display can be useful in improving users' trust and received high scores in terms of user acceptance, providing implications for the interaction design of self-driving cars and guiding future auditory display research [6]. Other research has described potential applications for monitoring passenger breathing to enhance the driving experience [16].

8.3   Unity Algorithmic Music

The project described in this section uses Unity as a game engine to create a 3D world from the perspective of a car driver, including a dirt road, grassy areas, and mountains. C# scripts control the game logic, allowing for acceleration, deceleration, and gear changes using key presses. The game sends OSC messages to a MaxMSP patch, which sends MIDI notes to Ableton to produce sounds. The sound mapping follows a cycle of fifths chord progression in C major, changing chords every measure at a constant tempo of 90 BPM in 4/4 time. The car's speed controls the subdivision speed of the chords, with faster speeds resulting in faster subdivisions. Turning the car or pressing the brakes produces additional musical effects, with a total of four instruments used in the sound design.

8.3.1   Technical Details

The project uses Unity as a game engine, rendering a three dimensional world viewed from the perspective of the driver's seat of a car. The terrain includes a dirt road surrounded by bumpy grass regions and mountains. During the videos, the car drives over both the road and grass regions. An example frame of the environment is shown in Figure 8.1.

Scripts controlling the game logic were coded in C#. The car accelerates while the gas key is held until reaching a maximum speed, and loses speed slowly when the gas key is not pressed until reaching zero speed. While the brakes key is pressed, the car loses speed quickly until reaching zero speed. Gravity affects the acceleration of the car, i.e. the car will roll down a hill in a case where there are no brakes or gas counteracting it, and accelerate slower up a steep hill when gas is pressed. The gear of the car can be toggled by a key press between reverse and drive. The left and right arrow keys turn the car in the corresponding direction, and the steering wheel visually turns to match.

FIGURE 8.1
Video frame from unity project.

The Unity game sends OSC messages to a patch coded in MaxMSP, a visual programming language for music, that details the speed, gear, turn direction, pressed state of the brakes, and pressed state of the gas. The MaxMSP patch processes these inputs and sends MIDI note messages to Ableton, a digital audio workstation, which produces the sounds. The Ableton project contains four instruments customized from the "Ambient and Evolving" presets.

8.3.2 Sound Mapping

The sounds during game-play follow the circle of fifths chord progression in C major, changing chords every measure at a constant tempo of 90 BPM in 4/4 time. Two instruments are assigned to the chords. One instrument plays the bass note of each chord, sustaining it for the full measure. The other plays randomized notes from two octaves of the chord (a choice of six notes total) at four different possible subdivision speeds: sixteenth notes, eight notes, quarter notes, or half notes. This subdivision is determined based on the speed of the car, where a faster car speed maps to a faster subdivision speed. The instruments were designed so that, when the car is in drive, the sustaining instrument has a longer note release time and more reverb than the instrument playing the subdivided notes. When the car is in reverse, these two instruments are swapped, giving the more sustained and reverb-containing instrument the faster-moving part.

Each time the steering wheel is used, a quarter note is played on a third instrument. A left turn plays a low octave C, and a right turn plays a higher octave C. While the brakes are pressed, a fourth instrument plays a sustained note. The brake instrument's sound was designed to be less strongly pitched, acting more as a stutter effect on top of the more musical sounds of the other instruments.

8.4   Unity Deep Learning

The next approach to robot sonification focused on using the same Unity simulation from the previous section with deep learning musical generation. A primary goal was to create varying, aesthetically pleasing music, while still representing the data from the car. For this reason we chose to map features from the car movement to broader musical features such as tempo and pitch contours.

8.4.1   Technical Details and Sound Mapping

At the basis of this sonification approach is a bidirectional LSTM to generate musical patterns. The network is inspired by Piano Genie [4], a neural network model that maps 8 buttons to 88 piano keys in a way that allows non-experts to play music by simply pressing combinations of the buttons. Both Piano Genie and our revised model were trained on the International Piano-e-Competition dataset, a collection of virtuoso performances on piano. Our model converts numbers in the range of 1 to 8 into MIDI sequences, where the important learning in the system is the contour and relation between each number. Constantly rising inputs from 1 to 8 will lead to rising patterns, and multiple notes input at the same time will result in chords.

   With this in mind our sonification process mapped different elements of the data to the range of 1 to 8, which created the pitches used in the piece, while the rhythms were placed based on events in the data. Table 8.1 shows each mapping. Tempo for the sonification is mapped directly to the speed of the car, with the range of the tempo between 80 and 140 BPM. To play back sounds we use Kontakt 7, and the 8Dio Century Strings virtual instruments, which provides a range of violin, viola, cello and double bass samples.Throughout the sonification there is an underlying accompaniment, varying between single quarter notes for parked, or chords when in drive and reverse. For park, drive or reversing different timbres are used for the strings. Park mode uses arco string staccato, while drive uses arco string legato. Reverse plays the same harmonic and rhythmic material as drive however uses string tremolo for playback. When the car is turning left or right, melodies are added in pizzicato strings, with the melody ascending for left and descending for right.

8.5   Unreal MetaSounds

The project described in this section engages with autonomous vehicles at the world-modeling stage, using dynamic audio processes as a means for conveying

TABLE 8.1
Summary of unity deep learning mappings.

Parameter	Output
Speed	Tempo, mapped between 80 to 140 BPM
Park	Play Park Music, Change Timbre
Drive/Reverse	Play Drive/Reverse, Change Timbre
Left/Right	Play Melody Sequence
Brakes	Double Notes on Flute

FIGURE 8.2
Video frame from unreal engine project.

information about the vehicle's environment to a human user/driver. Rather than replacing other streams of vehicle information, the process of sonifying the vehicle's relationship to its environment serves as a novel interface for audio synthesis, with parameters such as orientation, speed, and distance to obstacles mapped to synthesis parameters such as frequency, delay time, and amplitude. As the driver navigates the virtual space, a collection of sounds indicates both the vehicle's relationship to other objects as well as key pieces of information regarding the vehicle's current state. Some of these sonifications are constant (e.g. vehicle orientation and speed) while others occur only when corresponding events trigger them (e.g. approaching a wall or crossing the north-south axis). Figure 8.2 shows a screen capture of the Unreal gameplay.

8.5.1 Technical Details and Sound Mapping

The project employs Unreal Engine 5 to construct a virtual environment in which a test vehicle is driven around a simple obstacle course. Using MetaSounds (see Figure 8.3), Unreal Engine's built-in audio engine, several

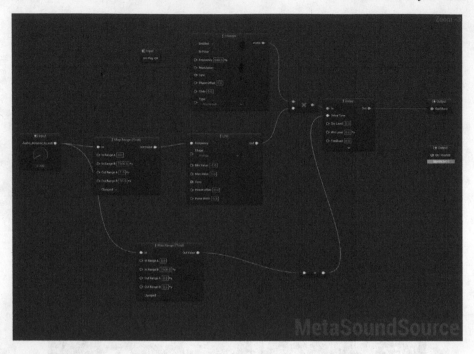

FIGURE 8.3
MetaSoundSource file from unreal engine project.

audio mappings are created to convey information about the player's vehicle and its relationship to the virtual environment:

8.5.1.1   Vehicle Orientation

North and South Markers: Two distinct sonic cues are used to signal that the driver/vehicle has crossed north or south during a rotation. The cue indicating a northern crossing is played an octave higher than the cue for the southern crossing.

Present Heading: Two constantly playing sine wave oscillators are used to indicate the current heading of the vehicle. Both oscillators play at the same frequency whenever the vehicle is facing due north or south. As the vehicle steers away from the north-south axis, the frequency of one of the oscillators changes accordingly. The resulting difference creates a beating pattern, which is used to indicate how far the vehicle has veered from the north-south axis.

8.5.1.2   Vehicle Speed

A square wave oscillator is used to modulate the amplitude of a triangle wave oscillator, which has its frequency mapped to the current speed of the vehicle.

As the vehicle accelerates, the frequency of the triangle wave rises, as does the modulation rate of the square wave. The result is a sound that is timbrally distinct from other indicators, and which the player can reference alongside orientation to determine vehicle speed and heading.

8.5.1.3 Object Identification

Two basic object types are defined in this prototype: walls and poles. Each wall is identified by a triangle wave oscillator set to a fixed frequency while each pole is identified by a pulsing burst of white noise.

8.5.1.4 Object Localization

Spatial: Walls and poles are treated as sound-emitting objects in Unreal Engine, with spatialization employed to place them at the corresponding location in the player's listening field. As the player changes the orientation of the vehicle, the perceived location of walls and poles changes accordingly. Likewise, the vehicle's distance from each sound-emitting object determines the attenuation of the sound source according to a falloff curve that extends out from the boundaries of the object. In the case of walls, this means that attenuation of the source follows a capsule-shaped spread rather than a spherical spread originating from the center point of the object. The falloff is designed to provide drivers with enough notice to avoid striking objects while also maintaining some degree of isolation from nearby objects.

Distance and acceleration: The distance between the vehicle and an object is mapped to both a low-frequency oscillator (modulating the amplitude of the wall's triangle wave oscillator) and the delay time of a delay module placed after the modulated triangle wave. When the vehicle is moving toward a wall, the delay time decreases, which leads to an increase in the perceived frequency being emitted by the wall. The opposite is true as the vehicle moves away from a wall, creating a Doppler effect from the player's perspective. The low-frequency oscillator likewise follows a direct relationship between frequency and distance between the vehicle and a wall. When the vehicle is positioned immediately adjacent to a wall, for example, the sound is heard as distinct pulses, whereas when the vehicle moves away from the wall the frequency of the amplitude modulation increases until the sound is no longer distinguishable as pulses.

8.6 Sonification with Beads

Computer vision and synthesized audio in autonomous vehicles are implemented to serve specific purposes, such as obstacle detection and user feedback. This project intends to employ data from autonomous vehicles within the scope of an interactive visual installation, where meaning is created through direct, real-time manipulation of audio samples as well as visual representation of the Berkley Deep Drive dataset. This is done using Processing software as well as an external library called Beads, which allows for sonification of the images.

8.6.1   Technical Implementation

The Berkley Deep Drive (BDD100K)[1] provides an extensive driving video and image dataset for studying and training autonomous vehicles within a wide range of diverse driving conditions [24]. The purpose of the construction of the 100k dataset was to improve on the limitations of existing visual content used for multitask learning for autonomous driving correlating to real-world computer vision applications required to execute tasks of various complexities [24].

This robust dataset, collected from over fifty thousand rides across New York, San Francisco Bay Area, and other regions [1] contains 40 second videos of diverse driving conditions with varying times of day and weather, ranging from city streets, residential areas and highways. "The frame at the 10th second of each video is annotated for image classification, detection, and segmentation tasks" [1]. These segmenations include; "lane marking, drivable area, full-frame semantic and instance segmentation, multiple object tracking, and multiple object tracking with segmentation" [24].

The concept for the author's project stemmed from a desire to provide humans a way to better understand computer vision in an interactive and artistic way. A single image from the dataset was used to explore the relationship between static images and dynamic interplay through data sonification and visual interaction. Only object bounding boxes were explored as a data source due to their simple coordinate system represented as floating point numbers, which were accessible for both, visual representation and audio synthesis.

The author chose to work with processing[2] as a software environment and the external Beads library[3] for sonifying the data. The audio is made up of five solo violin samples improvised and recorded by the author, reflecting the mood of the image. A looping bass line plays with the initiation of the processing sketch. The other samples respond to user cursor movement across the sketch display window. Five object bounding boxes (*box2d* object in JSON) object coordinates were chosen for visual display. When the user's cursor interacted with each "box", an audio sample assigned to those coordinates is initiated and manipulated based on the mouse movements (see Figure 8.4).

---

[1]https://www.bdd100k.com/
[2]https://processing.org/
[3]http://www.beadsproject.net/

FIGURE 8.4
User interaction with bounding boxes.

8.6.2 Beads

In order to facilitate playing multiple audio samples concurrently and one at a time, five sample players were created, one for each audio track, with all sample players continuously playing on a loop. A Glide object was then used to control the Gain values of each sample player. Since the bass line audio sample was not manipulated by the user, there was no need to create a Glide object to control its Gain value. When the user's mouseX and mouseY coordinates entered any of the displayed rectangle (object bounding box) values, the gain value of the corresponding audio sample would change from 0.0 to 0.5 and then back to 0.0 once the cursor went outside those values. Figure 8.5 shows an extract of Processing code using Beads library.

The x and y coordinates of each bounding box were then used to manipulate each corresponding track in real-time, using cursor movement. The first bounding box (x1: 751.147935, y1: 267.019365, x2: 771.111999, y2:285.735677) calculates the relationship between current vertical mouse location (mouseY) and previous vertical mouse location (pmouseY). When the cursor is moved downward within the boundaries of that rectangle, the audio sample is played in reverse with the playback rate slowed down by 285.735677 divided by the current vertical mouse coordinates (y2/mouseY, or box2d[3]/mouseY). This is shown in Figure 8.6.

Another bounding box shared coordinates with two other boxes. This overlapping data was sonified by triggering three audio samples simultaneously, representing the act of three distinct boundaries crossing and interacting.

```
if (mouseX > box2d1[0] && mouseX < width-10 && mouseY > box2d1[1] && mouseY < box2d1[1]+box2d1[3]) {

 if (pmouseY < mouseY) {
   player2.setLoopType(SamplePlayer.LoopType.LOOP_BACKWARDS);
   rateValue.setValue(box2d2[3]/mouseY);
 } else
 {
   rateValue.setValue(1);
 }
 tint(200);
 println(box2d1[0]);

 gainGlide2.setValue((float) 1.);
} else
{
 gainGlide2.setValue((float) 0.);
}
```

FIGURE 8.5
Extract of Beads code in processing.

```
if (pmouseY < mouseY) {
  player2.setLoopType(SamplePlayer.LoopType.LOOP_BACKWARDS);
  rateValue.setValue(box2d2[3]/mouseY);
```

FIGURE 8.6
Real-time audio processing using BDD data.

Rectangle x1:680.025953, y1:283.240168, x2:696.246757, y2:301.956478 was used with a granular sample player, allowing the user to turn the assigned audio track into grains of layering looping sounds (see Figure 8.7).

8.6.3 Visual representation

The chosen background image from the JSON dataset was 4bdd2193-546f8f2c.jpg. To avoid visual clutter, only the first four bounding boxes for that image were used. In order to provide visual feedback for the user, unfilled rectangles representing the object bounding boxes were drawn, with data directly from the JSON file appearing within each rectangle as the user entered their coordinates. A convolution effect [18] was used to alter the background image whenever the cursor was within the boundaries of any of the rectangles to signify the concept of clarity of focus on the context within the bounding box and blurriness of the peripheral vision outside its boundaries.

```
if (pmouseY < mouseY || pmouseX < mouseX) {
  grainSizeValue.setValue((float)box2d3[0]+box2d3[1]+box2d3[2]+box2d3[3]);
}
```

FIGURE 8.7
Granular processing using all four bounding box coordinates.

In this project, the author focused on sonifying only one parameter available within the extensive BDD dataset within a single image out of the available 100k. The coordinates of the object bounding boxes provided an abundance of room for musical and user interaction creativity. Building on the ideas explored in this project, a further expansion on the possibilities of sonifying the robust data from the BDD dataset could lend itself to a potentially large scale series of interactive works exploring the relationship between computer vision, sound synthesis, autonomous car training data and people's perception of this concept.

8.7 Discussion

8.7.1 Balancing Aesthetics with the Conveyance of Information

With any effort at sonifying data, there is inevitably some degree of information loss. Whether using vehicle speed to control an oscillator's frequency or mapping steering wheel movements to tempo and pitch contours, abstraction is a necessary step for translating sensor information into the audio domain. Even if the resulting mapping is direct (e.g. high speeds mapped to high frequencies), it is impossible to determine the exact speed of a vehicle by listening to an oscillator's output. Rather, we can follow the shape or outline of the speed over time, and in doing so gain an abstracted understanding of a vehicle's speed via sound.

This balance between fidelity and intelligibility can also be thought of as a trade-off between conveyance of information and aesthetics. An ideal sonification effort would maximize the amount of information conveyed while avoiding over-saturating the user's sensory input with a deluge of information. According to Csikszentmihalyi, the upper limit of human informational bandwidth is around 120 bits per second, or about double the amount of attention required to carry on a conversation [2]. To avoid exhausting the user's information processing faculties, a successful sonification effort would need to strike a balance between conveying the maximum amount of relevant information when necessary and allowing the user to relax their attention when possible.

Planned future development of the Unreal MetaSounds project includes incorporating this kind of attentional bandwidth management system through a background/foreground sonification model. Drawing from ambient music compositional strategies, this model will highlight important changes in sensor data through distinct sonic cues while otherwise relegating sonified vehicle data to a background process that is still present but only at the attentional periphery of the operator. In such a model, the trade-off between aesthetics and the conveyance of information is handled on a moment-by-moment basis, wherein normal operation is signified by a highly abstracted and attentionally undemanding soundscape while important events trigger a momentary shift in the sonification process to convey the maximum amount of relevant information as quickly as possible.

8.7.2   Adding Liveliness to Static Imagery

With the availability of such rich datasets as the BDDK, translating non human elements such as computer vision and robot sounds into an artistic work can provide new human insights into the world of robotics and machine learning. Although exploring each of the 100,000 images within the dataset is perhaps an unreasonable task, a set of multiple images across a variety of terrains, driving conditions and times of day would illustrate a representation of how a computer sees the world and what is significant for an autonomous vehicle. Multiple parameters from the data could be translated onto the static images in an interactive way, using both sonification and visualization of the accompanying data. In this way, a person interacting with the visual installation could hover over any area of an image seeing it react in a way that represents the robot car. Such approaches could be explored through mapping data to changes in volume or playback speed, depending on which elements of the image are being physically explored. Perhaps an obstacle could trigger a rise in volume or a night time drive would illuminate the road. Another approach would be to use multiple instances of the same image, focusing on specific objects within the data, such as lane markings or traffic lights. Finally, relating back to the entirety of the data, if such a task could be automated, then this would further bridge the gap between human and machine.

8.8   Conclusion

This chapter has explored four approaches to sonification of autonomous car data. While there are many other areas of research and potential mappings of car data to audio, our goal has been to describe just some of the possibilities of sonification and robotics broadly. Further research in this area can lead to the development of more sophisticated sonification systems that can adapt to changing driving conditions and driver preferences. While these systems may be used to augment a driver's awareness of their vehicle's surroundings, they may simultaneously function as generative compositional models that employ processes of abstraction to transform raw data into higher-level signifiers. In the case of the latter, it is possible to imagine a musical model in which both direct sensor inputs and sequences of events at the vehicle level may be mapped to musical gestures, rendering a vehicle's data streams as parametric inputs to an algorithmic composition system.

Likewise, the incorporation of autonomous vehicle training data sets into musical works offers the opportunity for human intervention into what is otherwise a largely machine-oriented endeavor (albeit one often aimed at transporting human passengers). By making the images and sensor information contained in these data sets the primary source material for interactive musical works, the performer and composer are reintegrated into fold. In doing so, they reiterate the uncertainty and improvisatory nature through which autonomous vehicles make sense of the world.

Bibliography

[1] PEDRO AZEVEDO. BDD100K to YOLOv5 Tutorial: https://medium.com/@pedroazevedo6/bdd100k-to-yolov5-tutorial-213e4a67d54b.

[2] CSIKSZENTMIHALYI, M. *Flow and the Foundations of Positive Psychology: The Collected Works of Mihaly Csikszentmihalyi.* Springer Netherlands, 2014.

[3] DEBASHI, M., AND VICKERS, P. Sonification of network traffic flow for monitoring and situational awareness. *PloS One 13*, 4 (2018), e0195948.

[4] DONAHUE, C., SIMON, I., AND DIELEMAN, S. Piano genie. In *Proceedings of the 24th International Conference on Intelligent User Interfaces* (2019), pp. 160–164.

[5] DOS SANTOS, F. L. M., DUBOZ, A., GROSSO, M., RAPOSO, M. A., KRAUSE, J., MOURTZOUCHOU, A., BALAHUR, A., AND CIUFFO, B. An acceptance divergence? media, citizens and policy perspectives on autonomous cars in the european union. *Transportation Research Part A: Policy and Practice 158* (2022), 224–238.

[6] FAGERLÖNN, J., LARSSON, P., AND MACULEWICZ, J. The sound of trust: sonification of car intentions and perception in a context of autonomous drive. *International Journal of Human Factors and Ergonomics 7*, 4 (2020), 343–358.

[7] FRID, E., AND BRESIN, R. Perceptual evaluation of blended sonification of mechanical robot sounds produced by emotionally expressive gestures: Augmenting consequential sounds to improve non-verbal robot communication. *International Journal of Social Robotics 14*, 2 (2022), 357–372.

[8] GROND, F., AND HERMANN, T. Interactive sonification for data exploration: How listening modes and display purposes define design guidelines. *Organised Sound 19*, 1 (2014), 41–51.

[9] HILDEBRANDT, T., MANGLER, J., AND RINDERLE-MA, S. Something doesn't sound right: Sonification for monitoring business processes in manufacturing. In *2014 IEEE 16th Conference on Business Informatics* (2014), vol. 2, IEEE, pp. 174–182.

[10] JANATA, P., AND EDWARDS, W. H. A novel sonification strategy for auditory display of heart rate and oxygen saturation changes in clinical settings. *Human Factors 55*, 2 (2013), 356–372.

[11] KRAMER, G., WALKER, B., BONEBRIGHT, T., COOK, P., FLOWERS, J., MINER, N., NEUHOFF, J., BARGAR, R., BARRASS, S., BERGER, J., ET AL. The sonification report: Status of the field and research agenda. report prepared for the national science foundation by members of the international community for auditory display. *International Community for Auditory Display (ICAD), Santa Fe, NM* (1999).

[12] LARSSON, P., MACULEWICZ, J., FAGERLÖNN, J., AND LACHMANN, M. Auditory displays for automated driving-challenges and opportunities.

[13] LATUPEIRISSA, A. B., PANARIELLO, C., AND BRESIN, R. Probing aesthetics strategies for robot sound: Complexity and materiality in movement sonification. *ACM Transactions on Human-Robot Interaction* (2023).

[14] LIN, P., ABNEY, K., AND JENKINS, R. *Robot Ethics 2.0: From Autonomous Cars to Artificial Intelligence.* Oxford University Press, 2017.

[15] MACDONALD, D. Designing adaptive audio for autonomous driving: an industrial and academic-led design challenge. Georgia Institute of Technology.

[16] MORIMOTO, Y., AND VAN GEER, B. Breathing space: Biofeedback sonification for meditation in autonomous vehicles. Georgia Institute of Technology.

[17] SAVERY, R., AYYAGARI, M., MAY, K., AND WALKER, B. Soccer sonification: enhancing viewer experience. In *International Conference on Auditory Display (25th: 2019)* (2019), Georgia Tech, pp. 207–213.

[18] SHIFFMAN, D. *Learning Processing: A Beginner's Guide to Programming Images, Animation, and Interaction.* Morgan Kaufmann, 2009.

[19] SILVA, Ó., CORDERA, R., GONZÁLEZ-GONZÁLEZ, E., AND NOGUÉS, S. Environmental impacts of autonomous vehicles: A review of the scientific literature. *Science of The Total Environment* (2022), 154615.

[20] THORVALD, P., KOLBEINSSON, A., AND FOGELBERG, E. A review on communicative mechanisms of external hmis in human-technology interaction. In *IEEE International Conference on Emerging Technologies and Factory Automation* (2022).

[21] TITTEL, C. Sound art as sonification, and the artistic treatment of features in our surroundings. *Organised Sound 14*, 1 (2009), 57–64.

[22] TÜNNERMANN, R., HAMMERSCHMIDT, J., AND HERMANN, T. Blended sonification–sonification for casual information interaction. Georgia Institute of Technology.

[23] WALKER, B. N., AND NEES, M. A. Theory of sonification. *The Sonification Handbook* (2011), 9–39.

[24] YU, F., CHEN, H., WANG, X., XIAN, W., CHEN, Y., LIU, F., MADHAVAN, V., AND DARRELL, T. Bdd100k: A diverse driving dataset for heterogeneous multitask learning. In *Proceedings of the IEEE/CVF Conference on Computer Vision and Pattern Recognition* (2020), pp. 2636–2645.

[25] ZAHRAY, L., SAVERY, R., SYRKETT, L., AND WEINBERG, G. Robot gesture sonification to enhance awareness of robot status and enjoyment of interaction. In *2020 29th IEEE International Conference on Robot and Human Interactive Communication (RO-MAN)* (2020), IEEE, pp. 978–985.

[26] ZHANG, R., JEON, M., PARK, C. H., AND HOWARD, A. Robotic sonification for promoting emotional and social interactions of children with asd. In *Proceedings of the Tenth Annual ACM/IEEE International Conference on Human-Robot Interaction Extended Abstracts* (2015), pp. 111–112.

9

Toward Improving User Experience and Shared Task Performance with Mobile Robots through Parameterized Nonverbal State Sonification

Liam Roy, Richard Attfield, Dana Kulić and Elizabeth A. Croft

9.1 Intro

As we continue to integrate collaborative machines within critical economic sectors including manufacturing, logistics, and healthcare, a growing number of users with diverse backgrounds will enter collaborative interactions with robots, increasing the need for seamless human–robot interaction (HRI). Nonverbal communication forms an essential component of human interactions and has accordingly been an important focus in the development of human–robot interactions [22]. Given the correct context, nonverbal interaction can express information with more universality, efficiency, and greater appeal than spoken words [1,6]. Nonverbal cues have been used for both active transmission of specific information (*explicit*) and passive conveyance of state information

DOI: 10.1201/9781003320470-9

(*implicit*). The combination of these two modalities has been shown to improve the mental models that humans develop for collaborative robots, thereby improving understanding and trust [6, 23, 36].

Despite it's advantages, nonverbal communication presents a trade-off related to the challenge of expressing complex ideas and the ambiguity of interpretation by human collaborators. This has been shown to occur even with seemingly intuitive nonverbal communication methods [8]. These disadvantages can be mitigated via human–robot familiarization. Improving user experience (UX) is an effective method for incentivizing users to spend more time with a nonverbal social robot, providing the opportunity to learn the mappings between it's nonverbal cues and corresponding states [28]. Thus, a positive feedback loop between UX and the learnability of a robot's nonverbal mapping could be realized, as studies show users who have a better awareness of a robot's internal state and intent are more likely to characterize interactions with that robot as interesting and enjoyable [2, 6].

Nonverbal cues including body language [15, 18, 30], gestures [1, 7, 20], facial expressions [5, 6], lights [2], sounds [9, 10] and color [21] have been widely implemented among collaborative machines deployed in real-world settings. In addition, the use of nonverbal communication in robotics has proven to be an effective method for improving user experience (UX) and shared task performance [2, 7, 10, 23]. The potential of expressive sounds for robot communication remains a relatively under-explored research area, with a general focus on emphatic emotions [29], such as joy, anger, and disgust. Less attention has been given to task-focused communicative expressions. Numerous developed state communication models map a single sound to a single state with minimal or no ability to incrementally transition between or vary within communicative states [9, 13]. The ability to incrementally adjust sound parameters would enable robots to achieve a greater degree of fluidity in social interactions, creating smoother and more enjoyable interactions.

The focus of this work is to develop a low-dimensional parameterized communication strategy based on NVS for conveying high-level robot information to facilitate human–robot collaboration. In robotics, using sound to represent robot states and information is referred to as state sonification [10, 35]. The proposed model functions by modulating a fixed number of parameters of a base sound in an attempt to communicate distinguishable high-level robot states. Our work builds on previous studies of this interaction modality [9, 10, 35] with the goal of reducing miscommunication and improving the user's mental model of a collaborative robot. In this work, we investigate the efficacy of using parameterized state sonification to convey high-level robot information relevant to human–robot collaboration, and how this communication modality affects the user's experience and shared task performance.

The developed communication method is validated using an online user study. This study tested participant understanding of the proposed nonverbal communication strategy using audio clips and the suitability of the communication strategy for HRI scenarios using video clips. Figure 9.1 shows a set

FIGURE 9.1
Key frames taken from video clips used in the online survey. (Top Left)
Human initiating a leader-follower interaction with a robot. (Top Right) Robot
attempting to communicate to a human that it needs assistance opening a
door. (Bottom Left) Robot approaching a curb alongside a human. (Bottom
Right) Human requesting assistance from robot while working on a bike.

of sample images from the video set. The results of this study present both
insight and direction concerning the use of simplified NVS communication for
human–robot collaboration.

9.2   Related Work

Implicit vs Explicit Communication. Nonverbal communication consists
largely of the implicit communication that can be observed in everyday interactions between people. While explicit communication, verbal or otherwise,
is effective for direct and explicit communication, supplementing information
implicitly can aid general understanding between collaborators. Using a collaborative task-based user study, Breazeal et al. showed that implicit nonverbal
robot communication can improve a user's mental model of a robot's internal
state, task efficiency, and error robustness [6]. Zinina et al. concluded that
people greatly prefer interacting with a robot that utilizes implicit communication [36].

Implicit Communication in Non-Humanoid Mobile Robots. Implicit communication is typically modeled on human gestures and facial
movements due to the ease with which people can pick up on these familiar

expressions. Mutlu et al. found that users were able to better predict robot intent using cues provided by a humanoid robot's gaze [23]. Breazeal et al. found similar results using gaze combined with shrugging motions [6]. While less attention has been paid to implicit communication for non-humanoid mobile robots, previous studies have explored expressive lights [2] and sonification [35].

Nonverbal sounds for communication. Nonverbal sounds (NVS) have been used to communicate robot information in numerous ways, with inspiration often being drawn from the world of science fiction. Jee et al. used musical theory to analyse the sounds developed by Ben Burtt for the cinematic robots R2-D2 and Wall-E, from the films Star Wars and Wall-E, respectively [16]. Amongst others, this work concludes that the intonation of an NVS for robot communication should correlate to human speech. Similarly, Schmitz et al. used the concept of affect bursts, defined as "very brief, discrete, nonverbal expressions of affect in both face and voice as triggered by clearly identifiable events", to produce synthetic NVS to represent human emotional states [29]. Komatsu showed that altering a sound using a continuous parameter, in this case pitch change, can influence a human's perception of an artificial agent's interactive state by asking participants to match sounds to the states agreement, hesitation, and disagreement [19]. Luengo et al. proposed a model for NVS generation that splits sounds into indivisible sonic terms, or quasons [9]. An automated version of this model could combine different configurations of these terms based on situational context to represent different interaction states of a robot. To construct these quasons, three sound parameter categories (amplitude, frequency, and time) were identified and validated using an online questionnaire.

Sonification Mapping. In robotics, sonification is the process by which sounds are used to represent robot states and information. Sonification mapping is the process by which these states and sounds are related, and can take the form of emotion and action representations. Different sonification techniques include juxtaposing rhythmic vs. continuous sounds [10], the use of auditory icons-earcons [25,31], and musical loops-based sonification [27]. Each technique has shown merit in accurately conveying specified actions, intent, or emotions. Recent publications have mapped music emotion [11], and more recently robot state sonification [10] using a 2D valence-arousal (VA) Figure [24]. A rendition of this graph can be seen in Fig. 9.2 and is further described in the *Methodology* section of this report.

9.3 Questions and Hypotheses

In this work, we investigated the efficacy of using parameterized state sonification to convey high-level robot information relevant to human–robot collaboration, and how this modality relates to UX and shared task per-

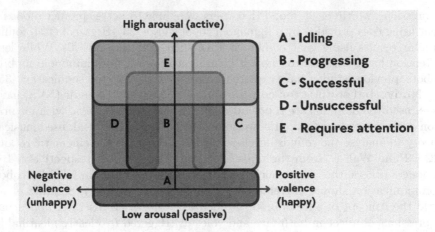

FIGURE 9.2
Visualization of five high-level robot states on a Valence-Arousal graph.

formance. This work seeks to realize a positive feedback loop present between a user's interaction enjoyment and the learnability of a robot's nonverbal mapping. Previous studies have affirmed the use of nonverbal modalities to facilitate enjoyable interactions with social robots [2, 6, 7, 26]. Informed by these prior works, we formulated the following research questions (Q1, Q2) and hypotheses (H1,H2):

Q1 How effective is a low-dimensional approach to parameterized state sonification for conveying high-level robot information relevant to human–robot collaboration?

Q2 How does user familiarization with this nonverbal communication strategy correlate to user experience (UX)?

H1 Linearly modulating two parameters of a sound will be a sufficient sonification strategy for communicating distinguishable high-level robot information relevant to human–robot collaboration.

H2 An appropriate sonification strategy will create a positive feedback loop, encompassing positive user experience, increased interest in subsequent interaction, and increased collaborative performance via familiarization.

The remaining sections of this chapter are structured as follows. The *methodology* and *user study* sections detail the approach we followed to answer the stated research questions and test our associated hypotheses. We then summarize our findings in the *results & analyses* section, followed by a *discussion* of these results. Finally, we close with final remarks and offer our thoughts on interesting avenues for relevant *future work*.

TABLE 9.1

Summary of five formulated robot states relevant to human–robot collaboration.

Collab. State	Description
Idling	At rest, awaiting human-initiated interaction
Progressing	On-track, actively processing an assigned task
Successful	Confident or able to complete an assigned task
Unsuccessful	Struggling or unable to complete an assigned task
Req. Attention	Attempting robot-initiated interaction

9.4 Methodology

Collaborative Robot States. In a related study, Baraka et al. [2] modeled three high-level robot states (on-task, stuck, requesting help) which most accurately represented their robot throughout their interaction scenario using onboard lights. Extrapolating from [2] and other studies [8,23,35] which focused on communicating robot states relevant to human–robot collaboration, we formulated five high-level robot states relevant to human–robot collaboration: *idling, progressing, successful, unsuccessful* and *requires attention*. Each of these states is briefly described in Table 9.1 and visualized as a continuous region along a two-dimensional valence-arousal (VA) graph in Figure 9.2.

The state *idling* is defined as a robot at rest, awaiting human-initiated interaction. This state is analogous to the standby mode common to personal electronics such as computers, televisions and mobile phones. In this state, the robot is not communicating auditory information. The state occupies the lower region of the VA graph, Figure 9.2 where arousal=0. Above *idling*, the state *progressing* occupies the central region of the graph. The state *progressing* overlaps with most other non-mutually exclusive states, as a robot can be actively processing a task with ranging degrees of confidence (correlated with valence) and urgency (correlated with arousal). On the right-hand region of the graph, the state *successful* is reached when the robot completes an assigned task. The overlapping region between *successful* and *progressing* refers to an on-task robot that is confidently progressing through an assigned task, whereas the right-most region refers to a robot which has successfully completed an assigned task.

The left-hand region of the graph *unsuccessful* is reached when a robot is unable to complete an assigned task. The overlapping region between *unsuccessful* and *progressing* refers to an on-task robot that is confused or struggling with an assigned task, whereas the left-most region refers to a robot which has failed or is unable to complete an assigned task entirely. Finally, the upper region of the graph refers to the state *requires attention*. In this state, a robot is attempting to get a human's attention and initiate an interaction. Similar to *progressing*, this state overlaps other states, as a robot may be actively seeking attention for positive or negative reasons.

Sonification Model Design. To develop our nonverbal state sonification model, we investigated sonification techniques proven to facilitate effective and enjoyable interaction with humanoid and social robots [6,35]. This investigation aimed to build a sonification model for a non-humanoid mobile robot based on sonification techniques which translate intuitively from humanoid and social robotics. We explored a sonification strategy in which we modulated a fixed number of parameters of a base sound in an attempt to communicate distinguishable high-level robot states. We focused on a low-dimensional model, strictly adjusting two parameters while keeping all remaining parameters constant, to reduce the number of variables and simplify the model validation process. By restricting the number of communication variables, we sought to reduce the required complexity of translation between the robot's state information and it's communication. Similar to [10], a 2D valence-arousal mapping [24] was used to characterize the continuous axes targeted by this 2D sound parameterization, as shown in Figure 9.2. Using a model initially developed for human emotional communication [3] presented the opportunity to explore different sound parameters that could map effectively to these axes, such as frequency to valence or volume to arousal.

To reduce the number of variables and simplify the model validation process, we structured our sonification model around one neutral sound set on a loop. This sound was selected based on it's neutral characteristics, namely, it's short duration and consistent pitch and volume. We situated this neutral sound at the origin of our VA graph. The sample we used was a 4-second synth sound file named *Infuction_F#.aif* from the Ableton Live 10 sample library, shown in Figure 9.3[1].

We experimented with mapping different sound parameters to each axis of our 2D VA graph. Parameters that were explored for the valence axis included the sound's relative harmonic key, pitch, and amplitude of pitch change within a single loop. For each experimental mapping, we used the digital audio workspace Ableton Live 10 to modulate the neutral sound situated at the origin. This output sound was used to gauge how the modulation of each candidate parameter correlated to perceived changes in the neutral sound's valence. The same process was done for the arousal axis, exploring parameters such as relative decibels (dB), the sound loop's duration (BPM), and the number of occurrences of the neutral sound within a single loop. Figure 9.4 shows a decision matrix used to identify two parameters from an initial list of six to use for the 2D sonification mapping. The criteria used to identify a suitable pair were the perceived changes to valence and arousal, the potential of the parameterization to be applied to any arbitrary base sound with previously stated neutral characteristics, and the simplicity of implementation. The performance measures shown in Figure 9.4 were estimated following an initial

---

[1] Sonification library: www.soundandrobotics.com/ch9

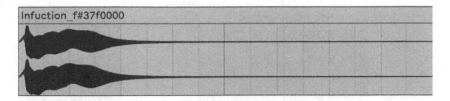

FIGURE 9.3
Visualization of the neutral sound file named *Infuction_F#.aif*. The dark-shaded region represents the audio file's sound wave. In this representation, the x-axis is the time domain, while the y-axis is the magnitude of the sound wave. Translation along the y-axis represents changes in the wave's frequency. The thickness of the wave is it's volume. The sound wave is shown twice as the audio file is stereo, meaning the file has both a unique sound wave for left and right channels in a 2-channel audio system. In this case, both left and right sound waves are identical.

analysis of the considered parameters.

From the decision matrix shown in Figure 9.4, the parameter selected for the valence axis of our 2D mapping was the amplitude of pitch change within a single loop. Changes in this parameter were estimated to have a high impact on the perceived valence, with a lower effect on arousal. This parameterization was also deemed relatively simple to implement, and easy to apply to different base sounds. The final parameter selected for the arousal axis of our 2D mapping was the number of occurrences of the neutral sound within a single loop. This

Parameter \ Criteria	Universal to Many Different Sounds	Simplicity to Implement	Perceived Change in Valence	Perceived Change in Arousal
Harmonic Key	Low	High	High	Low
Pitch	High	Low	Moderate	Low
Amplitude of Pitch Change	High	Moderate	High	Low
Relative Decibles (dB)	High	High	Low	Moderate
Sound Loop Durration (BPM)	Moderate	High	Low	Moderate
Occurrences of Sound in 1 Loop	High	Moderate	Low	High

FIGURE 9.4
Decision matrix used to identify appropriate parameters for the proposed sonification model.

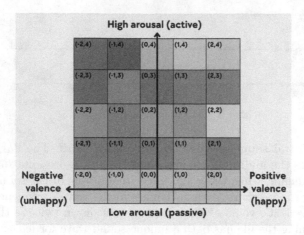

FIGURE 9.5

(Left) Visualization of 25 discretized regions on a valence-arousal graph. (Right) Sound files in Ableton Live 10 corresponding to each discretized region on the valence-arousal graph.

parameter was perceived to have a corresponding impact on arousal without affecting valence and was also estimated to be comparably simple to implement and apply to different base sounds. As well as being individually appropriate, these two parameters were estimated to be a suitable complement to one another.

Using the two selected parameters, the VA graph shown in Figure 9.5 was discretized into 25 regions with valance ranging from $[-2, 2]$ and arousal ranging from $[0, 4]$. At the region $(0,1)$, the neutral sound is played once within the communication loop, with no change in pitch through the loop. As shown in Figure 9.6, increases and decreases along the discretized valence axis represent weighted positive or negative pitch change throughout a single communication loop. As shown in Figure 9.7, increases and decreases along the discretized arousal axis represent adding or removing repetitions of the neutral sound within a single communication loop. No sound is produced in the grey regions in Figure 9.5 where arousal is zero.

The discretized regions shown on the VA graph in Figure 9.5 directly map to the selected high-level states represented as continuous regions in Figure 9.2. While regions at the center of the graph $(-1,1)$ through $(1,3)$ map to the state *progressing*, regions along the top of the graph $(-2,3)$ through $(2,4)$ map to *requires attention*. Similarly, regions $(1,1)$ through $(2,4)$ along the right-hand side map to *successful*, and regions $(-2,1)$ through $(-1,4)$ along the left-hand side map to *unsuccessful*. As previously discussed, *Idling* is represented by the greyed-out regions along the bottom of the graph $(-2,0)$ to $(2,0)$ where no sound is produced.

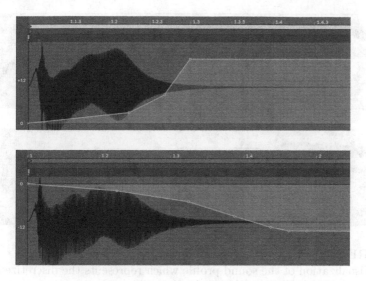

FIGURE 9.6
(Top) Visualization of the positive pitch modulation applied to the neutral sound wave to produce a valence = 1 sound. (Bottom) Visualization of the negative pitch modulation applied to the neutral sound wave to produce a valence = −1 sound. Note the axes of the left of both figures, representing the amplitude of pitch change for both pitch modulation examples.

9.5 User Study

Structure of Survey. To validate the developed sonification model, we carried out an online observational study using the platform Qualtrics. This user study was reviewed and approved by the Monash University Human Research Ethics Committee (MUHREC) with project ID 35703. This survey contained a series of audio samples from our sonification model, videos of HRI scenarios shot using a GoPro camera (all survey videos available at wwww.soundandrobotics.com/ch9), and associated interactive questions. This study was formatted as an observational survey to reduce possible stimulus variation between participants. As outlined by [33] and demonstrated by [14, 17, 32], screen-based methods are an effective way to fix the exact stimulus that each participant experiences. To this end, participants of our study were asked to listen to the same audio samples and observe the same videos. This study was conducted as an online survey, such to facilitate the recruitment of a large, diverse pool of participants. This also followed in the mould of previous similar studies, such as [9].

Before commencing the main sections of the survey, participants were asked to complete a preliminary data-collection consent form, along with a set of

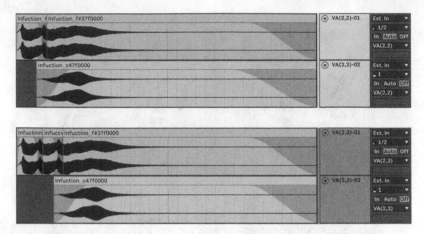

FIGURE 9.7
(Top) Visualization of the sound profile which represents the discretized region
VA = (2,2). Arousal = 2 is represented by two repetitions of the top wave
in each pair neutral sound wave. (Bottom) Visualization of the sound profile
which represents the discretized region VA = (2,3). Arousal = 3 is represented
by three repetitions of the shaded neutral sound wave. (Both) The bottom
wave in each pair sound wave represents the positive valence = 2 sound blended
with the repetitions of the neutral sound.

pre-survey questions. These questions were used to collect demographic data
along with self-scored experience and enthusiasm levels related to working
with collaborative robots. Upon completing this preliminary section, users
were brought to an introductory page outlining the concepts of valance and
arousal and their relation to robot states. This introduction was followed by
six questions in which users were presented with a blank VA graph discretized
into 25 regions as shown in Figure 9.5. For each of these six questions, users
were asked to listen to a sound selected from our sonification library and guess
which region this sound represented on the graph. The six sounds were chosen
to give a distributed representation of the VA graph. A visualization of this
selection process is shown in Figure 9.8.

   The next section of our survey was similar to the previous one, with the
addition of videos. For each question, users were presented with a video in
which a Jackal mobile robot was communicating with a human using the
designed sonification model. Each video captured a unique HRI scenario with
a unique outcome. The video was paused at a key point midway through the
interaction scenario, at which point users were asked which region the sounds
emitted from the Jackal represented on the discretized VA graph. Keyframes
from these video clips are shown in Figure 9.1. In addition to this selection,
users were asked to answer a multiple-choice question regarding which state
they perceived the robot to be in. The options for this question consisted of

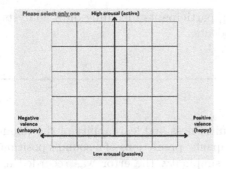 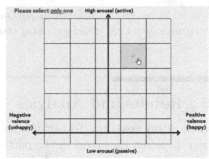

FIGURE 9.8
(Left) VA graph with 25 discretized selection regions presented to users. (Right) VA graph after users have made their selection.

each state outlined in Table 9.1 with the additional option *not sure.*

Unlike the previous section with a fixed number of questions (six), participants selected the number of video questions they wanted to complete. Upon completing an initial mandatory set of five video questions, users were presented with an option to view another video or proceed to the final video question. The number of video questions that users could complete within this section was capped at 15. Allowing users to select whether they would like to watch another video was used to gauge their interest in the robot interactions.

Following this set of video questions, users were presented with a final video question. This video was identical to the first video users watched in the previous section. The repeat video was used to gauge how contextual familiarization with the sonification model affected their ability to learn the sonification model and thus improve their robot state estimation accuracy. In addition, the variability in the number of video questions completed in the previous section presented an indication of the effects of extended familiarization with the sonification model.

Upon completing the repeated video question, users were presented with another section in which they were asked to listen to sounds without videos and guess which region these sounds represented on the discretized VA graph. Unique sounds from those presented at the beginning of the study were used in this section. Finally, participants were presented with a set of post-survey questions tailored from the BUZZ scale [12]. These questions probed for feedback on the audio communication model and user study overall.

Participants. This online survey was distributed as a single link over multiple Monash University social pages, along with several external networking platforms not affiliated with – the university. In all, our study received 37 complete responses. An analysis of our collected demographics data revealed an overall age spread of 18-60+, with 43% – of respondents in their 20s. In addition, 65% of respondents self-reported having little (2) to no (1) experience working with collaborative robots on a 1-5 Likert scale. To our surprise, there was a wide range of professions among participants including data analysts, government workers, teachers, and students.

9.6   Results and Analysis

Sonification testing. To test the intuitiveness and learnability of the sonification mapping, the errors in the participants' predictions of a sound's position on the VA graph were recorded. For simplicity, the error was recorded as the distance in grid squares of the prediction from the true position on the discretized VA grid. Figure 9.9 shows the trends of errors across the question set. There is a trend toward decreasing errors across the question set, although with a low statistical significance.

Further analysis of the errors in VA predictions is shown in Figure 9.10. These distributions show that the errors in arousal prediction were higher than those in valence, although with a large overlap. A 15% decrease in averaged overall participant error was observed in the second batch of sounds as opposed to the first, indicating a degree of learnability in the designed sonification method.

Interaction scenario testing. For each interaction video, participants were able to select up to two robot states that they believed to apply to the given scenario. The low arousal state, *Idle*, was mapped to a non-communicative state in which no sound was emitted from the robot. Thus, this state was not analyzed in the set of interaction videos. Figure 9.11 shows the percentage of participants who were able to correctly identify each high-level robot state, averaged over the question set. States *successful* and *requires attention* were

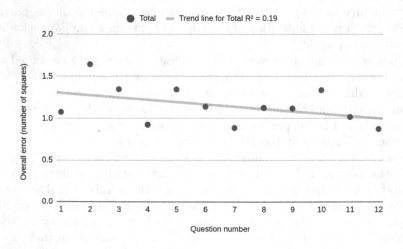

FIGURE 9.9
Error in VA predictions across sound question set.

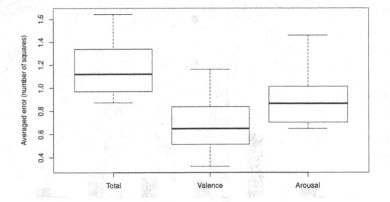

FIGURE 9.10
Error distributions across sound question set.

both identified correctly by over 70% of participants across the question set, while *progressing* and *unsuccessful* were both identified by roughly 50% of participants.

As participants were able to select multiple states for each scenario, further analysis is provided by examining the correct state selections in relation to the total selections made by participants. Figure 9.12 shows the participants' selection accuracies for each robot state using the *correct* and *incorrect* scores. These scores were calculated using the correct and incorrect selections as percentages of the total selections for each question (i) averaged over the question set (m) as shown in the following equations:

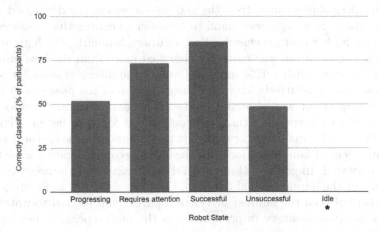

* No interaction scenarios were developed for this state

FIGURE 9.11
Participants' prediction accuracies for each robot state.

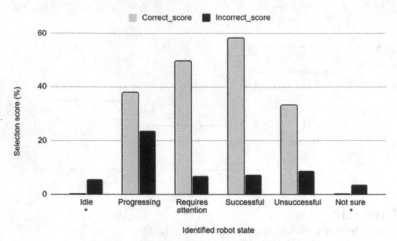

FIGURE 9.12
Correct and *incorrect* scores for each robot state.

$$correct\_score = \frac{1}{m} \sum_{i=1}^{m} \frac{correct\_selections_i}{total\_selections_i} * 100\%$$

$$incorrect\_score = \frac{1}{m} \sum_{i=1}^{m} \frac{incorrect\_selections_i}{total\_selections_i} * 100\%$$

Figure 9.12 shows some strengths and weaknesses of the developed sonification method. *Successful* was found to be easier to convey than *unsuccessful*, with a 25% higher average classification accuracy. Similarly, the high arousal state *requires attention* was correctly classified more than the lower arousal state, *progressing*, with a 12% margin. Overall, the incorrect selection scores of all classes were relatively low, with the exception of the *progressing* class.

No strong trend was observed in the classification accuracy across the question set, as the specific situational context was found to have a dominant effect. However, by analyzing the change in performance for the repeat question, a positive effect of familiarization with the Jackal's communication model may be substantiated. In general, there was little difference seen between the first question and the duplicate final question. However, as shown in Figure 9.13, a correlation between the number of videos a participant elected to watch and the classification accuracy improvement on the final repeated question was observed.

Figure 9.14 shows the effect of the participants' self-recorded feelings toward the communication model on the number of videos they chose to watch. In general, there is an observed relationship between fondness for the

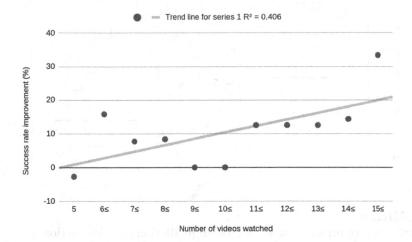

FIGURE 9.13
Effect of repeated interaction on duplicate scenario improvement.

communication model and a desire for repeated interaction via the survey videos. Paired with Figure 9.13, this points toward a positive relationship between user experience and robot understanding. However, this cannot be conclusively attributed to the communication method, as fondness for the robot's communication cannot be concretely separated from fondness for the robot itself.

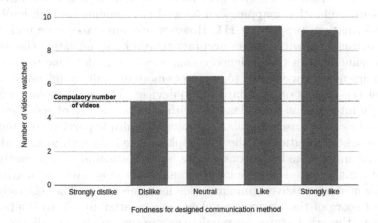

FIGURE 9.14
Average number of videos watched according to self-recorded feeling toward the communication model.

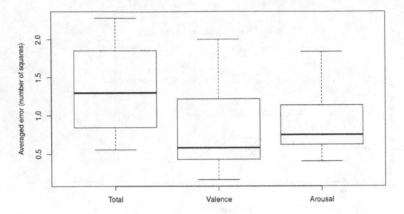

FIGURE 9.15
Distribution of errors in valence-arousal prediction of video section.

When analyzing the valence-arousal predictions made during the video section of the user study, a larger spread of errors was observed, as shown in Figure 9.15. There was no observable trend in these errors across the video set. This emphasizes the large role that situational context played in the participants' understanding of the robot during real-world interactions.

9.7 Discussion

The data presented in Figures 9.11 and 9.12 show that a low-dimensional continuous sonification mapping can be used to communicate high-level robot state information, supporting **H1**. However, certain states were understood by the participants with greater accuracy than others, indicating the need for further refinement in the designed communication model. Due to the use of lower frequencies, low-valence sounds were more difficult to differentiate than high-valence sounds on standard audio devices. This may explain why the *successful* interactions were classified with a higher degree of accuracy when compared against *unsuccessful*. Additionally, multiple participants reported that the *neutral* position on the VA graph sounded too high in arousal. This could explain why the high arousal state *requires attention* was correctly identified at a higher rate than the lower arousal state *progressing*, as participants regularly mistook passive communication for more active. The relatively high *incorrect* score of the *progressing* class may be attributable to it's position relative to the other states, as participants unsure of their predictions were perhaps likely to also select this more neutral state.

Overall, the majority of participants rated the communication as "Easy" or "Very easy" to understand, however, over a fifth rated the communication

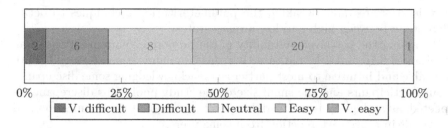

FIGURE 9.16

Participants' ease of understanding communication(absolute participant numbers shown on graph).

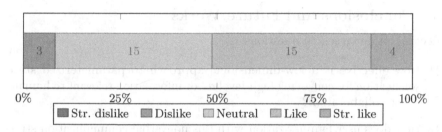

FIGURE 9.17

Participants' reactions to communication method (absolute participant numbers shown on graph).

as "Difficult" or "Very difficult", highlighting that the proposed strategy is not intuitive to all users. This breakdown is shown in Figure 9.16. The participants' feelings towards the communication model were determined using both a Likert scale question, shown in Figure 9.17, and a set of open-ended questions. Both mediums showed dominantly favorable reactions with a few negative responses. The correlation between likability and number of videos watched, shown in Figure 9.14, points to a potential relationship between user experience and interest in repeated interaction. Viewed in conjunction with Figure 9.13, the realized positive feedback loop between user experience, repeated interaction, and consequently improved robot understanding begins to support **H2**. A potential advantage key to this positive feedback loop is the improved user enjoyment resulting from a non-traditional, interesting communication form. Similar to **H1**, properly validating **H2** would require a comparison analysis between alternate nonverbal and verbal communication strategies.

With respect to the validation parameters, we found that participants were able to understand certain high-level robot states within different interaction scenarios. The relative difficulty observed in communicating certain states is believed to be due to features of the sonification model. From these findings, we believe that the model could be improved by improving sound clarity at low valence states and tweaking the parameterization techniques to better differentiate between high and low arousal states.

Regarding the perceived suitability of the proposed communications strategy, Figures 9.16 and 9.17 indicate that this communication strategy is both favourable and intuitive to most participants, acknowledging some discrepancy among participants and the small size of the study pool. This discrepancy is expected, as ambiguity is unavoidable in all non-explicit communication and may result in a negative reaction from some users.

9.8   Conclusions and Future Works

Reflecting on our research questions:

Q1 How effective is a low-dimensional approach to parameterized state sonification for conveying high-level robot information relevant to human–robot collaboration?

Q2 How does user familiarization with this nonverbal communication strategy correlate to user experience (UX) and shared task performance?

In our study, the proposed NVS model was able to effectively communicate the outlined set of robot states, with a degree of error due to the simplicity of the proposed 2D mapping. The findings also show variance between the different states; communicating the states *successful* and *requires attention* were less error-prone than *unsuccessful* and *progressing*. This finding presents an opportunity to experiment with alternate approaches for expressing mid-ranged arousal and negative valence. In the interest of making the sounds within our model more distinguishable, we plan to explore the use of different base sounds, blending sounds at the extremities of our VA mapping with auditory icons-earcons [31, 35], reducing the number of sounds emitted while *progressing*, correlating volume with arousal, rapidly oscillating the pitch of negative valence sounds rather than simply bending the pitch in the negative direction, and increasing our sonification model's dimensions of parameterization. Ultimately, we believe adding more complexity to our communication model while retaining structured parameterization would be an effective way to generate more meaningful and appealing sounds while further reducing human–robot miscommunication. To aid in characterizing the modulation of an additional sound parameter, we plan to explore the extension of a 2D valence-arousal (VA) mapping to a 3D valence-arousal-stance (VAS) mapping. This extension has proven effective for conveying robot emotion with a higher degree of complexity [4, 34]. An interesting avenue of future work would be to explore how the use of this higher-dimensional mapping might be useful for conveying high-level robot information relevant to human–robot collaboration.

Bibliography

[1] ADMONI, H., WENG, T., HAYES, B., AND SCASSELLATI, B. Robot nonverbal behavior improves task performance in difficult collaborations. Tech. rep., 2016.

[2] BARAKA, K., ROSENTHAL, S., AND VELOSO, M. Enhancing human understanding of a mobile robot's state and actions using expressive lights. *In 25th IEEE International Symposium on Robot and Human Interactive Communication, RO-MAN 2016* (2016), pp. 652–657.

[3] BRADLEY, M. M., AND LANG, P. J. Measuring emotion: The self-assessment manikin and the semantic differential. *Journal of Behavior Therapy and Experimental Psychiatry 25*, 1 (1994), 49–59.

[4] BREAZEAL, C. Emotion and sociable humanoid robots. *International Journal of Human Computer Studies 59*, 1-2 (2003), 119–155.

[5] BREAZEAL, C. Social robots that interact with people. *An Anthropology of Robots and AI* (2008), 72–88.

[6] BREAZEAL, C., KIDD, C. D., THOMAZ, A. L., HOFFMAN, G., AND BERLIN, M. Effects of nonverbal communication on efficiency and robustness in human-robot teamwork. *In 2005 IEEE/RSJ International Conference on Intelligent Robots and Systems, IROS* (2005), pp. 708–713.

[7] DRAGAN, A. D., BAUMAN, S., FORLIZZI, J., AND SRINIVASA, S. S. Effects of robot motion on human-robot collaboration. *In ACM/IEEE International Conference on Human-Robot Interaction 2015-March* (2015), pp. 51–58.

[8] FERNANDEZ, R., JOHN, N., KIRMANI, S., HART, J., SINAPOV, J., AND STONE, P. Passive demonstrations of light-based robot signals for improved human interpretability. *In RO-MAN 2018 - 27th IEEE International Symposium on Robot and Human Interactive Communication* (2018), pp. 234–239.

[9] FERNANDEZ DE GOROSTIZA LUENGO, J., ALONSO MARTIN, F., CASTRO-GONZALEZ, A., AND SALICHS, M. A. Sound synthesis for communicating nonverbal expressive cues. *IEEE Access 5* (2017), 1941–1957.

[10] FRID, E., AND BRESIN, R. Perceptual evaluation of blended sonification of mechanical robot sounds produced by emotionally expressive gestures: Augmenting consequential sounds to improve non-verbal robot communication. *International Journal of Social Robotics 14*, 2 (2022), 357–372.

[11] GREKOW, J. Music emotion maps in the arousal-valence space. *Studies in Computational Intelligence 747* (2018), 95–106.

[12] GRIER, P. Buzz. *Air Force Magazine 99*, 9 (2016), 87–92.

[13] HASTIE, H., DENTE, P., KÜSTER, D., AND KAPPAS, A. Sound emblems for affective multimodal output of a robotic tutor: A perception study. *In ICMI 2016 - Proceedings of the 18th ACM International Conference on Multimodal Interaction* (2016), pp. 256–260.

[14] HETHERINGTON, N. J., LEE, R., HAASE, M., CROFT, E. A., AND MACHIEL VAN DER LOOS, H. F. Mobile robot yielding cues for human-robot spatial interaction. *In IEEE International Conference on Intelligent Robots and Systems* (2021), pp. 3028–3033.

[15] INDERBITZIN, M., VALJAMAE, A., CALVO, J. M. B., VERSCHURE, P. F., AND BERNARDET, U. Expression of emotional states during locomotion based on canonical parameters. *In 2011 IEEE International Conference on Automatic Face and Gesture Recognition and Workshops, FG 2011* (2011), pp. 809–814.

[16] JEE, E. S., JEONG, Y. J., KIM, C. H., AND KOBAYASHI, H. Sound design for emotion and intention expression of socially interactive robots. *Intelligent Service Robotics 3*, 3 (7 2010), 199–206.

[17] KAMIDE, H., MAE, Y., KAWABE, K., SHIGEMI, S., HIROSE, M., AND ARAI, T. New measurement of psychological safety for humanoid. *In HRI'12 - Proceedings of the 7th Annual ACM/IEEE International Conference on Human-Robot Interaction* (2012), pp. 49–56.

[18] KNIGHT, H., AND SIMMONS, R. Laban head-motions convey robot state: A call for robot body language. In *2016 IEEE International Conference on Robotics and Automation (ICRA)* (5 2016), vol. 2016-June, IEEE, pp. 2881–2888.

[19] KOMATSU, T. LNCS 3784 - Toward Making Humans Empathize with Artificial Agents by Means of Subtle Expressions. Tech. rep.

[20] KWON, M., HUANG, S. H., AND DRAGAN, A. D. Expressing robot incapability. *In ACM/IEEE International Conference on Human-Robot Interaction* (2018), pp. 87–95.

[21] LÖFFLER, D., SCHMIDT, N., AND TSCHARN, R. Multimodal Expression of Artificial Emotion in Social Robots Using Color, Motion and Sound. *In ACM/IEEE International Conference on Human-Robot Interaction*, March (2018), pp. 334–343.

[22] MARIN VARGAS, A., COMINELLI, L., DELL'ORLETTA, F., AND SCILINGO, E. P. Verbal Communication in Robotics: A Study on Salient Terms,

Research Fields and Trends in the Last Decades Based on a Computational Linguistic Analysis, 2 2021.

[23] MUTLU, B., YAMAOKA, F., KANDA, T., ISHIGURO, H., AND HAGITA, N. Nonverbal leakage in robots. Association for Computing Machinery (ACM), p. 69.

[24] RUSSELL, J. A. Core affect and the psychological construction of emotion. *Psychological Review 110*, 1 (2003), 145–172.

[25] SAVERY, R., ROGEL, A., AND WEINBERG, G. Emotion musical prosody for robotic groups and entitativity. *In 2021 30th IEEE International Conference on Robot and Human Interactive Communication, RO-MAN 2021* (2021), pp. 440–446.

[26] SAVERY, R., ROSE, R., AND WEINBERG, G. Establishing human-robot trust through music-driven robotic emotion prosody and gesture. *In 2019 28th IEEE International Conference on Robot and Human Interactive Communication, RO-MAN 2019* (2019), pp. 4–10.

[27] SAVERY, R., ZAHRAY, L., AND WEINBERG, G. Before, between, and after: enriching robot communication surrounding collaborative creative activities. *Frontiers in Robotics and AI 8*, April (2021), 1–11.

[28] SAVERY, R., ZAHRAY, L., AND WEINBERG, G. Emotional musical prosody for the enhancement of trust: Audio design for robotic arm communication. *Paladyn 12*, 1 (2021), 454–467.

[29] SCHMITZ, M., FEHRINGER, B. C., AND AKBAL, M. Expressing emotions with synthetic affect bursts. In *CHI PLAY 2015 - Proceedings of the 2015 Annual Symposium on Computer-Human Interaction in Play* (10 2015), Association for Computing Machinery, Inc, pp. 91–96.

[30] SRIPATHY, A., BOBU, A., LI, Z., SREENATH, K., BROWN, D. S., AND DRAGAN, A. D. Teaching Robots to Span the Space of Functional Expressive Motion.

[31] STERKENBURG, J., JEON, M., AND PLUMMER, C. Auditory emoticons: Iterative design and acoustic characteristics of emotional auditory icons and earcons. *Lecture Notes in Computer Science (including subseries Lecture Notes in Artificial Intelligence and Lecture Notes in Bioinformatics) 8511 LNCS*, PART 2 (2014), 633–640.

[32] TAN, J. C. A., CHAN, W. P., ROBINSON, N. L., CROFT, E. A., AND KULIC, D. A Proposed Set of Communicative Gestures for Human Robot Interaction and an RGB Image-based Gesture Recognizer Implemented in ROS.

[33] VENTURE, G., AND KULIĆ, D. Robot expressive motions. *ACM Transactions on Human-Robot Interaction 8*, 4 (2019), 1–17.

[34] WAIRAGKAR, M., LIMA, M. R., BAZO, D., CRAIG, R., WEISSBART, H., ETOUNDI, A. C., REICHENBACH, T., IYENGAR, P., VASWANI, S., JAMES, C., BARNAGHI, P., MELHUISH, C., AND VAIDYANATHAN, R. Emotive response to a hybrid-face robot and translation to consumer social robots. *IEEE Internet of Things Journal 9*, 5 (2022), 3174–3188.

[35] ZAHRAY, L., SAVERY, R., SYRKETT, L., AND WEINBERG, G. Robot gesture sonification to enhance awareness of robot status and enjoyment of interaction. In *29th IEEE International Conference on Robot and Human Interactive Communication, RO-MAN 2020* (8 2020), IEEE, pp. 978–985.

[36] ZININA, A., ZAIDELMAN, L., ARINKIN, N., AND KOTOV, A. Nonverbal behavior of the robot companion: A contribution to the likeability. *Procedia Computer Science 169*, 2019 (2020), 800–806.

10

How Happy Should I be? Leveraging Neuroticism and Extraversion for Music-Driven Emotional Interaction in Robotics

Richard Savery, Amit Rogel, Lisa Zahray and Gil Weinberg

DOI: 10.1201/9781003320470-10

10.1   Introduction

Emotion in robotics covers a broad range of uses, from enhanced social interaction [35] to improved survivability and performance [2]. Personality has also been utilized in human robotic interaction research, such as in works that embed human personality in a robot to drive certain reactions and uses [21]. Another common approach is using human personality to understand robot perception, such as the overall impact of the uncanny valley [28]. While emotion is considered a critical feature of personality and is intertwined with the definition of personality itself [40], less research has been conducted addressing the interaction of personality, emotion, and robotics. We contend that sound and music, is intrinsically emotional and tied to human personality, and an effective medium to explore the relationship between each area.

In this work, we consider links between two of the Big Five personality types, Neuroticism and Extraversion, their impact on human emotional responses, and how these traits can be leveraged for HRI. The Big Five is the most common measure of personality in psychology [12,38] and is considered cross-cultural [30] with each trait representing discrete areas of the human personality [58]. The personality traits in the Big Five, also known by the acronym OCEAN, are Openness to experience, Conscientiousness, Extraversion, Agreeableness, and Neuroticism. Here, we focus on Neuroticism and Extraversion, which have shown robust and consistent findings in regards to their role in emotion regulation for a human's personality [3]. These personality traits lead to emotion strategies such as the human process of exerting control over the intensity and type of emotion felt and how that emotion is displayed [16].

We contend that human personality strategies for emotion can be used to drive design choices for robot emotional responses. A human's personality traits can lead to unique approaches to emotion; for example, individuals with low levels of extraversion are much more likely to outwardly display lower valence emotions. We believe that by mimicking human emotion strategies that are drawn from personality models, we can create varying versions of robotic responses to stimuli. By modelling human responses to emotional stimuli we can create robotic personas that are perceived differently by human collaborators.

In this work we developed two separate emotional responses to positive and negative stimuli, projected through audio and gestures. These personality types are based on human emotion strategies for different levels of Neuroticism and Extraversion. We believe that emotion strategies can be leveraged to portray varying robotic personas, with each persona receiving different ratings from human participants. We propose that through duplicating consistent emotion strategy from the Big Five framework, robots will achieve higher likeability and improved collaboration metrics than a control group.

For the study, we embedded custom emotional gestures and emotional musical prosody (EMP) in an industrial robotic arm. We believe robotic arms are especially well positioned to benefit from increased social interaction through audio and gesture, as they lack facial features and other communication methods often present in social robotics. The robotic gestures used were based on human body language poses and were validated before use. The audio system was based on an emotional musical prosody engine that has been shown effective for robotic arm interaction [45]. Avoiding speech and language has many advantages when it is not required for the interaction, such as reduced cognitive load [56] and improved trust [43].

This study aimed to address two key questions, firstly how a robot's personality type, as portrayed through emotion regulation strategies alter the perception of the robot. The second question aims to understand if a users' personality alters their preference for robot emotion regulation strategies. The study found that ultimately all users prefer robots with low neuroticism and high extraversion and that music and gestures is an effective medium to portray emotion in a robotic arm.

10.2   Background

10.2.1   Emotion and Robotics

Emotions can be classified in a variety of manners. The most common discrete categorization as proposed by Ekman [13] includes fear, anger, disgust, sadness, happiness and surprise. Emotions can also be classified by a continuous scale such as the Circumplex model; a two-dimension model using valence and arousal [39]. Research in robotics and emotion has seen continued growth across the last 20 years [44] and can be divided into two main categories – emotion for social interaction, and emotion for improved performance and "survivability" [2]. For social interaction, emotion can improve general expressiveness and interaction metrics [31]. For improved performance or survivability, robots can use emotion to reinforce or correct actions such as improved navigation [57].

10.2.2   Personality and Robotics

There are a variety of frameworks for the analysis of human personality in psychology literature, with the most common categorizations classifying personality between three and seven traits [23]. In human robot interaction literature, the term personality is not always used consistently and often lacks an agreed upon framework [41]. It is relatively common for HRI researchers to describe robot personality based on distinctive responses to stimuli, without basing their work on any specific personality model [6, 32]. Some studies have shown the potential of embedding psychologically driven personality models in human robot interaction [48]. These include aligning human and robot actions based on human personality [51], predicting the acceptability of a robot in a teaching environment [9], and understanding the impact of personality on understanding robot intentionality [7].

The Big 5 has been used previously in robotics, such as work focusing on extraversion and introversion in medical settings [49]. Other work has demonstrated human participants could accurately identify whether a robot was acting as an introvert or extrovert [27]. General attitudes based on a human's personality traits to robots has also been studied [34]. Likewise, past studies have shown that humans identify personality traits on robots, with general preferences emerging for positive traits [53]. Emotion modeling has been incorporated into some robotic personality models. For example, [1] use custom, subjective variations in emotional response to create nine unique personalities. [47] and [36] developed a robotic personality based on the Big Five, while using emotional responses based on possible relations between each class of the Big Five and emotion.

10.2.3   Emotion Regulation Strategies for Robotics

Emotion regulation is the process of attempting to modify both an internal feeling of emotion and our external expression of an emotion [15]. There are three core features of emotion regulation that separate regulation from common approaches to emotion in robotics. The first is regulation relies on an intrinsic or extrinsic activation of a goal to modify emotions [17]. The second feature emphasizes attempting to mentally engage with the cause of the emotion and changing one's internal reaction [18]. The third feature relies on varying the length and intensity of an emotional reaction [50].

Emotion regulation is a key element of emotion in humans and has direct links to personality, and has been hardly addressed in HRI research. Research has begun to cover potential deep learning applications for creating emotion regulation [19], strategies, however these have focused on generative processes and not human applications. A meta-analysis of emotion regulation and the Big Five found 32,656 papers including reference to regulation strategies linked to personality [4]. These findings are not always consistent however both Extraversion and Neuroticism had robust findings across the survey.

Overall, the literature in human psychology strongly indicates that emotion regulation strategies can be linked to personality traits for high Neuroticism and low Extraversion or low Neuroticism and high Extraversion in humans. [37] in particular, describe contrasting response types for positive and negative emotion. High Neuroticism and low Extraversion (HighN-LowE) personalities are consistently more likely to respond to positive stimuli with lower valence emotions, such as relief, whereas low Neuroticism and high Extraversion (LowN-HighE) are much more likely to respond directly with Joy or Happiness. For negative stimuli, HighN-LowE have a much higher likelihood to show disgust, fear, or guilt, while LowN-HighE are more likely to express sadness. In this paper, we utilize these approaches to present a LowN-HighE robot and a HighN-LowE robot, each capable of responding with a different range of emotions to stimuli. This creates personality models that are able to respond to positive or negative stimuli, with varying response types, allowing a positive response to take multiple forms.

10.3 Stimulus

To present models of emotional strategies, we developed and embedded gesture and audio based interactions in an industrial robotic arm. Our experiment design consisted of emotional robot gestures and responses to tagged image stimuli, followed by text questions. These responses were emulated from a study of response to visual stimuli with human personality types in existing research [37].

We chose to use a robotic arm due to its rapid expansion in use, with expected growth continuing into the foreseeable future, largely due to factory and industry settings. Research has also shown that embedding emotion driven gestures and audio in non-anthropomorphic robotic arms is more effective in portraying affect, than embedding such gestures and audio in social robots [45]. Our stimulus was designed as arm gestures that would respond to emotion tagged images in a manner derived from the personality traits.

10.3.1 Emotional Musical Prosody

We utilized an existing music-driven vocal prosody generator designed to represent emotions in audio [46]. Emotional musical prosody contains audio phrases that do not have semantic meaning, but are tagged with an emotion. They are useful in environments where sentiment or alerts are required from sound, without semantic meaning. The model we used included validated emotional phrases that use a voice-like synthesized processed sound. The dataset was labeled using the Geneva Emotion Wheel (GEW) [42], which combines both a continuous classification approach based on valence and arousal as well as discrete labels. The model lists 20 distinct emotions over a circle, with positions corresponding to the circumplex model of affect [39]. Each quadrant of the GEW corresponds to a different high/low valence-arousal pair, with arousal on the vertical axis and valence on the horizontal axis. The GEW emotions also correspond to the emotions linking HighN-LowE and LowN-HighE, allowing us to use the classification directly in a personality model.

10.3.2   Gestures

To physically display emotion strategies we used the generative system described in Chapter 13, mapping human gestures to a 7-joint robotic arm. The movements for each joint were created by hand, the guidelines and matching our emotion driven musical prosody engine. The gesture system was designed by studying traditional human body language postures. Human gestures were broken down into their fundamental movements based on [52] and [11]. These motions were then mapped to various joints on the robot. Most of these mappings involved designing erect/collapsed positions for the robot as well as forward/backward leaning motions to create a linear profile of the robot that matched human gestures.

While human gestures informed the robotic arm's movement speed, rest times between movements and number of movements were designed to synchronize with the audio phrases to create a connection between the emotional prosody and the physical movements of the robot. After primary joint movements were established, smaller, subtle movements were added to some of the remaining joints to increase the animacy of the robot.

Video samples with audio are available at https://soundandrobotics.com/ch10

10.3.3   Validation

Human perception of the robotic gestures and sounds used in the experiments was validated in a user study. Each participant completed a survey containing 30 videos. Each video was approximately 8 seconds long and depicted a robot gesture and sound corresponding to a particular emotion. 17 different emotions were represented among the videos, chosen due to the emotions used in personality based-responses. After each video, participants were asked to identify the emotion they perceived, along with its intensity on a scale of 1-5, using the Geneva Emotion Wheel. One video was used as an attention check, which showed a robot gesture along with audio instructing the participant to select a particular choice. The validation used a total of 20 participants from Amazon Mechanical Turk. One participant was eliminated due to failing the attention check, leaving a total of 19 valid participants. Of these, there were 11 from the United States, 6 from India, 1 from Thailand, and 1 from Malaysia. 17 identified as male, and 2 as female. The mean age was 36.5. The gestures and audio had previously been validated independently (only audio and only gestures), which we believe allowed us to test only a small group.

We utilized two metrics to analyze the validity of the videos, based on [10] – the mean weighted angle of the emotions reported by participants and the respective weighted variance. Both of these metrics were weighted according to reported intensity, and were converted to units of emotions on the wheel. The average emotion error (absolute difference between weighted reported emotion and ground truth emotion) was 1.7 with a standard deviation of 1.1. The average variance was 2.8. All emotion errors were below 3.5 except for one video, which represented admiration and had an error of 5.0. These results show that participants were able to interpret the expressed emotions within a small range of error, making the videos suitable for use in the experiments.

We believed the emotion error rate was well within a reasonable rate for this study. The error of 1.5, with a standard deviation of 1.1 showed that even when participants did mistake an emotion for another, they were usually only one emotion off, which was within the range of a possible response in our personality model.

10.4   Experiment

10.4.1   Method

Research question 1 examines how the robot's personality alters its perception amongst all participants. This question does not consider the participants' personality type and instead aims to identify broad trends amongst all interactions. We considered the traits or anthropomorphism, animacy, likeability, and perceived intelligence for each robot.

Research Question 1) How does a robot's personality type as portrayed through emotion regulation strategies alter anthropomorphism, animacy, likeability, and perceived intelligence?

We hypothesized that the robot with LowN-HighE will achieve greater ratings for likeability and perceived intelligence, while we will see no difference in anthropomorphism and animacy across all participants combined. We believed that emotion regulation strategies matching LowN-HighE are conducive to immediate likeability in a short term experiment as they show less unpredictability. We believed predictability will also contribute to an increase in perceived intelligence.

Our second research question considered the effect a users' personality will have on how they interact with the robotic arm.

Research Question 2) How does a users' personality type impact their ratings of different emotion regulation strategies for anthropomorphism, animacy, likeability and perceived intelligence?

We hypothesized that each category will have a preference for the emotion regulation strategy that matches their own personality type for likeability and perceived intelligence, while there will be no difference for anthropomorphism and animacy. While the previous question described our belief that LowN-HighE would achieve better results, overall we believe that would occur largely to the addition of LowN-LowE or HighN-HighE, whereas each group individually will show significant variation in results.

Participants first read a consent form and entered their names to confirm consent. They then completed the Ten Item Personality Measure (TIPI) [14], which gives the users' personality with the Big Five emotion model. TIPI was chosen as it has shown strong convergence with widely used longer measures, and has been shown to effectively gather personality in online platforms such as Mturk [8].

The main section of the experiment involved participants seeing a photo followed by a robotic response. We used photos from the open effective standardized image set (OASIS) [26], which features a range of images tagged with valence and arousal ratings. We chose photos that clearly showed positive or negative sentiment but also with a high standard deviation still within the bounds of positive or negative, implying a range of emotional response. We used a between experiment design, with participants randomly split into two groups, either seeing a robot responding to the stimuli with LowN-HighE or a robot responding with HighN-LowE. The responses were based on the response type described in Section 10.2.3, with each image returning an emotion based on the varying emotion regulation strategies. The same images were used for each robot personality type.

Figure 10.1 shows a sample sad image with a still of the robotic response. For each photo participants were asked to identify if the accompanying emotional reaction matched the image with a yes, no, or "other" option. This was inserted to force participants to watch, as every expected response was yes. Stimuli were randomly ordered for each participant with an attention check also appearing randomly. The attention check involved a related image as well as audio requiring the participant to type a specific phrase in the selection box "other".

FIGURE 10.1
Sample stimulus and still of robot response.

Following reviewing the emotion stimuli participants were shown three text questions with an accompanying emotional response. The responses to each question were matched to expected responses by personality as found in work by [37].

1. How stressful was the task you just completed?

2. To what extent did you experience positive emotions?

3. To what extent did you experience negative emotions?

After viewing all stimuli, participants completed the Godspeed Questionnaire. Participants were asked to complete the survey while considering the robot across all videos shown for each image. Godspeed is a commonly used human–robot interaction standard for measuring anthropomorphism, animacy, likeability, perceived intelligence, and perceived safety of robots [5]. We chose not to ask participants about perceived safety as felt it was not relevant to the research question or reliably observed given the experiment design. The Godspeed Questionnaire involves 28 questions (22 without perceived safety), rating users' impression of a robot for terms such as Artificial to Lifelike, which combine to give the broader metrics. Following the Godspeed test, we collected participant demographic information including year of birth, country of origin, and gender. The combined study took no more than 15 minutes, with the average time to completion of 11 minutes. The survey form was hosted on Qualtrics.

We had 100 participants complete the study on MTurk, of which 8 were eliminated due to failing an attention check, leaving a pool of 92. Of the 92 participants, the mean age was 42 with a standard deviation of 10 and a range of 22 to 69. 36 participants identified as female and 57 as male. Each participant was paid $2.00. 21 participants' country of origin was India, with the other 71 from the United States. We found no significant variation in responses from differences in countries of origin, gender or age.

This study was performed online using pre-recorded videos instead of live interaction or video watching in person. We believe that for this experiment this was an acceptable experimental design as ultimately our analysis focused on external viewing and analyzing a group of robots. Multiple past papers have shown no significant variation in results when a participant is watching a robot on video compared to in person [54, 55]. We also believe the use of MTurk and Prolific has some advantages over in person studies, allowing us a far larger and more diverse participant pool than possible in person. It has also been shown that compared to university pools, MTurk participants are more careful [20]. When combined with our multiple point attention check we are confident that our results would be replicated in person.

10.4.2   Results

We first analyzed the participants' personality results and found the break down between Neuroticism and Extraversion as HighN-HighE $n=11$, LowN-LowE $n=13$, HighN-LowE $n=27$, and LowN-HighE $n=36$. For the Godspeed test, we first calculated Cronbach's Alpha for each category. The results for each category were: Animacy *0.83*, Anthropomorphism *0.88*, Likeability *0.92*, and Intelligence *0.91*. This indicates a high internal consistency across all survey items.

10.4.2.1   Research Question 1

The robot personality with LowN-HighE emotion responses had a higher mean for both likeability and perceived intelligence. After conducting pair-wise t-tests the results were significant for both categories; for likeability $(p = 0.011)$ and for perceived intelligence $(p = 0.015)$.

For likeability LowN-HighE the results were *(M = 4.191, SD = 0.684)*, with a confidence interval of *(3.903, 4.480)*. LowN-HighE had a high effect size of 0.856. HighN-LowE had *(M = 3.606, SD = 0.924)* and a confidence interval *(3.272, 3.940)*. For the intelligence statistics LowN-HighE had *(M = 3.992, SD = 0.790)* and the confidence interval *(3.658, 4.325)*. LowN-High had a high effect size of *0.741*. For intelligence HighN-LowE had *(M = 3.406, SD = 0.919)* and the confidence interval *(3.074,3.737)*. For anthropomorphism and animacy the results were not significant $(p{\geq}0.05)$. These results proved our hypothesis and showed that the robotic personality type did alter the general populations' ratings for likeability and perceived intelligence. Figure 10.2 shows a box-plot of the results.

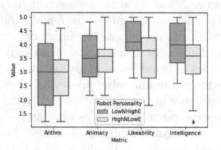

FIGURE 10.2

Comparing robot personality across all participants.

10.4.2.2 Research Question 2

Both human personalities rated the robot with LowN-HighE higher for likeability, with a pair-wise t-test giving significant results for LowN-HighE $(p=0.025)$ but not for HighN-LowE $(p=0.147)$. Figure 10.3 shows an overview of these results. This partly supported the hypothesis with LowN-HighE preferring LowN-HighE, but without significant results for HighN-LowE. Likewise perceived intelligence rating was higher from both for LowN-HighE, but again only with significant results for LowN-HighE human personalities $(p=0.049)$, and for HighN-LowE $(p=0.78)$.

Contradicting our hypothesis both animacy and anthropomorphism showed ratings for robot personality that matched that of the human personality. Users with LowN-HighE rated the robot with LowN-HighE better for both animacy and anthropomorphism although neither was significant $(p \geq 0.05)$. HighN-LowE also rated animacy and anthropomorphism higher for the robot with HighN-LowE, with a significant result for anthropomorphism $(p = 0.004)$.

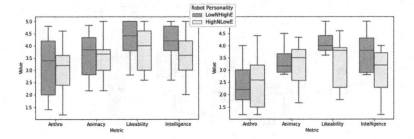

FIGURE 10.3
Comparing human personality across platform. Left indicates humans with
LowN-HighE, right HighN-LowE.

Further discussion of these results is available in Section 10.5, including comparisons with the results from our second experiment.

10.4.3 Supplementary Results: Openness, Conscientious and Agreeableness

Our research questions focused on collecting and analyzing the personality traits
Neuroticism and Extraversion, however standard personality measures for the
Big-5 also include Openness, Conscientiousness and Agreeableness. Openness is
linked to levels of curiosity and willingness to try new things; conscientiousness
is considered a efficiency and organization, while agreeableness is related to
friendliness and compassion. As previously described these traits do not have
consistent findings in relation to emotion regulation, nevertheless we believe
analyzing the links between human's ratings for openness to our other variables
is worth consideration to guide future work.

Our results for human Openness to experience matched expectations, with
the more open a participant the more likely they were to rate both robot
personalities as likable and intelligent. Comparing openness and intelligence
gave a Pearson's correlation coefficient of *0.4* with *p=0.002*, indicating a
moderate positive relationship. Figure 10.4 shows the high and low openness
trait for each metric.

While [8] found Mturk personality surveys gave accurate results, we believe
TIPI was insufficient for measuring conscientiousness and could not draw any
conclusions on the trait. TIPI includes two questions for measuring conscientiousness, asking for a self-rating of participants' dependability and carefulness.
For Mturk we believe participants would be wary to mark either rating too
low and risk their rating on the platform. This lead to a distribution with 88
participants rating themselves as highly conscientious and 5 giving themselves
a low conscientious rating.

We found no relation between agreeableness and preferences for emotion
regulation or robotic personalities. The Pearson correlation coefficient for each

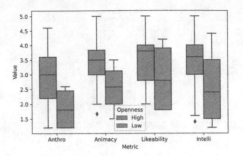

FIGURE 10.4
Openness to experience personality trait rating for each metric.

metric was: animacy (0.136, *p=0.195*), anthropomorphism (0.46, *p=0.661*), likeability (0.195, *p=0.062*), and perceived intelligence (0.190, *p=0.069*).

This replicates common psychology findings, that find agreeableness plays a part in emotion regulation near exclusively in social emotion settings [24,25,29].

10.5 Discussion

10.5.1 Human and Robot Personality

We found LowN-HighE consistently more likable for all users, with significant results for the LowN-HighE human with LowN-HighE robot. While we can not conclude why this is the case, we believe it may be due to the nature of short-term interaction. Especially in a single encounter, it is reasonable to assume that a robotic agent that shows higher extraversion and more emotional stability (through lower neuroticism) is more immediately likable regardless of a user's personality.

LowN-HighE also received higher ratings for perceived intelligence across both personality classes. This indicates that perceived intelligence is much more than just the ability to accurately complete a task. All users almost unanimously rated the robot as correctly identifying the emotion, yet still found a significant difference in perceived intelligence. As for likeability, we believe this reduced intelligence rating is due to higher levels of emotional instability.

Contradicting our hypothesis anthropomorphism and animacy ratings corresponded to human personality types, with HighN-LowE and LowN-HighE both rating their matching robotic personality higher. While we did not predict this, we believe this does make sense as users who see emotion regulation strategies closer to their own may be more likely to see anthropomorphic characteristics in a robot and more lifelike behavior.

10.5.2 LowN-LowE, HighN-HighE

Our core personality design involved HighN-LowE and LowN-HighE, however in our participant pool we had users with these personality traits. For this reason we include some preliminary findings on the group. Our sample size from experiment one was significantly smaller for both these groups (n=11 and n=13). Figure 10.5 shows the results for all personality types. LowN-LowE and HighN-HighE personalities are less common and less easily grounded in literature, so any conclusions from this data are not easily verified. However, there are some clear distinctions between comparisons of each human personality. HighN-HighE has almost no variation between robot personality with no significant results. This implies either that emotion regulation strategies do not impact this personality type, or that neither of our emotion regulation strategies strongly impacted HighN-HighE personalities. LowN-LowE personalities however did not have significant results for the LowN-HighE robot, for perceived intelligence ($p=0.48$) and likeability ($p=0.49$). This matches the results achieved for the general population and the LowN-HighE group. Despite these results, there is still future work required to draw any conclusions about LowN-LowE and HighN-HighE personalities and robotics.

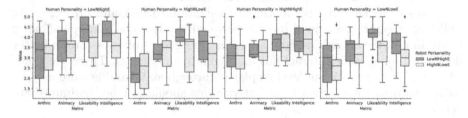

FIGURE 10.5
Comparing LowN-LowE and HighN-HighE.

10.5.3 Limitations

While attempting to control for all weaknesses in the study, there are several limitations that are worth describing. We did not collect information on participants on how they perceived the personality of each robot, so do not have a firm metric that the robot was believed to be a certain personality. This however was a considered decision; it has been repeatedly shown that untrained humans are inaccurate at predicting other human's personality types through observation, especially over short interactions [22,33], so there is no reliable way to gauge whether a personality type was perceived by a human user. Additionally emotion, while a strong part of personality, is just one component; we do not make a claim that we fully captured any personality trait, instead our goal was primarily to leverage features of personality traits and emotion. Nevertheless, future work attempting to identify how emotion regulation in robotics portrays a personality type to users would be of benefit.

Our study used videos of the robots interacting instead of in person participation. We believe for this experiment this did not alter the end results and improved overall outcomes as we were able to recruit many more participants than would be possible in person. Multiple past papers have shown no significant variation in results when a participant is watching a robot on video compared to in person [55]. In future work, we expect to apply lessons learned from these studies to in person experiments and interactions and believe lessons learned from video will apply to in person studies.

10.5.4   Future Work

This research will enable three new directions in future robotic research. The first is extended research in robot customization, based on a human's personality type. This can not only include audio features as described in this paper but also consideration of all areas of robotic design. We envision future studies where robot personalities are adapted in the short-term and over longer use, to the personality traits of individual users.

The second key area is robotic customization to task and project goals, building through the lens of robotic personality. By embedding personality traits in robots through design variations we believe robotics can be better developed for specific interactions and human experiences. There are times when a higher neuroticism level displayed through audio and gesture may be useful, such as times when robot interactions should not be the immediate action.

Finally, we believe this research outlines the need for further consideration and research of personality traits and their links to human robot interaction. While this was only an early step into the role of personality traits and potential links, future steps focusing on embedding specific personalities into robotics can lead to many enhanced outcomes. Using human personality preferences to design robotic emotional responses can have multiple broader implications. Emotion regulation strategies for a robot, whereby the strength of emotional response to stimuli are altered by a personality based design framework, provides the opportunity to drive new areas in human–robot interaction and develop new knowledge regarding the mechanisms that underlie affect based interaction. Developing an understanding of the potential of personality and emotion informed design can lead to the creation of deeper interactions between humans and robots and inform the development of a new framework for emotion driven interaction

10.6 Conclusion

The paper presents a new framework for developing emotional regulation and personality strategies for human–robot interaction through the use of sound and gesture. It explores how the Big Five personality traits can inform future designs of emotion-driven gestures and sound for robots. In particular, it studies the interplay between human and robotic Neuroticism and Extraversion and their effect on human perception of robotic personality. Key contributions include the development and implementation of novel affect and personality models for non-anthropomorphic robotic platforms. Other contributions include a groundwork understanding of emotion regulation strategies in human–robot interaction and novel insights regarding the underlying mechanism of emotion and affect in robotics.

Bibliography

[1] AHN, H. S., BAEK, Y. M., NA, J. H., AND CHOI, J. Y. Multi-dimensional emotional engine with personality using intelligent service robot for children. In *2008 International Conference on Control, Automation and Systems* (2008), IEEE, pp. 2020–2025.

[2] ARKIN, R. C., AND ULAM, P. An ethical adaptor: Behavioral modification derived from moral emotions. In *2009 IEEE International Symposium on Computational Intelligence in Robotics and Automation-(CIRA)* (2009), IEEE, pp. 381–387.

[3] BARAŃCZUK, U. The five factor model of personality and emotion regulation: A meta-analysis. *Personality and Individual Differences 139* (2019), 217–227.

[4] BARAŃCZUK, U. The five factor model of personality and emotion regulation: A meta-analysis. *Personality and Individual Differences 139* (2019), 217–227.

[5] BARTNECK, C., KULIĆ, D., CROFT, E., AND ZOGHBI, S. Measurement instruments for the anthropomorphism, animacy, likeability, perceived intelligence, and perceived safety of robots. *International Journal of Social Robotics 1*, 1 (2009), 71–81.

[6] BUI, H.-D., DANG, T. L. Q., AND CHONG, N. Y. Robot social emotional development through memory retrieval. In *2019 7th International Conference on Robot Intelligence Technology and Applications (RiTA)* (2019), IEEE, pp. 46–51.

[7] CHEVALIER, P., MARTIN, J.-C., ISABLEU, B., AND TAPUS, A. Impact of personality on the recognition of emotion expressed via human, virtual, and robotic embodiments. In *2015 24th IEEE International Symposium on Robot and Human Interactive Communication (RO-MAN)* (2015), IEEE, pp. 229–234.

[8] CLIFFORD, S., JEWELL, R. M., AND WAGGONER, P. D. Are samples drawn from mechanical turk valid for research on political ideology? *Research & Politics 2*, 4 (2015), 2053168015622072.

[9] CONTI, D., COMMODARI, E., AND BUONO, S. Personality factors and acceptability of socially assistive robotics in teachers with and without specialized training for children with disability. *Life Span and Disability 20*, 2 (2017), 251–272.

[10] COYNE, A. K., MURTAGH, A., AND MCGINN, C. Using the geneva emotion wheel to measure perceived affect in human-robot interaction. In *Proceedings of the 2020 ACM/IEEE International Conference on Human-Robot Interaction* (New York, NY, USA, 2020), HRI '20, Association for Computing Machinery, pp. 491–498.

[11] DAEL, N., MORTILLARO, M., AND SCHERER, K. Emotion expression in body action and posture. *Emotion (Washington, D.C.) 12* (11 2011), 1085–1101.

[12] DIGMAN, J. M. Personality structure: Emergence of the five-factor model. *Annual Review of Psychology 41*, 1 (1990), 417–440.

[13] EKMAN, P. Basic emotions. *Handbook of Cognition and Emotion 98*, 45-60 (1999), 16.

[14] GOSLING, S. D., RENTFROW, P. J., AND SWANN JR, W. B. A very brief measure of the big-five personality domains. *Journal of Research in Personality 37*, 6 (2003), 504–528.

[15] GROSS, J. J. Emotion regulation: Past, present, future. *Cognition & Emotion 13*, 5 (1999), 551–573.

[16] GROSS, J. J. Emotion regulation. *Handbook of Emotions 3*, 3 (2008), 497–513.

[17] GROSS, J. J., SHEPPES, G., AND URRY, H. L. Emotion generation and emotion regulation: A distinction we should make (carefully). *Cognition and Emotion (Print) 25*, 5 (2011), 765–781.

[18] GYURAK, A., GROSS, J. J., AND ETKIN, A. Explicit and implicit emotion regulation: a dual-process framework. *Cognition and Emotion 25*, 3 (2011), 400–412.

[19] HAO, M., CAO, W., LIU, Z., WU, M., AND YUAN, Y. Emotion regulation based on multi-objective weighted reinforcement learning for human-robot interaction. In *2019 12th Asian Control Conference (ASCC)* (2019), IEEE, pp. 1402–1406.

[20] HAUSER, D. J., AND SCHWARZ, N. Attentive turkers: Mturk participants perform better on online attention checks than do subject pool participants. *Behavior Research Methods 48*, 1 (2016), 400–407.

[21] HENDRIKS, B., MEERBEEK, B., BOESS, S., PAUWS, S., AND SONNEVELD, M. Robot vacuum cleaner personality and behavior. *International Journal of Social Robotics 3*, 2 (2011), 187–195.

[22] JOHN, O. P., DONAHUE, E. M., AND KENTLE, R. L. Big five inventory. *Journal of Personality and Social Psychology* (1991). Behavior research methods, 48(1), 400–407.

[23] JOHN, O. P., SRIVASTAVA, S., ET AL. The big five trait taxonomy: History, measurement, and theoretical perspectives. *Handbook of Personality: Theory and Research 2*, 1999 (1999), 102–138.

[24] KIM, B.-R., CHOW, S.-M., BRAY, B., AND TETI, D. M. Trajectories of mothers' emotional availability: Relations with infant temperament in predicting attachment security. *Attachment & Human Development 19*, 1 (2017), 38–57.

[25] KOCHANSKA, G., FRIESENBORG, A. E., LANGE, L. A., AND MARTEL, M. M. Parents' personality and infants' temperament as contributors to their emerging relationship. *Journal of Personality and Social Psychology 86*, 5 (2004), 744.

[26] KURDI, B., LOZANO, S., AND BANAJI, M. R. Introducing the open affective standardized image set (oasis). *Behavior Research Methods 49*, 2 (2017), 457–470.

[27] LEE, K. M., PENG, W., JIN, S.-A., AND YAN, C. Can robots manifest personality?: An empirical test of personality recognition, social responses, and social presence in human–robot interaction. *Journal of Communication 56*, 4 (2006), 754–772.

[28] MACDORMAN, K. F., AND ENTEZARI, S. O. Individual differences predict sensitivity to the uncanny valley. *Interaction Studies 16*, 2 (2015), 141–172.

[29] MANGELSDORF, S., GUNNAR, M., KESTENBAUM, R., LANG, S., AND ANDREAS, D. Infant proneness-to-distress temperament, maternal personality, and mother-infant attachment: Associations and goodness of fit. *Child Development 61*, 3 (1990), 820–831.

[30] McCRAE, R. R., AND COSTA JR, P. T. Personality trait structure as a human universal. *American Psychologist 52*, 5 (1997), 509.

[31] MICHAUD, F., AUDET, J., LETOURNEAU, D., LUSSIER, L., THEBERGE-TURMEL, C., AND CARON, S. Experiences with an autonomous robot attending aaai. *IEEE Intelligent Systems 16*, 5 (2001), 23–29.

[32] MIWA, H., UMETSU, T., TAKANISHI, A., AND TAKANOBU, H. Robot personalization based on the mental dynamics. In *Proceedings. 2000 IEEE/RSJ International Conference on Intelligent Robots and Systems (IROS 2000)(Cat. No. 00CH37113)* (2000), vol. 1, IEEE, pp. 8–14.

[33] MOUNT, M. K., BARRICK, M. R., AND STRAUSS, J. P. Validity of observer ratings of the big five personality factors. *Journal of Applied Psychology 79*, 2 (1994), 272.

[34] MÜLLER, S. L., AND RICHERT, A. The big-five personality dimensions and attitudes to-wards robots: A cross sectional study. In *Proceedings of the 11th Pervasive Technologies Related to Assistive Environments Conference* (2018), pp. 405–408.

[35] OGATA, T., AND SUGANO, S. Emotional communication robot: Wamoeba-2r emotion model and evaluation experiments. In *Proceedings of the International Conference on Humanoid Robots* (2000).

[36] PARK, J.-C., KIM, H.-R., KIM, Y.-M., AND KWON, D.-S. Robot's individual emotion generation model and action coloring according to the robot's personality. In *RO-MAN 2009-The 18th IEEE International Symposium on Robot and Human Interactive Communication* (2009), IEEE, pp. 257–262.

[37] PENLEY, J. A., AND TOMAKA, J. Associations among the big five, emotional responses, and coping with acute stress. *Personality and Individual Differences 32*, 7 (2002), 1215–1228.

[38] PERVIN, L. A. *The Science of Personality*. Oxford university press, 2003.

[39] POSNER, J., RUSSELL, J. A., AND PETERSON, B. S. The circumplex model of affect: An integrative approach to affective neuroscience, cognitive development, and psychopathology. *Development and Psychopathology 17*, 3 (2005), 715–734.

[40] REVELLE, W., AND SCHERER, K. R. Personality and emotion. *Oxford Companion to Emotion and the Affective Sciences 1* (2009), 304–306.

[41] ROBERT, L., ALAHMAD, R., ESTERWOOD, C., KIM, S., YOU, S., AND ZHANG, Q. A review of personality in human–robot interactions. *Available at SSRN 3528496* (2020).

[42] SACHARIN, V., SCHLEGEL, K., AND SCHERER, K. Geneva emotion wheel rating study (report). geneva, switzerland: University of geneva. *Swiss Center for Affective Sciences* (2012).

[43] SAVERY, R., ROSE, R., AND WEINBERG, G. Establishing human-robot trust through music-driven robotic emotion prosody and gesture. In *2019 28th IEEE International Conference on Robot and Human Interactive Communication (RO-MAN)* (2019), IEEE, pp. 1–7.

[44] SAVERY, R., AND WEINBERG, G. A survey of robotics and emotion: Classifications and models of emotional interaction. In *Proceedings of the 29th International Conference on Robot and Human Interactive Communication* (2020).

[45] SAVERY, R., ZAHRAY, L., AND WEINBERG, G. Emotional musical prosody for the enhancement of trust in robotic arm communication. In *29th IEEE International Conference on Robot & Human Interactive Communication* (2020), Trust, Acceptance and Social Cues in Human-Robot Interaction.

[46] SAVERY, R., ZAHRAY, L., AND WEINBERG, G. Prosodycvae: A conditional convolutionalvariational autoencoder for real-timeemotional music prosody generation. In *2020 Joint Conference on AI Music Creativity* (2020), CSMC + MUME.

[47] SOHN, K., KRISHNAMOORTHY, S., PAUL, O., AND LEWIS, M. A. Giving robots a flexible persona: The five factor model of artificial personality in action. In *2012 12th International Conference on Control, Automation and Systems* (2012), IEEE, pp. 133–139.

[48] SYRDAL, D. S., DAUTENHAHN, K., WOODS, S. N., WALTERS, M. L., AND KOAY, K. L. Looking good? appearance preferences and robot personality inferences at zero acquaintance. In *AAAI Spring Symposium: Multidisciplinary Collaboration for Socially Assistive Robotics* (2007), pp. 86–92.

[49] TAPUS, A., ȚĂPUȘ, C., AND MATARIĆ, M. J. User—robot personality matching and assistive robot behavior adaptation for post-stroke rehabilitation therapy. *Intelligent Service Robotics 1*, 2 (2008), 169–183.

[50] THOMPSON, R. A. Emotion and self. *Socioemotional Development 36* (1990), 367.

[51] VOLLMER, A.-L., ROHLFING, K. J., WREDE, B., AND CANGELOSI, A. Alignment to the actions of a robot. *International Journal of Social Robotics 7*, 2 (2015), 241–252.

[52] WALBOTT, H. G. Bodily expression of emotion. *European Journal of Social Psychology 28*, 6 (1998), 879–896.

[53] WHITTAKER, S., ROGERS, Y., PETROVSKAYA, E., AND ZHUANG, H. Designing personas for expressive robots: personality in the new breed of moving, speaking, and colorful social home robots. *ACM Transactions on Human-Robot Interaction (THRI) 10*, 1 (2021), 1–25.

[54] WOODS, S., WALTERS, M., KHENG LEE KOAY, AND DAUTENHAHN, K. Comparing human robot interaction scenarios using live and video based methods: towards a novel methodological approach. In *9th IEEE International Workshop on Advanced Motion Control, 2006.* (2006), pp. 750–755.

[55] WOODS, S., WALTERS, M., KOAY, K. L., AND DAUTENHAHN, K. Comparing human robot interaction scenarios using live and video based methods: towards a novel methodological approach. In *9th IEEE International Workshop on Advanced Motion Control, 2006.* (2006), IEEE, pp. 750–755.

[56] YIN, B., CHEN, F., RUIZ, N., AND AMBIKAIRAJAH, E. Speech-based cognitive load monitoring system. In *2008 IEEE International Conference on Acoustics, Speech and Signal Processing* (2008), IEEE, pp. 2041–2044.

[57] ZHANG, H., LIU, S., AND YANG, S. X. A hybrid robot navigation approach based on partial planning and emotion-based behavior coordination. In *2006 IEEE/RSJ International Conference on Intelligent Robots and Systems* (2006), IEEE, pp. 1183–1188.

[58] ZILLIG, L. M. P., HEMENOVER, S. H., AND DIENSTBIER, R. A. What do we assess when we assess a big 5 trait? a content analysis of the affective, behavioral, and cognitive processes represented in big 5 personality inventories. *Personality and Social Psychology Bulletin 28*, 6 (2002), 847–858.

11

Augmenting a Group of Task-Driven Robotic Arms with Emotional Musical Prosody

Richard Savery, Amit Rogel and Gil Weinberg

DOI: 10.1201/9781003320470-11

11.1   Introduction

Research at the intersection of music, emotion, and robotics has focused on work in robotic musicianship and human–robot interaction (HRI) studies. In robotic musicianship, robots are designed to perform and compose music acting as a musical collaborator [35]. Research in music and HRI however, has focused on methods for music or sound to improve how humans interact with robots [22]. In this work, we focus on how embedding musical features and gestures into robotic systems can alter the interaction and improve key HRI collaboration metrics, including trust, warmth and competence. We expand past work in the field by looking at the intersection of two rarely addressed areas, large groups of robots and the impact of sound on interaction. We contend that music, as one of the most emotionally driven human forms of communication, can play a key role in HRI.

Emotional musical prosody (EMP), where short musical phrases are used to convey an emotion, has been shown to improve trust, likeability and perceived intelligence in HRI [32]. The use of musical phrases to improve interaction in robotics still has many future avenues for research, in particular we believe group robotic environments, where multiple sources of speech may lead to higher cognitive loads and distraction, could be improved through musical phrases. The role of emotional contagion, where the emotion of a robot alters the emotion the a human collaborator has also not been addressed.

In this paper we explore how EMP can be used in HRI in a large group containing 10 robotic arms and one human collaborator (see Figure 11.1). We explore this area using an existing generator for emotional musical prosody [33] in combination with new custom gestures on a group of Ufactory Xarms. We aimed to explore the role of emotion contagion in a group of robotic arms performing a task, in this case moving an object between robotic arms that is passed from a human collaborator.

To develop these findings we conducted two studies, the first study was online through Amazon Mechanical Turk, and had 111 participants rating video footage of the robots interaction. We first analyzed whether EMP can lead to emotional contagion in human participants. We then considered the impact on HRI metrics for trust, competence, warmth and discomfort, as well as the relation between these metrics and the level of contagion. Our findings suggested that participants preferred robots using EMP, and emotional contagion could occur from robots to human participant. The second study was conducted with 16 participants in-person, primarily aiming to build on conclusions from the first study, while gathering extensive qualitative data to direct future research both using EMP, as well as music and HRI.

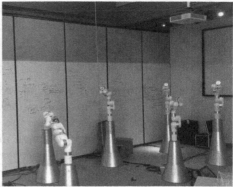

FIGURE 11.1
User passing ring to robot and robots passsing rings between each other.

11.2 Background

11.2.1 HRI and Emotion

Research in HRI often explores how different variables can alter human-based metrics. One of the most widely used survey in HRI is the Godspeed Questionnaire Series, which measures anthropomorphism, animacy, likeability, perceived intelligence, and perceived safety of robots [4,39]. Each metric in the Godspeed survey is measured with 4-5 bipolar sub-questions. Other surveys are often created for a specific metric with more extended questions such as ratings for self-efficacy [23] or willingness to interact [12]. It is also very common for psychology and social studies metrics to be adopted within the field of robotics, such as the mind attribution scale [16]. In HRI a reoccuring issue is the development of trust, as low levels of trust can lead to under-utilization in work and home environments [17]. Trust is generally categorized into either cognitive trust or affective trust [14]. Perceiving emotion is a crucial for the development of affective trust in human-to-human interaction [24], as it increases the willingness to collaborate and expand resources bought to the interactions [15].

Research focusing on the role of emotion in robotics has seen continual growth over the last thirty years, spanning many applications and platforms [31]. This research can primarily be divided into two main categories, emotion for improved performance (called "survivability") and emotion for social interaction [2]. Survivability invokes the belief that emotion is central to animals' ability to survive and navigate the world and can likewise be used in robots. This includes situations such as an internal emotion based on external danger to a robot [1]. The second category – social interaction – addresses anyway emotion is used to improve interaction, such as analyzing a humans emotion, or portraying emotion to improve agent likeability and believability [19].

11.2.2   Sound and HRI

There is only limited work in sound and HRI outside the use of speech systems, with research on the impact of sound relatively rare [42]. Studies have been conducted to analyze whether the use of a beep improves the perception of a robot with positive results, although more considered application of the range of possible audio sounds has not been conducted [11]. Consequential sounds are the sounds made by a robot in normal operation, such as motors and physical movement. The sound from motors has been used as a communication tool through modification of gesture [13], as well as used to improve localization [7]. Overall, consequential sounds have been analyzed for their impact on interaction with primarily negative results [18, 38]. Sonification of robotic movements has been examined, such as in relation to emotions for children with Autism Spectrum Disorder (ASD) [43], or for general movement of robots [5]. While there are multiple attempts to incorporate sound beyond spoken language into robotics, it is ultimately very limited in scope with broad potential for further research. There has not been the same sound tested on multiple platforms, or even the same platform in different interaction types, and each sound implementation is very rarely explored outside single one-off studies.

11.2.3   Emotional Musical Prosody

EMP was developed in previous work by the authors [29] to leverage the emotional power of musical improvisation, combined with the musical features of language, to create a new method of non-semantic communication called Emotional Musical Prosody (EMP). In linguistics, prosody refers to the parts of speech that are not the words, including the intonation, stress, and rhythm of the words. These features have parallels in music and the way a performer expresses music. EMP offers many advantages for HRI, by not using language, it can lower cognitive load while adding an emotion-driven personality, which has been repeatedly shown to improve collaboration metrics [33].

The first phase of developing EMP relied on gathering a new dataset from three vocalists, Mary Carter, Ella Meir and Aya Donahue. Emotions can be classified in countless ways, such as Ekman's six categories of anger, surprise, disgust, enjoyment, fear, and sadness or the cirumplex model, which places emotion on a two-dimensional circular space. We chose to use the Geneva Emotion Wheel (GEW), which uses 20 different emotion types, such as love and admiration. GEW allowed us to capture a wide range of emotions, within a range we felt would be possible for the performers. The final dataset contained 12 hours of audio, with around 450 phrases for each emotion. After collecting the dataset, the authors developed a new method using deep learning to generate audio phrases. This method focused on real-time generation, capable of rapidly responding to a human's emotion (more details on the underlying generation technology are available in [33]).

After the dataset and generation method were established, multiple studies aimed to understand how EMP could work in dyadic HRI (between one human and one robot) across multiple robotics platforms [10, 25–30, 33], as well as in Chapter 10 for portraying personality. EMP was compared between social robots, who are designed primarily for social interaction with humans, with a humanoid robot, which aims to replicate human behavior, and an industrial robotic arm [34]. EMP significantly improved trust, likeability and perceived intelligence for the industrial arm and social robot but not the humanoid robot. Emotional Contagion and EMP has not yet been explored in depth however, nor has large scale groups with interaction with human participants. Emotional contagion refers to the process of emotion transferring from one agent to another, commonly in the form of shared group laughter. In group human interactions, emotional contagion has been shown to improve cooperation and trust in team exercises [3, 21, 37]. Music contagion has been studied extensively, with a complex relationship developed between the emotion portrayed and the effect on music [8], nevertheless music has consistently lead to emotional contagion in many listeners [9].

11.3   Study 1: Fundamentals of Emotion Contagion, Music and Robotics

The primary goals of our first study were to engage a diverse audience, who would view groups of robots using EMP and establish some fundamental principles. Our goal was to understand participant perceptions of emotional contagion, rather than aiming to actually measure emotional contagion in the participants. We developed three research questions aiming to understand the role of emotional contagion, warmth, competence, discomfort and trust in task-basked group robotic activities. We analyzed the variation in robots when using EMP compared to performing a task alone, with tasks sometimes being performed successfully and sometimes failing.

RQ1 Does EMP improve the ratings for warmth, competence, discomfort and trust compared to the task alone?

RQ2 Does EMP lead to emotional contagion when compared to a task alone?

RQ3 How do levels of emotional contagion compare to human ratings for warmth, competence, discomfort and trust?

Research Question 1 was designed to reconfirm that using EMP in a group of robots would improve the metrics of warmth, competence, discomfort and trust over the performance of a task alone. In previous work EMP has been measured in multiple use cases, but not in combination with a task with options for success or failure. Robots involved in specific tasks however is a much more real world scenario, than idle robots. It is possible that the addition of a task will override any emotional or other reaction that may have be drawn from prosody. We hypothesized however that ratings for warmth, competence and trust would all increase in the EMP version, with a lower rating for discomfort. We believed this would replicate previous studies reactions to EMP [34], despite the change of scenario and addition of success and failure in the task.

Research Question 2 explored whether the emotional reactions from robots would actually lead to participants believing a human user would show a different emotional response. We hypothesized that participants would be influenced by the emotional content of the audio, and change their results when compared to people who view the task alone, expanding on simpler previous findings about the role of interaction [28].

Research Question 3 aimed to identify how, if at all, the ratings for warmth, competence, discomfort and trust correlate with the likelihood a participant believing an emotional response would occur from the user. We conducted this question as an exploratory study, aiming to broadly see if any correlation existed. Our hypothesis was that emotional contagion would be linked to higher ratings of warmth, competence and trust, as participants would relate better to the robots when they identify with an emotional response.

11.3.1   Method

To answer these research questions we developed an online study, where participants would view video footage of the group of robotic arm. The stimuli for this experiment was a group of robots tasked with passing a ring from a human to a box. This task required interaction between the participant, robot, and group of robots. We used a ring passing task to ensure that the person's reactions to each emotion are a results of the emotion, and not from robots moving. The robots would fail to deliver the ring 50 % of the time. After the robot would succeed or fail in its task, the rest of the robotic group would react in response to the ring passers. If the task fails, the robots can react with either valance (happy or sad). If the task succeeds, the robots would have different arousal levels (happy or calm). This resulted in four possible conditions for EMP, either Task Failure Anger, Task Failure Joy, Task Success Calm, Task Success Joy. The robots emotional reactions were designed based on a set of emotional musical prosody phrases, previously validated [25] and described in section 11.2.3. The gestures were designed using the rule-based system described in Chapter 13.

Video samples can be seen at https://www.soundandrobotics.com/ch11

Participants first completed a consent form, followed by watching the video stimuli. The study was a within design, with all participants viewing every video. For each robot video participants rated the emotion a "human interacting with the robot will most likely feel". This was presented as a multiple choice question with the choices, "happy or excited", "angry or disgusted", "sad or depressed", and "relaxed or calm", with one option from each quadrant of the circumplex model. Participants also had an option to enter free text, or answer "None". Following the video we measured warmth, competence, and discomfort using The Robotic Social Attributes Scale (RoSAS) [6]. RoSAS is an 18 item scale that requires participants to identify how closely associated certain words are with the robot (such as "reliable" and "scary"). To measure trust we used the 14 point version of the Trust Perception Scale-HRI [36], which asks participants to rate the percentage of time a robot will act in a certain way, such as "dependable"; and "provide appropriate information". The questions for RoSAS and the Trust Perception Scale were randomly ordered for each participant. We also optionally asked for participant demographic information and allowed an open general text response to discuss the study, and for other comments on the robot.

The study was conducted online using Amazon Mechanical Turk (MTurk) to gather participants, with the survey hosted online using Qualtrics. We recruited 118 participants, of whom we used the responses of 111. Through the study there were a range of attention checks, including a video overdubbed with audio asking for a specific response, and a survey question requiring a specific response. We also tracked time spent on each question, and overall time on the survey, as well as participants IPs. In past Mturk studies we have found multiple participants working from the same IP, which is allowed on the platform, but prevents us from knowing if there was any collaboration. Overall, we recruited 118 participants, and we used the responses of 111 who all passed the attention checks. Participants were all based in the United States, with ages *(M = 44, STD = 10.4, max = 70, min = 20)* and 46 identify as female, 64 as male and one non-binary.

11.3.2 Results

11.3.2.1 RQ 1

For Research Question 1 we first calculated Cronbach's Alpha for the combined metrics in the trust survey. This gave a result of *0.91*, indicating high internal reliability for the questions. A pair-wise T-test showed did not find a significant result *(F=1.7, p =0.08)* after Holm–Bonferroni considering the four variables. The effect size, measured using Cohen's D was *0.32*.

For each question in RoSAS we first calculated Cronbach's Alpha, with the results: Competence *0.92*, Warmth *0.93*, and discomfort 0.81 indicating high internal reliability for each metric. A pair-wise indicated that competence was not significant, with an effect size of *0.06*. Warmth was significant *(F=5.5, p¡0.001* with an effect size of *0.94*. Discomfort was also significant, *(F3.6, p¡0.001* with an effect size of *0.61*. Figure 11.2 shows a box plot of the results.

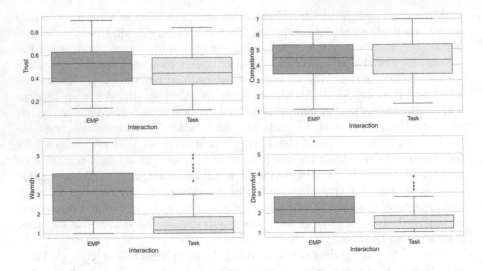

FIGURE 11.2
EMP compared to task for HRI metrics.

11.3.2.2   RQ 2

To analyze whether participants in an online study would recognize the potential for emotional contagion we compared responses between task alone and EMP conditions. Considering task failure, we found that with the addition of anger-tagged EMP participants, were much more likely to choose "sad or depressed" and likewise for task failure with Joy were much more likely to choose "happy or excited". For successful tasks we also found joy-tagged EMP increased the chance of choosing "happy or excited", however for success calm, participants were more likely to choose "sad or depressed". Figure 11.3 presents a bar plot of the results.

11.3.2.3   RQ 3

To answer Research Question 3 we conducted linear regression on each of the four metrics compared against the contagion rating. A result was as a positive contagion whenever the emotion matched the emotion of the robot. We found a significant result for Trust, Competence and Warmth, but not discomfort, indicating some correlation between each result and the level of contagion (see Table 11.1. Figure 11.4 displays a regression plot of the results.

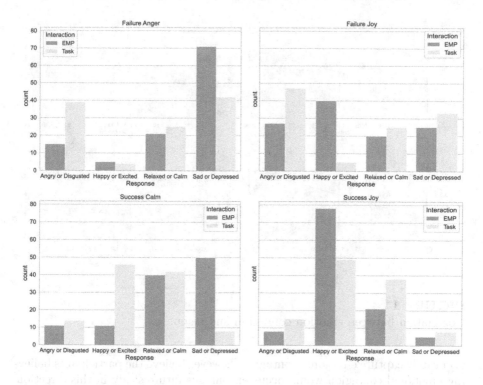

FIGURE 11.3
EMP compared to task for emotional response.

TABLE 11.1
Linear regression of interaction between emotion contagion and HRI metric.

	Slope	Intercept	r	p	Standard Error
Trust	0.065	0.446	0.339	0.011	0.024
Competence	0.351	3.945	0.265	0.048	0.174
Warmth	0.577	2.408	0.375	0.004	0.194
Discomfort	0.069	2.137	0.066	0.63	0.142

11.3.3   Discussion

Research Question 1 confirmed that results from past EMP studies were replicable and carried across to larger groups of robots in a different setting. We believe confirming these findings helps to strengthen past work in EMP, while justifying the continued exploration. Research Question 2 supported broad findings that emotion contagion could occur from robots human participants. We certainly make no claim that an online study in this manner can

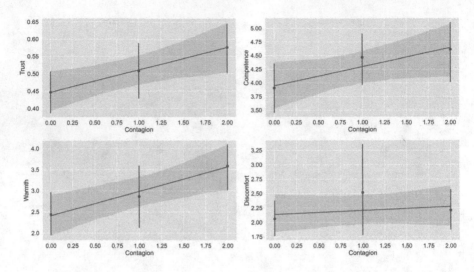

FIGURE 11.4
Correlation between emotion contagion and response.

accurately capture emotional contagion, however believe the participants believing emotional contagion would occur encourages future study in this direction. Research Question 3 further encouraged that the participants who believed there would be emotional contagion.

11.4   Study 2: Exploration Research into the Role of EMP in Groups

The results of our first study confirmed past studies and that EMP is able to change ratings for HRI Metrics. Additionally it suggested that emotional contagion could correlate with a robots emotional reaction. To continue exploring the role of EMP beyond past research we decided to conduct an in-person study with a participant interacting with many robots simultaneously. We were primarily interested in gathering a wide range of qualitative data that could help inform future understandings of EMP and sound in group robotics more broadly.

Our in-person study focused on an extended interaction with the robots that was video recorded, followed by a 20 minute semi-structured interview. Our stimuli includes three robots that people could interact with. Each robot had a unique color to help participants differentiate the three robots. In addition, participants had three sets of four rings (12 total). The sets of rings matched the colors on the robots. The colors on the rings signified a participant which robot to give the ring to. This ensures that each robot gets an even number of rings to interact with. After a robot receives a ring, it will pass it along to another robot, that places it in a box. Afterward, all the robots will react to the robots success/failure. The success and failure patterns were the same as the first study. We placed a camera in the group of robots that monitored the participants facial expressions and interactions.

As a participant first walked into the lab, they were asked to read and complete a consent form, that outlined the basic interaction with the robots and that they were being recorded on video. After signing the consent form, participants were given 12 rings and instructed to hand the top ring to the robot with a matching color. We instructed the users that the robots "Are tasked to pass the ring to a back robot, that must put the ring in a box". The participant continues to observe the robot pass the ring, place it into the box, and observe the robots reactions. While we did not initiate some information, we told participants that the robots were aware of task results if they asked. At the end of the 12 rings, participants were given 4 white rings that they were allowed to place on any robot. We documented the white ring placement. The task result and robot reactions are randomized for the white rings. The white rings served as base questions for the extended interview. Participants proceeded to complete a qualtrics survey about the robots and then spoke with a lab member for an extended interview.

Participants were undergraduate student volunteers at Georgia Institute of Technology. Each one was rewarded 20 dollars for participating. Of the 20 students who participated in the study, 16 results were used due to technical malfunctions in the environment.

After completing the study we conducted a constant comparison analysis on the results of the interviews to develop themes and ideas that emerged [20]. Constant comparison analysis begins by building categories and subcategories from the analyzed speech, then organizing codes and continual refinement of categorization. These themes split into three categories, firstly, perception of robot emotion, musical reactions and gesture, which focused on manners in which EMP shaped participant responses. The second category, robotic groups, featured responses specific to groups of robots. The third category HRI metrics, refers to comments from participants that relate to common HRI studies.

11.4.1 Perception of Robot Emotion, Musical Reactions and Gesture

11.4.1.1 Robot Emotion

During the study process we did not mention emotion at any stage. Nevertheless, the vast majority of participants interpreted the robots sound and movement as emotion-driven. This was not necessarily expected, as the embodiment of emotion into a robotic arm is rare [32] and no participants had interacted with arms previously. One participant described that the "Start was very unexpected" as they didn't expect an emotional response', while another participant echoed this general feeling, although also adding the "emotion was a bit scary at first, but then cool towards the end".

There was a general mixed perception about whether the gestures, the music, or a combination of both was driving the perception of emotion. Some descriptions of how the robots conveyed emotion focused entirely on the gesture, such as, "looking down and not making eye contact is kind of considered to be sad and just looking up and jumping around is happier", "I thought it was sad is because it was shrinking into a corner and turning away from me and "when it was like more like pessimistic they would like bow their heads more and like it was like kind of it felt kind of dispirited". However other participants believed the gestures alone did not convey emotion, "but the audio made it an emotion, otherwise it will just be like just arms moving up and down". Descriptions of the emotional content of music included "it was music designed for robots to specific emotions" and that the music "showed characteristic things humans associate with emotions".

Participants generally interpreted the emotions as binary, primarily as happy or sad but also used terms such as "more optimistic or more pessimistic" or "positive" and "negative". While the robots were designed to show four emotions, no participant consciously noted a difference between the level of emotion displayed by the robots. One participant did disagree with the notion of including emotion in a robotic system, stating "they don't if they should be happy, so couldn't draw emotion from anything". No other participants however questioned the reason for including emotional responses.

11.4.1.2   Musical Reaction

There was fairly limited reaction to the music as a separate feature of the robots, outside of the emotional role. Participants agreed that the music showed emotions, but only one participant commented on the aesthetic decisions of the music. One participant notes that it was interesting that the robots were using a human-like voice, but they believed it fit the robot. Ultimately all participants viewed the music as an embedded part of the robotic system, and generally not a separate process.

11.4.1.3   Gesture

Many participants commented on the auxiliary gestures of the robots, and it was common for the participants to be intrigued by the relations of the music to gesture. "I felt like that was some kind of like abstract dance, but it was also cool to see them like move around in different directions". No participants questioned the utility of the additional gestures, as we had expected may occur, with participants instead focusing on understanding what the gestures where trying to convey, as described in Section 11.4.1.1.

11.4.1.4   Language Choices

Participants often questioned their own choice of phrases and language in describing the robot. We did not give the robot any name, or imply any gender during the studies, allowing the participants to develop their own understanding of the platform. Multiple participants paused when assigning a gender to the robot, including changing direction mid-sentence, such as: "It was dropped by him – I don't know why I'm calling in to him, but it dropped it and they were happy anyway". Throughout the interviews other participants would use phrases such as "the robot tried" and "the robot wanted" before then self-correcting to remove the agency from the robot.

While many participants did not their use of language in describing robots, others comfortably personified the robots. Statements included language such as, "It is sometimes like one started partying and the other was like joining in the party and the third is like, oh, I see, uh party, let me join in and then finally it was just the last one partying" or "Sometimes it was just funny that the board dropped it and it just was like, yeah, congratulations or something like that". Many participants also independently described the robots as animals, such as an "octopus" or "spider". Other participants note that they seemed like "a weird animal", or "a group of a foreign species".

11.4.2   Robotic Groups

11.4.2.1   Group Appearance and Interaction

It was extremely common for participants to be taken aback by the group size and number of robots in the room. None of our participants had previously interacted or seen a single robot arm in person, possibly adding to the sometimes jarring experience of encountering a group of robots. Participants noted that they were "initially confused by why there were so many" and "very surprised by look of so many robots". Nevertheless, by the end of the experiment, and after interacting with the robots, all participants noted they felt comfortable with the group of robotic arms.

They were multiple perspectives on the concept of entitativity arising from the interviews, with participants generally divided about whether the arms were a single entity or a group of individuals. One participant described the arms as a robot "passing between itself", while others saw the robots as "10 robots all doing their own thing". Another participant noted "So like they yeah they function together apart like rather than having their like individual responses". A different participant noted that there were "three leader robots". Ultimately the perception of the group was extremely varied, ranging across many ideas without any clear conclusions between all participants.

11.4.2.2 Group Interaction

The form of interaction amongst the group was also widely discussed by participants. There were many references to group coordination and synchronization such as "I do think the group coordinated, but like they each had their own little changes", "a lot of them were synchronized as well" and "the robots did coordinate as a group, which was like memorable like they would do the same dance movements and they would have like their tunes would harmonize with each other, which was cool".

11.4.3 HRI Metrics

11.4.3.1 Self-Efficacy

An important component of the interviews was developing an understanding of participants self-efficacy with the group of robots. Self-efficacy is an individuals confidence that they will be able to successfully interact with a robot, and has implications for how often people want to interact with a robot and the sort of tasks they are comfortable completing [23]. From our interviews all participants felt "very confident in my own tasks", with no participants mentioning any concern over their ability to interact with the robots or to recognize how they could interact.

11.4.3.2 Trust and Confidence

In comparison to self-efficacy, trust and confidence refers to the users perception that the robot will behave reasonably, while self-efficacy refers to the humans' confidence in their own role. We found multiple lines of thoughts amongst the participants. A small minority stated their confidence and trust as a binary, either believing the system is trustworthy or not. The majority of participants stated they trust the robot to some degree but would be uncomfortable with tasks, with statements such as, "if it was like a hot beverages I wouldn't use it. I would be like scared".

Additionally most participants asked for extra information before they would be confident and trust the robots in a wider range of circumstances. This included requests for more details on how exactly the robots worked, to help improve understandings of when the robots were performing as expected. Statements include "I'd need to understand how they work better to use them more, but comfortable overall and think I would be happy to use". Multiple participants also noted that would be confident if they robots underwent some further troubleshooting.

Lastly, it is worth mentioning that one participant consciously noted that they felt "safe as I believed it was a controlled experiment". As for any study, the effect of environment should be considered before assuming real-world implications.

11.4.3.3 Intimidation

In our interview process, no participants described any level of intimidation toward the arms. One participant noted, "robotic arms are not very scary as such because they can't move other than the rotation, so I wasn't very intimidated, intimidated about". We believe this may however be a direct result of our participant pool, who were all undergraduate students at ANON. While no participants had experience with robotic arms before, considering the general university environment they are much less likely to be intimidated by the robots than a general population.

11.4.3.4 Reasons for Failure

When asked why the robots were failing the vast majority of participants assigned the error to human programmers and not the robot itself. Participants went as far as to suggest the human programmer "miscalculated the angles" or that the task was "incorrectly designed for the robot". Many participants also noted they were very surprised when the robot first dropped a ring, and then assumed there was a bug that led to the errors, instead of independent robot errors.

11.5 Discussion

Throughout both studies we found multiple areas arise that are worthy of further study and description. We contend that our results in previous sections imply a vast range of potential interactions between musical study and HRI.

11.5.1 Memorable Moments around Musical Interaction

A key idea that occurred through the interview process was that the most memorable and interesting moments were based on musical interactions. From the participants point of view these interactions were not always about the music, but could be what the music implied through the robot platform. For example, a participant noted their favorite moment was "When they [the robots] laughed at one of the robots failing". Even in a such a unique setting where a participant was seeing a group of many robots for the first time, music has the potential to enhance and change the interaction.

11.5.2 Music Adds Emotion

We found that music was clearly able to add a perceived emotion to the robot systems. While in some ways this is an expected result, the importance of almost unanimous description of robotic emotion from the participants should not be overlooked. Robotic arms themselves are inherently non-emotional, and in robotics literature are very rarely used to display emotion. In this way, music can add an entire new range of approaches to interacting with this technology.

11.5.3   Embodied Music

In our setup, music was coming directly from underneath each robot, with each robot having it's own speaker. From all feedback it was clear that participants heard the music as from each robot itself, and not as a generalised sound. In early prototypes we had considered having four speakers or other arrangements, but chose to use a speaker in each arm, despite the extra system complexity. Before the study there was the potential for the musical phrases to not be directly associated with the robot and instead as background music. While we did not test this directly, we believe speaker placement, where it was embodied within the robot, was important to the perception of the voices as coming directly from each robot and drastically alters the interaction.

11.5.4   Music in Groups

We found very mixed perspectives on the role of music in groups and how it altered the perception of the group dynamics. There is a range of future work that could be done in this area. One potential direction is more variability in the timbre of the sound, or trying different melodic approaches for each robot. For this initial research we consciously choose to reduce the variables and use the same voice for each robot.

11.5.5   Music and HRI

The intersection of musical phrases and HRI is very much an under researched area, with only minimal work addressing the areas. With this in mind, it is worth continuing to describe the important role that music could have. We confirmed in our first study that the addition of music can increase trust and likeability. Music however could foreseeably have many other roles in robotics, such as extra dissonance reducing trust when a system should be avoided. This work is ultimately only one of many potential musical approaches to HRI.

11.5.6   Limitations

Both the studies presented in this paper were carefully designed to avoid limitations when possible. However, as in any study there limitations within each study. The online studied used pre-recorded videos instead of live interaction. We believe that for this was an acceptable experimental design as ultimately and allowed us to initially gather a wide range of data from a very diverse participant pool. Multiple past papers have shown no significant variation in results when a participant is watching a robot on video compared to in person [40,41]. By combining an online and in-person study we aimed to collect a very diverse range of opinions on robotic systems and develop a foundation for future research.

A further limitation was the length of each study, which was conducted over a single session. One participant even noted that "I mean, if you're doing that the whole day everyday, like I mean that's gonna get boring". In the future we aim to study longer interactions over multiple sessions to identify the variation that occurs during repeat encounters.

11.6 Conclusion

This research reiterates the important role music can have in communication and HRI. We were able to demonstrate that EMP is capable of changing HRI metrics in an online study, confirming past results, while suggesting the possibility of emotional contagion. Our second study focused on broad qualitative results, aiming to emphasize the perception of music in robotic systems and develop further avenues for research. We intentionally focused on robotic arms as this is both a world-wide growth area but also a platform well suited to gain from the addition of music. To robotic arms, music can add emotion and an entire range of communication options. We believe this form of interaction can enhance collaboration across many settings, ranging from robotic arm interface testing, to large scale factory use. Overall, we hope this research helps expand broader ideas about the possibilities of sound and robotics

Bibliography

[1] ARKIN, R. C., FUJITA, M., TAKAGI, T., AND HASEGAWA, R. An ethological and emotional basis for human–robot interaction. *Robotics and Autonomous Systems 42*, 3-4 (2003), 191–201.

[2] ARKIN, R. C., AND ULAM, P. An ethical adaptor: Behavioral modification derived from moral emotions. In *2009 IEEE International Symposium on Computational Intelligence in Robotics and Automation-(CIRA)* (2009), IEEE, pp. 381–387.

[3] BARSADE, S. G. The ripple effect: Emotional contagion and its influence on group behavior. *Administrative Science Quarterly 47*, 4 (2002), 644–675.

[4] BARTNECK, C., KULIĆ, D., CROFT, E., AND ZOGHBI, S. Measurement instruments for the anthropomorphism, animacy, likeability, perceived intelligence, and perceived safety of robots. *International Journal of Social Robotics 1*, 1 (2009), 71–81.

[5] BELLONA, J., BAI, L., DAHL, L., AND LaVIERS, A. Empirically informed sound synthesis application for enhancing the perception of expressive robotic movement. Georgia Institute of Technology.

[6] CARPINELLA, C. M., WYMAN, A. B., PEREZ, M. A., AND STROESSNER, S. J. The robotic social attributes scale (rosas) development and validation. In *Proceedings of the 2017 ACM/IEEE International Conference on Human-Robot Interaction* (2017), pp. 254–262.

[7] CHA, E., FITTER, N. T., KIM, Y., FONG, T., AND MATARIĆ, M. J. Effects of robot sound on auditory localization in human-robot collaboration. In *Proceedings of the 2018 ACM/IEEE International Conference on Human-Robot Interaction* (2018), pp. 434–442.

[8] DAVIES, S. Infectious music: Music-listener emotional contagion.

[9] DAVIES, S., COCHRANE, T., SCHERER, K., AND FANTINI, B. Music-to-listener emotional contagion. *The Emotional Power of Music: Multidisciplinary Perspectives on Musical Arousal, Expression, and Social Control* (2013), 169–176.

[10] FARRIS, N., MODEL, B., SAVERY, R., AND WEINBERG, G. Musical prosody-driven emotion classification: Interpreting vocalists portrayal of emotions through machine learning. In *18th Sound and Music Computing Conference* (2021).

[11] FISCHER, K., LOHAN, K., SAUNDERS, J., NEHANIV, C., WREDE, B., AND ROHLFING, K. The impact of the contingency of robot feedback on hri. In *Collaboration Technologies and Systems (CTS), 2013 International Conference on* (2013), IEEE, pp. 210–217.

[12] FRAUNE, M. R., NISHIWAKI, Y., SABANOVIĆ, S., SMITH, E. R., AND OKADA, M. Threatening flocks and mindful snowflakes: How group entitativity affects perceptions of robots. In *Proceedings of the 2017 ACM/IEEE International Conference on Human-Robot Interaction* (2017), pp. 205–213.

[13] FREDERIKSEN, M. R., AND STOEY, K. Augmenting the audio-based expression modality of a non-affective robot. In *2019 8th International Conference on Affective Computing and Intelligent Interaction (ACII)* (2019), IEEE, pp. 144–149.

[14] FREEDY, A., DeVISSER, E., WELTMAN, G., AND COEYMAN, N. Measurement of trust in human-robot collaboration. In *Collaborative Technologies and Systems, 2007. CTS 2007. International Symposium on* (2007), IEEE, pp. 106–114.

[15] GOMPEI, T., AND UMEMURO, H. Factors and development of cognitive and affective trust on social robots. In *International Conference on Social Robotics* (2018), Springer, pp. 45–54.

[16] KOZAK, M. N., MARSH, A. A., AND WEGNER, D. M. What do i think you're doing? action identification and mind attribution. *Journal of Personality and Social Psychology 90*, 4 (2006), 543.

[17] LEE, J. D., AND SEE, K. A. Trust in automation: Designing for appropriate reliance. *Human Factors 46*, 1 (2004), 50–80.

[18] MOORE, D., MARTELARO, N., JU, W., AND TENNENT, H. Making noise intentional: A study of servo sound perception. In *2017 12th ACM/IEEE International Conference on Human-Robot Interaction (HRI* (2017), IEEE, pp. 12–21.

[19] OGATA, T., AND SUGANO, S. Emotional communication robot: Wamoeba-2r emotion model and evaluation experiments. In *Proceedings of the International Conference on Humanoid Robots* (2000).

[20] ONWUEGBUZIE, A. J., AND FRELS, R. *Seven Steps to a Comprehensive Literature Review: A Multimodal and Cultural Approach.* Sage, 2016.

[21] OSOSKY, S., SCHUSTER, D., PHILLIPS, E., AND JENTSCH, F. G. Building appropriate trust in human-robot teams. In *2013 AAAI Spring Symposium Series* (2013).

[22] PELIKAN, H., ROBINSON, F. A., KEEVALLIK, L., VELONAKI, M., BROTH, M., AND BOWN, O. Sound in human-robot interaction. In *Companion of the 2021 ACM/IEEE International Conference on Human-Robot Interaction* (New York, NY, USA, 2021), HRI '21 Companion, Association for Computing Machinery, pp. 706–708.

[23] ROBINSON, N. L., HICKS, T.-N., SUDDREY, G., AND KAVANAGH, D. J. The robot self-efficacy scale: Robot self-efficacy, likability and willingness to interact increases after a robot-delivered tutorial. In *2020 29th IEEE International Conference on Robot and Human Interactive Communication (RO-MAN)* (2020), IEEE, pp. 272–277.

[24] ROUSSEAU, D. M., SITKIN, S. B., BURT, R. S., AND CAMERER, C. Not so different after all: A cross-discipline view of trust. *Academy of Management Review 23*, 3 (1998), 393–404.

[25] SAVERY, R. Emotional musical prosody: Validated vocal dataset for human robot interaction. In *2020 Joint Conference on AI Music Creativity,* (2020).

[26] SAVERY, R. *Machine Learning Driven Emotional Musical Prosody for Human-Robot Interaction.* PhD thesis, Georgia Institute of Technology, 2021.

[27] SAVERY, R. Machine learning driven musical improvisation for mechanomorphic human-robot interaction. In *Companion of the 2021 ACM/IEEE International Conference on Human-Robot Interaction* (2021), pp. 559–561.

[28] SAVERY, R., ROGEL, A., AND WEINBERG, G. Emotion musical prosody for robotic groups and entitativity. In *2021 30th IEEE International Conference on Robot & Human Interactive Communication (RO-MAN)* (2021), IEEE, pp. 440–446.

[29] SAVERY, R., ROSE, R., AND WEINBERG, G. Establishing human-robot trust through music-driven robotic emotion prosody and gesture. In *2019 28th IEEE International Conference on Robot and Human Interactive Communication (RO-MAN)* (2019), IEEE, pp. 1–7.

[30] SAVERY, R., ROSE, R., AND WEINBERG, G. Finding shimi's voice: fostering human-robot communication with music and a nvidia jetson tx2. In *Proceedings of the 17th Linux Audio Conference* (2019), p. 5.

[31] SAVERY, R., AND WEINBERG, G. A survey of robotics and emotion: Classifications and models of emotional interaction. In *Proceedings of the 29th International Conference on Robot and Human Interactive Communication* (2020).

[32] SAVERY, R., AND WEINBERG, G. Robots and emotion: a survey of trends, classifications, and forms of interaction. *Advanced Robotics 35,* 17 (2021), 1030–1042.

[33] SAVERY, R., ZAHRAY, L., AND WEINBERG, G. Before, between, and after: Enriching robot communication surrounding collaborative creative activities. *Frontiers in Robotics and AI 8* (2021), 116.

[34] SAVERY, R., ZAHRAY, L., AND WEINBERG, G. Emotional musical prosody for the enhancement of trust: Audio design for robotic arm communication. *Paladyn, Journal of Behavioral Robotics 12,* 1 (2021), 454–467.

[35] SAVERY, R., ZAHRAY, L., AND WEINBERG, G. Shimon sings-robotic musicianship finds its voice. In *Handbook of Artificial Intelligence for Music.* Springer, Cham, 2021, pp. 823–847.

[36] SCHAEFER, K. E. Measuring trust in human robot interactions: Development of the "trust perception scale-hri". In *Robust Intelligence and Trust in Autonomous Systems*. Springer, 2016, pp. 191–218.

[37] STOCK, R. M. Emotion transfer from frontline social robots to human customers during service encounters: Testing an artificial emotional contagion modell.

[38] TENNENT, H., MOORE, D., JUNG, M., AND JU, W. Good vibrations: How consequential sounds affect perception of robotic arms. In *2017 26th IEEE International Symposium on Robot and Human Interactive Communication (RO-MAN)* (2017), IEEE, pp. 928–935.

[39] WEISS, A., AND BARTNECK, C. Meta analysis of the usage of the godspeed questionnaire series. In *2015 24th IEEE International Symposium on Robot and Human Interactive Communication (RO-MAN)* (2015), IEEE, pp. 381–388.

[40] WOODS, S., WALTERS, M., KHENG LEE KOAY, AND DAUTENHAHN, K. Comparing human robot interaction scenarios using live and video based methods: towards a novel methodological approach. In *9th IEEE International Workshop on Advanced Motion Control, 2006.* (2006), pp. 750–755.

[41] WOODS, S., WALTERS, M., KOAY, K. L., AND DAUTENHAHN, K. Comparing human robot interaction scenarios using live and video based methods: towards a novel methodological approach. In *9th IEEE International Workshop on Advanced Motion Control, 2006.* (2006), IEEE, pp. 750–755.

[42] YILMAZYILDIZ, S., READ, R., BELPEAME, T., AND VERHELST, W. Review of semantic-free utterances in social human–robot interaction. *International Journal of Human-Computer Interaction 32*, 1 (2016), 63–85.

[43] ZHANG, R., BARNES, J., RYAN, J., JEON, M., PARK, C. H., AND HOWARD, A. Musical robots for children with asd using a client-server architecture. In *International Conference on Auditory Display* (2016).

Part III

Robotic Musicianship and Musical Robots

12

Musical Robots: Overview and Methods for Evaluation

Emma Frid

Musical robots are complex systems that require the integration of several different functions to operate successfully. These processes range from sound analysis and music representation to mapping and modeling of musical expression. Recent advancements in Computational Creativity (CC) and Artificial Intelligence (AI) have added yet another level of complexity to these settings, with aspects of Human–AI Interaction (HAI) becoming increasingly important. The rise of intelligent music systems raises questions not only about the evaluation of Human–Robot Interaction (HRI) in robot musicianship but also about the quality of the generated musical output. The topic of evaluation has been extensively discussed and debated in the fields of Human–Computer Interaction (HCI) and New Interfaces for Musical Expression (NIME) throughout the years. However, interactions with robots often have a strong social or emotional component, and the experience of interacting with a robot is therefore somewhat different from that of interacting with other technologies. Since musical robots produce creative output, topics such as creative agency and what is meant by the term "success" when interacting with an intelligent music system should also be considered. The evaluation of musical robots thus expands beyond traditional evaluation concepts such as usability and user experience. To explore which evaluation methodologies that might be

DOI: 10.1201/9781003320470-12

appropriate for musical robots, this chapter first presents a brief introduction to the field of research dedicated to robotic musicianship, followed by an overview of evaluation methods used in the neighboring research fields of HCI, HRI, HAI, NIME, and CC. The chapter concludes with a review of evaluation methods used in robot musicianship literature and a discussion of prospects for future research.

12.1   Background

The history of musical automata predates digital technology. Archimedes invented the first known humanoid musical automaton, an elaborate clepsydra (water clock) combined with a Byzantine whistle, in the 3rd century BC and attempts to mechanize musical instruments in the form of mechanically wind-fed organs were done as early as in the 4th century BC [43, 88]. Mechanical automatic musical instruments that play pre-programmed music with negligible human intervention can be traced back at least to the 9th century [55, 87]. The algorithmic thought in Western music composition goes even further back in time, to the beginning of notation [129]. In more recent years, advances in computational power, sound processing, electrical engineering, as well as Artificial Intelligence (AI) and Virtual/Augmented Reality (VR/AR), have paved the way for new interaction possibilities with robots that go beyond physical corporality. Today, technological developments have blurred the line between robots as tangible entities and robots as abstract intelligent agents. The emergence of such musical systems introduces a need to understand and evaluate robotic systems in the musical and socio-cultural context in which they are used. But how should these systems be evaluated, and which properties should be considered important, when pursuing such an activity? This is the focus of this chapter.

Before diving deeper into a discussion of evaluation methods, it is first important to define the term *"musical robot"*; which properties of a system are required to be considered a musical robot and – more importantly – what is *not* a musical robot? To be able to answer these questions, we may refer to standards and robot taxonomies. It has been suggested that the concept of a *"robot"* predates the word by several centuries, and that the history of robots has been intertwined with the arts [149]. In ISO Standard 8373:2012, a robot is defined as *"a programmed actuated mechanism with a degree of autonomy to perform locomotion, manipulation or positioning"* [66]. Autonomy in this context refers to the *"ability to perform intended tasks based on current state and sensing, without human intervention"*.

Given that music, or musicking [140], is an activity that is embedded in a social context, it is worth reviewing taxonomies from the field of Human–Robot Interaction (HRI) in this context. Several different taxonomies have been

TABLE 12.1

Overview of Onnasch and Roesler's taxonomy to structure and analyze Human–Robot Interaction (adapted from [115]). Abbreviations: a = anthropomorphic, z = zoomorphic, t = technical, N_H = number of humans, N_R = number of robots.

Interaction context	**Field of application**	**Exposure to**
	Industry, service, military & police, space expedition, therapy, education, entertainment, none	<u>Robot</u>: embodied, depicted <u>Setting</u>: field, laboratory
Robot **Robot task specification**	**Robot morphology**	**Degree of robot autonomy**
Information exchange, precision, physical load reduction, transport, manipulation, cognitive stimulation, emotional stimulation, physical stimulation	<u>Appearance</u>: a/z/t <u>Communication</u>: a/z/t <u>Movement</u>: a/z/t <u>Context</u>: a/z/t	<u>Information acquisition</u>: − to + <u>Information analyses</u>: − to + <u>Decision making</u>: − to + <u>Action implementation</u>: − to +
Team **Human role**	**Communication channel**	**Proximity**
Supervisor, operator, collaborator, cooperator, bystander	<u>Input</u>: electronic, mechanical, acoustic, optic <u>Output</u>: tactile, acoustic, visual	<u>Temporal</u>: synchronous, asynchronous <u>Physical</u>: following, touching, approaching, passing, avoidance, none
Team composition $N_H = N_R$ $N_H > N_R$ $N_H < N_R$		

proposed (see e.g. [169, 170]). A recent example is the taxonomy introduced by Onnasch and Roesler in [115], which divides HRI work into three clusters with different foci: (1) *interaction context* classification, (2) *robot* classification, and (3) *team* classification. An overview of the three different clusters, and their corresponding categories to specify an HRI scenario, is presented in Table 12.1. The *interaction context* cluster involves, for example, the field of application. For musical robots, relevant examples include entertainment, education, and therapy. The interaction context also relates to how you are exposed to the robot; exposure can be embodied, which is the case for a physical robot, or depicted, which is the case for a virtual agent. This exposure can be in a field versus laboratory setting. The *robot classification* cluster focuses on the robot's work context and design; robot task specification, robot morphology, and degree of robot autonomy. In this context, robot morphology refers to the appearance of the robot, among other factors. For example, a robot can be classified as anthropomorphic (human-like) or zoomorphic (animal-like). It could also be more task-driven than human, i.e. technical. Finally, the third cluster, *team classification*, focuses on the human role (supervisor, operator, collaborator, co-operator, or bystander), the team composition (number of robots versus humans), the communication channel (e.g. tactile or acoustic communication), and proximity (temporal or physical).

A framework focused on classification of *social robots* was presented in [8]. This classification characterizes robots along seven dimensions (somewhat overlapping with the categories discussed in [115]): *appearance, social capabilities, propose and application area, relational role, autonomy and intelligence,, proximity,* and *temporal profile*. Although musical robots may find themselves on different points along these dimensions, some broader themes can be identified. For example, musical robots often have artifact-shaped or bio-inspired *appearance*. In other words, the design of musical robots is often inspired by acoustic instruments or features of, or even the entire, human body.[1] Different robots have different levels of *social capabilities*. For example, musical robots usually communicate using non-verbal modalities, producing sounds. They may also use motion, gestures, and lights. Some musical robots can model and recognize social aspects of human communication and respond accordingly. For example, they may interpret musical phrases played by a musician and adopt their musical response. The *purpose and application areas* of musical robots span across a wide range of different domains. Musical robots can be used for personal empowerment, to expand on human abilities, and to empower people to enhance creativity on an individual level. The *relational role* of the robot, i.e. the role that the robot is designed to fulfill within an interaction, can also take many forms. For example, musical robots can act as co-players in an ensemble, solo performers, and music teachers, among other roles. They may greatly vary when it comes to their *autonomy and intelligence*, for example, in terms of their ability to perceive environment-related and human-related factors such as physical parameters (speed, motion), non-verbal social cues (gestures, gaze, facial expression), and speech. They may also differ in terms of planning of actions and how much they can learn through interaction with humans and the environment, over time. When it comes to the spatial *proximity* of the interaction, the most common scenario for a musical robot is that the robot exists in a shared space, interacting directly with a human (but there are, of course, also other scenarios). Finally, the *temporal profile* of musical robots can vary when it comes to time span, duration, and frequency of the interactions.

To further narrow down what we mean by the term *musical robot*, we may turn to literature on *machine musicianship*, and *robot musicianship*, in particular. An influential book in this context is *"Machine Musicianship"* by Robert Rowe [129]. Rowe describes that the training of a musician begins by teaching basic musical concepts that underlie the musical skills of listening, performance, and composition. Such knowledge is commonly referred to as *musicianship*. Computer programs that are designed to implement any of these skills, for example, the skill to make sense of the music that is being heard, will benefit from a musician's level of musicianship. Another influential example from the literature is the work *"Robotic Musicianship – Embodied*

---

[1]Consider, for example, *"The Prayer"*, a singing mouth robot by Diemut Strebe [151], versus full-body humanoids such as the "Waseda Saxophonist Robot No. 2 (WAS2)" [148].

Artificial Creativity and Mechatronic Musical Expression" by Gil Weinberg et al. [163]. In this book, the authors describe robot musicianship research as work focused on *"the construction of machines that can produce sound, analyze and generate musical input, and interact with humans in a musically meaningful manner"* [162]. They define two primary research areas in this field: *Musical Mechatronics* [81] and *Machine Musicianship* [129]. The first relates to the study and construction of physical devices that generate sound through mechanical means, whereas the latter refers to the development of algorithms and cognitive models of music perception, composition, performance, and theory. The two disciplines are said to be brought together by *Robotic Musicianship*. Weinberg et al. describe that the ultimate goal of robotic musicianship is to design robots that can demonstrate musicality, expression, and artistry, while stimulating innovation and creativity in other musicians [162]. Rather than imitating or replacing human creativity, the goal of robotic musicianship is to supplement human creativity, and to enrich musical experiences for humans [162]. In this way, robot musicianship may advance music as an art form by creating novel musical experiences that can encourage humans to create, perform, and think of music in new ways.

Robot musicianship brings together perceptual and generative computation with physical sound generators to create systems capable of (1) *rich acoustic sound production*, (2) *intuitive physics-based visual cues from sound producing movements*, and (3) *expressive physical behaviors through sound accompanying body movements* [133]. Robotic musicians make use of various methods for music generation. This includes generative functions such as composition, improvisation, score interpretation and accompaniment, which in turn can rely on statistical models, predefined rules, abstract algorithms, or actuation techniques [162]. Going beyond sound-producing ability, an important aspect of robotic musicianship is the cognitive models and algorithms that enable the machines to act like skilled musicians. A robotic musician should have the ability to extract information relevant to the music or performance and be able to apply this information to the musical decision process. This is something that Weinberg et al. refer to as *Musical Intelligence*. As stressed by Ajay Kapur in [81], a robot must be able to sense what the human is doing musically, and the machine must deduce meaningful information from all its sensor data and then generate a valid response. Moreover, as described in [145], an idealized musical robot should integrate musical representation, techniques, expression, detailed analysis and control, for both playing and listening. Musical robots usually put emphasis on *Machine Listening*, i.e, on extracting meaningful information from audio signals using sensing and analysis methods [131]. To provide the robot with information about other musicians (for example, to be able to synchronize musical events) visual sensing and computer vision techniques, as well as multimodal analysis focused on inertia measuring units capturing acceleration and orientation of limbs, may also be used [162].

12.2   Musical Robots

The terms *"musical robots"* and *"robotic musical instruments"* can refer to a wide range of different types of musical machines [84]. Ajay Kapur [81] defines a robotic musical instrument as *"a sound-making device that automatically creates music with the use of mechanical parts, such as motors, solenoids and gears"*. Steven Kemper [84] suggested that although approaches lacking autonomy could more accurately fall under the term *"musical mechatronics"* (see e.g. [81, 171]), the popular conception of robots rooted in mythology includes any machines that can mimic human actions (citing [72,153]). As such, he considers any approach in which an electromechanical actuator produces a visible physical action that models the human act of making music as *"musical robotics"*, regardless of level of autonomous control.

Several overviews of the history of musical robots have been published throughout the years. A review of musical automata from classical antiquity to the 19th century was provided by Krzyzaniak in [88]. Kapur published a comprehensive overview of piano robots, percussion robots, string robots, wind robots, and turntable robots in [81]. Weinberg et al. [162] provided an overview of musical robots designed to play traditional instruments, with examples of robots playing percussive instruments, stringed instruments, and wind instruments. An introduction to research trends for musical performance robots was given by Solis and Takanishi in [146]. Finally, foundations of musical robotics and how such systems experienced a rebirth even in the face of loudspeaker technology dominance, thanks to their ability to serve as uniquely spatialized musical agents, was discussed in [97, 107].

Some of the earliest examples of musical robots include different types of musical automatons and automatic musical instruments driven by water and air, such as water clocks, wind-fed organs, and systems involving whistles or flutes, including mechanical birds (see [88, 109]). When it comes to programmable music machines, notable examples include Ismail ibn al-Razzaz al-Jazari's mechanical boat with four musical automata, as well as his early examples of percussion robots. Two programmable humanoid music robots that often reoccur in the literature are Jacques de Vaucanson's *"Flute Player"* automaton from 1738 [34], and Pierre Jaquet-Droz's *"Musical Lady"* from the 1770s [149,168].

Kapur describes the *"Player Piano"* as one of the first examples of mechanically played musical instruments. On the topic of piano robots, he mentions the *"Pianista"* piano by Henri Fourneaux and *"Pianola"* by Edwin Scott Votey. Kapur also discusses humanoid techniques in which the entire human body is modeled to play the piano, for example, the *"WABOT-2"* from Waseda University [125]. Today, there are electronic systems for control of mechanical pianos; automated pianos controlled by MIDI data can be purchased for example from Yamaha (Disklavier) and QRS Music (Pianomathon). When it comes to percussion robots, Kapur categorizes them into three subcategories: membranophones, idiophones, and extensions[2]. An example of a membranophone robot is *"Cog"*, which can hit a drum with a stick [167]. Idiophone examples include Gerhard Trimpin's robotic idiophones [157] and the LEMUR[3] *"TibetBot"* by Eric Singer, which plays on three Tibetan singing bowls using six arms [138]. The extension category includes, for example, combinations of many instruments, e.g. the LEMUR *"ModBots"*,[4] which are modular robots that can be attached anywhere. Kapur divides string robots into subcategories based on if they are plucked verus bowed. Examples from the plucked category include *"Aglaopheme"* by Nicolas Anatol Baginsky [7], the electric guitar robot from Sergi Jordà's *"Afasia"* project (see [73]), and *"GTRBOT666"* from Captured By Robots [19]. Two bowed robot examples are the *"MUBOT"* by Makoto Kajitani [80] and Jordà's *"Afasia"* violin robot [73]. A more recent example that would fall under this category is Fredrik Gran's cello robot [52]. Kapur defines wind robots as mechanical devices performing wind instruments like brass, woodwinds, and horn-type instruments. Examples mentioned include *"MUBOT"* [79] which performs on the clarinet, the Waseda University anthropomorphic robot playing the flute [142, 143] and robotic bagpipes such as *"McBlare"* by Roger Dannenberg [30]. Humanoid woodwind robots, and challenges in designing such systems, are discussed more in detail in [147].

A classification framework based on the ways in which musical robots express creativity was introduced by Kemper in [84]. The framework distinguishes between six musical robot categories, see Table 12.2. Category 1 generally prioritizes versatile, humanoid robots engaging in quintessentially "human" activities over novel musical output. One example is the *"Toyota Partner Robot"* which can play trumpet, violin, and an electronic drum kit [154]. Category 2 is different from 1 in the sense that these robots model human actions, for example, by replicating humanoid organs such as lips or oral cavity, which in turn may affect the efficiency. They often involve complex mechanical models that can limit sonic possibilities (e.g. the ability to play at super virtuosic speed). Examples include pioneering work from Waseda University, such as piano keyboard - [125], flute - [142, 143], and saxophone robots [148], as well as a robotic finger for harp plucking [20]. Robots in category 3 assume an anthropomorphic form but do not model the specific actions of human performance. They are generally focused more on musical output and appearance, with an anthropomorphic nature highlighted by their look, rather than an attempt to model the human actions of performance. Examples include robotic bands

---

[2]These are percussion robots that do not fall into the other two categories.

[3]LEMUR stands for "League of Electronic Musical Urban Robots". It is a group of artists and technologists who create robotic musical instruments, founded by Eric Singer (https://lemurbots.org/index.html).

[4]The *"ModBots"* were, for example, used in the multi-armed percussion Indian God-like robot *"ShivaBot"* [139].

TABLE 12.2
Summary of Kemper's musical robot classification system [84].

No	Category	Examples
1	Nonspecialized anthropomorphic robots that can play musical instruments	[154]
2	Specialized anthropomorphic robots that model the physical actions of human musicians	[20, 125, 142, 143, 148]
3	Semi-anthropomorphic robotic musicians	[33, 130, 141, 165]
4	Non-anthropomorphic robotic instruments	[128, 138]
5	Cooperative musical robots	[55, 163]
6	Individual actuators used for their own sound production capabilities	[117, 175]

such as *"The Trons"* [141] and *"Compressorhead"* [33], the robotic drummer *"Haile"* [165], and the robotic marimba player *"Shimon"* [130]. Category 4 are non-anthropomorphic instruments that are either mechatronic augmentations of acoustic instruments, e.g. Disklaviers, or new acoustic analog instruments. Such robots tend to focus more on sonic nuance than on modeling human performance actions. Examples mentioned by Kemper include the *"Expressive Machines Musical Instruments (EMMI)"* see e.g. (128) and LEMUR's musical robots [138]. Category 5 focuses on cooperative musical robots, i.e. systems that combine human performance and robotic actuation in a shared interface. Kemper mentions *"String Trees"* [55] and Georgia Tech's *"Robotic Drumming Prosthetic Arm"*, which robotically augments the capabilities of the human body [163]. Finally, category 6 includes projects focused on sound and movement of individual actuators, such as the arrangement of *"Imperial March"* from Star Wars by Pawel Zadrożniak played on floppy disc drives [117], and large-scale sound sculptures featuring individual motors actuating resonant objects by Zimoun [175].

Although most of the musical robots described in the literature are physical robots, there are also examples of virtual robot musicianship. Some HRI taxonomies (e.g. [61, 169]) have explicitly distinguished between exposure to embodied versus depicted robots, since there is a growing body of research suggesting that physically embodied robots are perceived differently than virtual agents (see e.g. [86]). The ethical dimensions of virtual musicians and machine (or robo) ethics were explored in [22]. Topics discussed were, for example, *"vocaloids"*[5] such as *"Hatsune Miku"*, *"Kagamine Rin"* and *"Kagamine Len"*, as well as *"Utatane Piko"*. Hatsune Miku [29] has gained widespread success as a virtual musician, performing in concerts as a 3D hologram (for an analysis of the recent popularity of three-dimensional holographic performances in

---

[5]Vocaloids make use of the Yamaha Vocaloid software [158] for speech and singing sound synthesis.

popular music, see [103]). Miku was also available as a voice assistant hologram from the company Gatebox [51], a product somewhat similar to Amazon Alexa, although that product is now discontinued. Going beyond Japanese vocaloid characters, attempts have also been made to recreate representational simulations of celebrity musicians. Such efforts include avatar simulations of Kurt Cobain and resynthesis of Freddie Mercury's singing voice [22], as well as holograms of Gorillaz, Mariah Carey, Beyoncé, Michael Jackson, Old Dirty Bastard of Wu-Tang Clan, Eazy-E of N.W.A., and Tupac Shakur[6] [103]. Yet another example of virtual robot musicianship is the attempt to recreate Jerry Garcia from The Grateful Dead using Artificial Intelligence, as mentioned in [21].

Certain musical robots focus specifically on improving access to music-making. Such robots can, for example, take the form of wearable technology or prosthetic devices that support people with disabilities in their musicking. These musical robots can be considered a subcategory of the type of Digital Musical Instruments (DMIs) that are called *Accessible Digital Musical Instruments (ADMIs)*. ADMIs can be defined as *"accessible musical control interfaces used in electronic music, inclusive music practice and music therapy settings"* [48]. Examples include the *"Robotic Drumming Prosthetic Arm"*, which contains a drumming stick with a mind of its own; the *"Third Drumming Arm"*, which provides an extra arm for drummers[7]; and the *"Skywalker Piano Hand"*, which uses ultrasound muscle data to allow people with an amputation to play the piano using dexterous expressive finger gestures [164]. Another example is *"TronS"*, a prosthetic that produces trombone effects; *"Eleee"*, a wearable guitar prosthetic; and *"D-knock"*, a Japanese drum prosthetic [60].

The above section has shed light on the breadth of work carried out within the fields of musical robotics and robot musicianship. Given the broad range of different interfaces that may be considered musical robots, it is to be expected that no unified framework for evaluation of robot musicianship exists. A strategy that has been used to find appropriate evaluation methods proposed in the fields of new musical interfaces [42] and creativity support tools [121] is to look into other disciplines to identify existing methods, and to adapt those to the specific use case (if required). Building on this idea, the following sections review evaluation methods used in neighboring research fields to inform the method selection strategy for evaluation of musical robots. More specifically, the succeeding sections will explore methods used in Human–Computer Interaction (HCI), Human–AI Interaction (HAI), New Interfaces for Musical Expression (NIME), and Computational Creativity (CC).

---

[6]The latter performing with Dr. Dre and Snoop Dogg in a famous concert in 2012.

[7]Although this robot was primarily designed to be used by people without disabilities.

12.3   Evaluation in Human–Computer Interaction

The term *"evaluation"* is commonly used to describe a range of different activities and goals in the field of Human–Computer Interaction (HCI). Traditionally, considerable focus has been on evaluating *usability*. The ISO Standard 9241-11:2018 defines usability as the *"extent to which a system, product or service can be used by specified users to achieve specified goals with effectiveness, efficiency and satisfaction in a specified context of use"* [65]. Usability evaluation encompasses methodologies for measuring the usability aspects of a system's user interface and identification of specific problems (see e.g. [110]). Many different evaluation methodologies have been proposed, each with different limitations and disadvantages. Evaluation processes can roughly be divided into two broad categories: *formative* evaluations, which take place during a design process, and *summative* evaluations, which take an already finished design and highlights its suitability for a specific purpose [136]. In other words, some evaluation methods can be applied already at an early stage of the design process, whereas others are intended to be used when the final interface design has been implemented.

Nielsen describes four main ways of conducting a user interface evaluation: *automatically, empirically, formally,* and *informally* [110]. *Automatic* usability evaluation involves using programs that can compute usability measures based on user interface specifications. An overview of the state of the art in usability evaluation automation was presented in [68]. *Empirical* evaluation involves testing an interface with real users. Many different methods may be employed, ranging from questionnaires to observations. A commonly used empirical method is *usability testing* [39]. Usability tests have the following six characteristics: the focus is on usability; the participants are end users or potential end-users; there is some artifact to evaluate (a product design, a system, or a prototype); the participants think aloud as they perform tasks; the data are recorded and analysed; and the results of the test are communicated to appropriate audiences. *Formal (model-based)* evaluation uses a model of a human to obtain predicted usability measures by either calculation or simulation. The goal of this procedure is to get some usability results before implementing a prototype and testing it with human subjects. A notable example is GOMS (Goals, Operators, Methods, Selection) modelling, which involves identifying methods for accomplishing task goals and calculating predicted usability metrics [85]. Finally, *informal evaluation* methods are non-empirical methods for evaluating user interfaces. Nielsen uses the term *usability inspection* to describe such methods that are based on having evaluators inspect the interface [110]. Commonly used methods include heuristic evaluation [111] and cognitive walk-through [94].

It should be noted that the topic of evaluation has been extensively debated and reviewed in the field of HCI over the years (see e.g. [10,100] for an overview) and that opinions about who to best evaluate interfaces greatly differ depending on who you ask. Some have even suggested that usability evaluations may be harmful [53] and proposed to not use such traditional methods to validate early design stages or culturally sensitive systems, advocating instead for the use of more reflective and critical methods. Others have emphasized that the increased interest in experience-focused (rather than task-focused) HCI brings forward a need for new evaluation techniques [83]. As mentioned in [105], discussions about user-orientated quality assessment of technology have lately moved away from a focus on usability, satisfaction, efficiency, effectiveness, learnability, and usefulness; instead, attempts have been made to shift focus to the user experience and the wider relationship between people and technology, exploring concepts such as engagement, pleasure, presence, and fun (see [13, 102, 126] for further reading). In [105], the authors suggest that three aspects should be considered simultaneously when designing and evaluating technology: *functionality, usability*, and *user experience*. In this context, *functionality* relates to technical issues (for example, which features that should be provided by the device, and aspects of performance, reliability, and durability). *Usability* relates to user issues. Finally, *user experience* focuses on the wider relationship between the product and the user, i.e. the individual's personal experience of using it [104, 105].

Although the word *user experience* (UX) has been around since the 1990s, there is still no widely accepted definition of the term (see e.g. [59, 93]). The ISO Standard 9241 defines user experience as *"a person's perceptions and responses resulting from the use and/or anticipated use of a product, system or service"* [65]. In a side note, this is said to include *"all the users' emotions, beliefs, preferences, perceptions, physical and psychological responses, behaviors and accomplishments that occur before, during and after use"*. MacDonald and Atwood [100] suggest that a major challenge for UX evaluators is the lack of shared conceptual framework, despite that multiple models have been proposed (see e.g. [45, 113]). An example of a tool for evaluation of user experience is the *User Experience Questionnaire*, which consists of 26 items arranged into six scales: attractiveness, perspicuity, efficiency, dependability, stimulation, and novelty (a shorter version of the UEQ with only 8 items is displayed in Table 12.3) [92, 156]. A model of UX introduced by Hassenzahl suggested that products have both *pragmatic* (e.g. an ability to help users achieve goals) and *hedonic* attributes (e.g. an ability to evoke feelings of pleasure and self-expression) [58]. The authors of [100] stressed that UX methods tend to focus solely on hedonic attributes, while usability evaluation methods mostly are used to capture pragmatic attributes, voicing the need for methods that can seamlessly integrate both hedonic and pragmatic feedback.

Apart from being evaluated from a usability and user experience perspective, musical robots may also benefit from accessibility evaluation. This is particularly important for musical robots that find themselves at the intersection of robot musicianship and ADMIs. Accessibility can be conceptualized as usability for a population with the widest range of user needs, characteristics,

TABLE 12.3
Items in the shorter version of the User Experience Questionnaire (UEQ) [156].

1	Obstructive	☐☐☐☐☐☐☐	Supportive
2	Complicated	☐☐☐☐☐☐☐	Easy
3	Inefficient	☐☐☐☐☐☐☐	Efficient
4	Confusing	☐☐☐☐☐☐☐	Clear
5	Boring	☐☐☐☐☐☐☐	Exciting
6	Not interesting	☐☐☐☐☐☐☐	Interesting
7	Conventional	☐☐☐☐☐☐☐	Inventive
8	Usual	☐☐☐☐☐☐☐	Leading edge

and capabilities (see [65]). This fits within the *universal design* [26] or *design for all* [40] philosophies, which may be used as starting point for accessibility evaluations. The seven design principles of universal design include: *equitable use, flexibility in use, simple and intuitive use, perceptible information, tolerance for error, low physical effort,* and *size and space for approach and use* [26]. For an overview of methods for evaluating accessibility, usability, versus user experience, as well as a discussion of strengths and weaknesses of these concepts, see [119].

12.4   Evaluation in Human–Robot Interaction

Researchers have stressed that the experience of interacting with robots is different from interacting with other technologies and that such experiences often involve a strong social or emotional component [174]; people tend to treat robots similar to how they treat living objects and ascribe them life-like qualities [46, 152]. Robots' physical and social presence, and their tendency to evoke a sense of agency, i.e. a capacity to act that carries the notion of intentionality [37], creates a complex interaction different from the one involving other artifacts and technologies [174]. This poses certain challenges when it comes to evaluation. Even if interactions with robotic technology and themes discussed in HCI research have things in common [44], HRI researchers have emphasized that HRI is different from Human–Computer Interaction [31], and that evaluation methods from HCI should be applied to HRI with care [174].

Young et al. [174] published a summary of methodologies, techniques, and concepts from both HCI and HRI research, focusing on strategies that they deemed useful for the unique and deep social component of interactions between a person and a robot. The following evaluation approaches were discussed: (a) *task completion and efficiency,* (b) *emotion,* (c) *situated personal experience,* and (d) *frameworks for exploring social interactions with robots.* The authors suggest that (a) can be used as a wider part of evaluation of social

HRI but that other techniques are required for a more comprehensive view of the entire HRI experience. For (b), one suggestion was to monitor biological features (e.g. heart rate, blood pressure, and brain activity, number of laughs, and so on). However, given the holistic, rich, and multi-faceted nature of social interactions, such simplifications of emotion into quantities and discrete categories will have limitations. Other methods proposed within category (b) included self-reflection using think-aloud and interviews. For (c), the authors emphasized that the situated holistic experience of interacting with a robot includes aspects of social structure, culture, and context. Researchers have stressed the importance of accepting the complex nature of interactions [28,137] and proposed to focus on uncovering themes and in-depth descriptions of such complexities [11, 62, 67]. Young et al. (173) provide examples of how this can be tackled using qualitative techniques such as participant feedback and interviews, grounded theory, cultural- or technology probes, contextual design, and in situ context-based ethnographic and longitudinal field studies. They stress that context sensitive evaluation should value that individuals have unique, culturally grounded experiences; that one should take care when generalizing across people (citing [15,28]); and that evaluators themselves carry culturally rooted personal biases toward robots, participants, and scenarios (see [28]). Complementary to above-described methods, evaluators can use specific frameworks (d) when exploring social interaction with robots. For example, Norman's three-level framework for analysing how people interact with and understand everyday objects may be used [112]. This framework highlights different temporal stages of interaction with a product: *the initial visceral impact, the behavioral impact during use,* and *the reflective impact after* interacting with a product.

Based on above discussion, Young et al. (173) present an attempt to classify the rich interaction with a robot into articulated concepts: *visceral factors, social mechanisms,* and *social structures.* These perspectives can be integrated into existing HCI and HRI evaluation methods and provide a new vocabulary that encourages investigators to focus more on emotional and social aspects of interaction. The *visceral factors* of interaction involve the immediate and automatic human responses on a reactionary level that is difficult to control, for example, instinctual frustration, fear, joy, and happiness. The *social mechanics* involve the application of social language and norms. This includes higher-level communication and social techniques; for example, gestures such as facial expressions and body language, spoken language, and cultural norms such as personal space and eye-contact. Finally, *social structures* refer to macro-level social structures, i.e. the development and changes of the social relationships and interaction between two entities over a longer period of time. This can be seen as the trajectory of the two other perspectives, and relates to how a robot interacts with, understands, and modifies social structures. As an illustrative example, they mention how cleaning-robot technology in homes may shift who is responsible for cleaning duties [46]. The authors describe how these three perspectives can serve as tools throughout the evaluation process at various

TABLE 12.4
The USUS Evaluation Framework, as presented in [166].

Factor	Indicator	Methods
Usability	Effectiveness Efficiency Learnability Flexibility Robustness Utility	Expert evaluation, user studies Questionnaires, interviews
Social Acceptance	Performance expectancy Effort expectancy Self efficacy Forms of grouping Attachment Attitude toward using technology Reciprocity	Questionnaires, focus groups Questionnaires
User Experience	Emotion Feeling of security Embodiment Co-experience Human-oriented perception	Questionnaires, physiological measurements, focus groups Questionnaires, focus groups Questionnaires
Social Impact	Quality of life Working conditions Education Cultural context	Questionnaires, focus groups, interviews

stages, from designing a study to conducting it, to analysing collected data.

A comprehensive overview of evaluation methods in HRI was presented in [78]. This book discusses questionnaires for HRI research, processes for designing and conducting semi-structured interviews, standardized frameworks for evaluation, evaluation of user experience of human–robot interactions (see [96]), evaluations based on ethology and ethnography, as well as recommendations for reliable evaluations. One of the discussed frameworks is the USUS Evaluation Framework [166], which is described in Table 12.4. It consists of a theoretical framework based on a multi-level model involving factors split into indicators, extracted and justified by literature review, and a methodological framework consisting of a mix of methods derived from HRI, HCI, psychology, and sociology. The USUS Evaluation Framework was later reformulated by Wallström and Lindblom into the USUS Goals Evaluation Framework [159], after the authors identified a lack of evaluation methods that included UX goals, i.e. high-level objectives driven by the representative use of the system, in social HRI.

12.5   Evaluation in Human–AI Interaction

Researchers in the HCI community have proposed several guidelines and recommendations for how to design for effective human interaction with AI systems. Several large companies have also published white papers aiming to serve as guidance for development of AI systems (see e.g. [116] and [64]). Amershi et al. proposed 18 generally applicable design guidelines for Human–AI Interaction (HAI) in [3]. These guidelines are intended for AI-infused systems, i.e. systems that have features harnessing AI capabilities that are exposed directly to an end user. The AI design guidelines in [3] are separated into categories depending on when during the user's interaction that they are applied: *initially, during interaction, when something goes wrong*, and *over time* [3]. A summary of the guidelines is presented in Table 12.5. Although described as guidelines rather than evaluation criteria, these points can be used for evaluation purposes (see e.g. [50]). A research protocol for evaluating Human-AI Interaction based on the guidelines was recently published in [95].

Eight of the guidelines presented in [3] overlap with the principles for Mixed-Initiative Systems by Horvitz [63]. A taxonomy for Mixed–Initiative Human–Robot Interaction was presented in [70][8]. Related areas of research that may be interesting to explore for the purpose of identifying evaluation methodologies suitable for musical robots include work on Mixed-Initiative Co–Creative Systems [172], Mixed–Initiative Creative Interfaces [36], and Human–AI Co–Creativity [122]. To the author's knowledge, there is still no unified framework for evaluation of Human–AI co-creative systems (although some attempts have been made, see e.g. [82]). Potential pitfalls when designing such systems were discussed in [17]. The pitfalls were identified starting from three speculation prompts: issues arising from (a) *limited AI*, (b) *too much AI involvement*, and (c) *thinking beyond use and usage situations*. The first category includes issues related to invisible AI boundaries, lack of expressive interaction, and false sense of proficiency; the second to conflicts of territory, agony of choice, and time waste; and the third to AI bias, conflict of creation and responsibility, and user and data privacy.

Another relevant area of research in this context is the field of EXplainable AI (XAI) [54]. XAI aims to interpret or provide a meaning for an obscure machine learning model whose inner workings are otherwise unknown or non-understandable by the human observer [4]. This relates to the notion of *"explainable agency"*, which refers to autonomous agents such as robots explaining their actions, and the reasons leading to their decisions [91]. In the context of robots, a range of other related terms are also used to explore similar concepts, e.g. understandability, explicability, transparency, and predictability (see [4] for an overview). A systematic review on explainable agency for robots and agents was presented in [4], highlighting a considerable lack of evaluations in the reviewed papers. A framework and paradigm for evaluation of explanations of AI was provided in [47]. This framework suggests that an explanation needs

---

[8] *"Mixed Initiative Interaction HRI (MI-HRI)"* is defined as *"A collaboration strategy for human–robot teams where humans and robots opportunistically seize (relinquish) initiative from (to) each other as a mission is being executed, where initiative is an element of the mission that can range from low-level motion control of the robot to high-level specification of mission goals, and the initiative is mixed only when each member is authorized to intervene and seize control of it"*.

TABLE 12.5

The 18 Human-AI interaction design guidelines proposed by Amershi et al. [3], categorized by when they likely are to be applied during interaction with users.

No	AI Design guideline	Description	Timepoint
1	Make clear what the system can do.	Help the user understand what the AI system is capable of doing.	Initially
2	Make clear how well the system can do what it can do.	Help the user understand how often the AI system may make mistakes.	
3	Time services based on context.	Time when to act or interrupt based on the user's current task and environment.	During interaction
4	Show contextually relevant information.	Display information relevant to the user's current task and environment.	
5	Match relevant social norms.	Ensure the experience is delivered in a way that users would expect, given their social and cultural context.	
6	Mitigate social biases.	Ensure the AI system's language and behaviors do not reinforce undesirable and unfair stereotypes and biases.	
7	Support efficient invocation.	Make it easy to invoke or request the AI system's services when needed.	When wrong
8	Support efficient dismissal.	Make it easy to dismiss or ignore undesired AI system services.	
9	Support efficient correction.	Make it easy to edit, refine, or recover when the AI system is wrong.	
10	Scope services when in doubt.	Engage in disambiguation or gracefully degrade the AI system's services when uncertain about a user's goals.	
11	Make clear why the system did what it did.	Enable the user to access an explanation of why the AI system behaved as it did.	
12	Remember recent interactions.	Maintain short term memory and allow the user to make efcient references to that memory.	Over time
13	Learn from user behavior.	Personalize the user's experience by learning from their actions over time.	
14	Update and adapt cautiously.	Limit disruptive changes when updating and adapting the AI system's behaviors.	
15	Encourage granular feedback.	Enable the user to provide feedback indicating their preferences during regular interaction with the AI system.	
16	Convey the consequences of user actions.	Immediately update or convey how user actions will impact future behaviors of the AI system.	
17	Provide global controls.	Allow the user to globally customize what the AI system monitors and how it behaves.	
18	Notify users about changes.	Inform the user when the AI system adds or updates its capabilities.	

to (1) *provide knowledge*, (2) *be trustworthy*, (3) *be useful*, (4) *update the receiver's estimation about the probability of events occurring*, and (5) *change the receivers mental model.*

12.6 Evaluation of New Interfaces for Musical Expression

Evaluation of acoustic musical instruments was discussed by Campbell in [18]. The author posed the question *"if we are to optimize a musical instrument, who determines success?"* The author suggests that this should either be the musician playing the instrument, or the listener who hears the sound. Campbell points out that musicians and physicists approach evaluation differently, and they often lack a common vocabulary to discuss these differences. He emphasized the importance of understanding cross-modal interference and how this may influence judgment of instrument quality, mentioning for example differences in perceived tone quality of the piano (a phenomenon caused by cross-modal interference between auditory and haptic channels) and the importance of distracting visual cues when evaluating the quality of violins and brass instruments.

When it comes to the digital domain, the topic of evaluation has also been extensively debated in the fields dedicated to New Interfaces of Musical Expression (NIME) and Digital Musical Instruments (DMIs) for quite some time. Attempts have been made to explore what the word *"evaluation"* means for the NIME community, with findings suggesting that there are different understandings of the term [9]. A review of papers published in the NIME conference proceedings revealed that the word often is used to denote the process of collecting feedback from users to improve a prototype. Others use the term to assess the suitability of existing devices for specific tasks, or to compare devices. In addition, evaluation is sometimes used to describe emerging interaction patters when using devices. As pointed out by Barbosa et al. [9], complicating factors involved in NIME evaluation include that several stakeholders often are involved in the design of the instruments, and the requirements of one stakeholder may not necessarily intersect those of another. Barbosa et al. also stress that the time window of the evaluation[9], as well as the level of expertise, can influence the evaluation results.

Traditionally, evaluation methods for NIMEs and DMIs have largely been based on frameworks used in the field of Human–Computer Interaction (HCI). Wanderley and Orio [160] discuss the application of such methodologies in the evaluation of input devices for musical expression. In particular, they

---

[9]The experience of playing a musical instrument usually changes the more you play on the instrument.

focused on specific tasks used in HCI to measure the performance of an input device in the music domain. More specifically, Wanderley and Orio propose to use a set of musical tasks for *usability evaluation*. For a musical instrument, such tasks could focus on the production of musical entities and include the generation of, for example *isolated tones*, i.e. pitches at different frequencies and loudness levels; *basic musical gestures* like glissandi, trills, vibrato, and grace notes; and *musical phrases*, such as scales and arpeggios, as well as more complex contours with different speeds and articulations. In addition, tasks could focus on reproducing continuous timbral changes or different rhythms for such musical entities, given a specific loudness. Wanderley and Orio also propose a set of relevant features to be tested in usability evaluations of controllers used in the context of interactive music: *learnability, explorability, feature controllability,* and *timing controllability.*

In the framework for musical instruments proposed by Kvifte and Jensenius, the authors discuss three perspectives of instrument description and design: that of *the listener, the performer,* and *the instrument constructor* [89]. Barbosa et al. [9] highlight that, for example, playability might be important for a performer, but not for an audience. The idea that there are many perspectives from which one can view the effectiveness of an instrument was also stressed by O'Modhrain [114], who suggested that if performance is considered a valid means of evaluating a musical instrument, a much broader definition than what is typically used in HCI is required. O'Modhrain emphasizes that in addition to players and audiences, there are also composers, instrument builders, component manufacturers, and perhaps also customers, to consider. The different stakeholders can have different views of what is meant by the term evaluation. A complicating factor in this context is that the boundaries between roles usually are blurred in DMI design (this is usually not the case for design of acoustic musical instruments). Since DMI designs can be evaluated from multiple perspectives, different techniques and approaches are required. O'Modhrain aims to provide a structure to these competing interests by providing a framework for evaluation that enables performers, designers, and manufacturers to more readily identify the goal of an evaluation, and to view their methods in the light of prior work. Her framework includes a summary of methods that a given stakeholder might use to evaluate a DMI against a given design goal. Possible evaluation goals include *enjoyment, playability, robustness,* and *achievement of design specifications.*

Young and Murphy [173] suggest that the evaluation of DMIs should focus on *functionality, usability,* and *user experience*. The functionality testing should aim to highlight potential issues before longitudinal studies are carried out, i.e. the usability and user experience studies. The initial stages of a device's evaluation should focus on capturing low-level device characteristics, thus creating a generalized device description (for example, through evaluation of the musical tasks proposed in [160]). This should be followed by a process in which a device is reduced to its physical variables in terms of a taxonomy of input. In this step, you contextualize a device's evaluation in terms of stakeholders, questioning who is evaluating the device and why. Several HCI paradigms exist that can be augmented to fit these processes.

Useful tools for evaluation and classification of NIMEs include, for example, the phenomenological dimension space for musical devices introduced by Birnbaum et al. [12]. This framework can be used to describe a musical instrument along a set of seven axes: *required expertise, musical control, feedback modalities (outputs), degrees of freedom (input), inter-actors, distribution in space*, and *role of sound*. Another example is the epistemic dimension space for musical devices presented by Magnusson in [101], which includes eight parameter axes: *expressive constraints, autonomy, music theory, explorability, required foreknowledge, improvisation, generality*, and *creative simulation*. Yet another useful framework was provided by Jordà in [74]. This framework focuses on the musical output diversity of the instrument and how the performer can control and affect this diversity, dividing instrument diversity into *macro-diversity, mid-diversity*, and *micro-diversity*. Macro-diversity determines the flexibility of an instrument to be played in different contexts, music styles or assuming varied roles; mid-diversity refers to how different two performances or compositions played with the instrument can be; and micro-diversity to how two performances of the same piece can differ.

When it comes to evaluation of user experience for NIMEs, a recent review was published by Reimer and Wanderley [120]. Findings suggested that UX-focused evaluations typically were exploratory and that they were limited to novice performers. The authors propose to use the *"Musicians Perception of the Experiential Quality of Musical Instruments Questionnaire (MPX-Q)"* to compare UX for different instruments [135]. This questionnaire is based on psychometric principles and consists of three interrelated subscales: (1) *experienced freedom and possibilities*, (2) *perceived control and comfort*, and (3) *perceived stability, sound quality, and aesthetics*. Referring to [173] and building on ideas previously discussed in [42], Reimer and Wanderley also propose that standardized frameworks to evaluate UX in other fields could be adapted for NIME evaluation (see e.g. [38] for an overview), mentioning for example the *"Gaming Experience Questionnaire (GEQ)"* from ludology[10].

The suitability of HCI evaluation tools, which put emphasis on technological aspects of musical instruments and describe them as *"devices"* with properties viewed from a *"usability"* and *"accessibility"* perspective, has been widely debated in the NIME community. Some have even questioned the use of the word *"evaluation"*, proposing to instead employ the term *"user experience study"*, thereby broadening the scope of such work to acknowledge that while ergonomics and efficiency are important, they are not the primary determinants of the quality of a musical interface [71]. Stowell et al. [150] suggested that while the framework proposed by Wanderley and Orio [160] is useful, it has drawbacks. For example, the reduction of musical interaction to simple tasks

---

[10]The study of gaming.

may compromise the authenticity of the interaction. Since musical interactions involve creative and affective aspects, they cannot simply be described as tasks for which aspects such as completion rates are measured [150]. Task-based methods may be suited to examine usability, but the experience of the interaction is subjective and requires alternative approaches for evaluation. Stowell et al. propose that the following questions should be considered when evaluating interactive music systems: (1) *Is the system primarily designed to emulate the interaction provided by a human, or by some other known system? (2) Is the performer's perspective sufficient for evaluation? (3) Is the system designed for complex musical interactions, or for simple/separable musical tasks? (4) Is the system intended for solo interaction, or is a group interaction a better representation of its expected use pattern? (5) How large is the population of participants on which we can draw for evaluation?* They also present two methods for evaluation of musical systems: a qualitative approach using structured discourse analysis and a quantitative musical Turing-test method. Finally, they suggest that the design of evaluation experiments should aim to reflect the authentic use context as far as possible.

Rodger et al. [127] questioned the adoption of tools from traditional HCI to understand what constitutes a good musical instrument. The implication of viewing musical activities as something compromised of a "device" and a "user" is that the instrument is considered an entity with a set of intended functional behaviors, known to the designer and employed by the user, for the purpose of a specific goal. This is a limiting view of how musicians interact with instruments. There are many examples in which musical instruments are used in manners that differ from the original intended design. The idea that the instrument should be assessed by how readily it supports an intended design function can also be questioned by what Rodger et al. call "instrument resistance", i.e. that the effortfulness of playing an instrument may serve as a source of creativity. Viewing musicians as users of musical devices results in conceptual issues, since musicians vary in their capabilities and histories of embodied knowledge. As such, the idea of a "prototypical user" doesn't suit this context. The functional properties of an instrument can only be meaningfully understood relative to the capabilities of specific musician at a specific period in her musical development. Moreover, it is hard to make sense of what a musician does with an instrument if divorced from both the immediate and extended socio-cultural context. Rodger et al. therefore propose an evaluation approach in which instruments are understood as processes rather than devices, and musicians are viewed as agents in musical ecologies, i.e. a system compromising an agent and environment, rather than users. Evaluations of instruments should align with the specificities of the relevant processes and ecologies concerned. In this context, a specifity is defined as *"the effective components of the musician-instrument system relative to the relevant musical activities and contexts of interest"*. In other words, the evaluation should be relative to its environmental context, and not focus on a generalizable methodology based on a prototypical user. The consequence of this stance is that instruments may mean different things to different musicians.

Also El-Shimy and Cooperstock [42] stressed that the nature of musical performance requires that designers re-evaluate their definition of user "goals", "tasks", and "needs". They stress the importance of creativity and enjoyment rather than efficiency [42]. El-Shimy and Cooperstock reviewed literature focused on user-driven evaluation techniques offered by HCI, ludology, interactive arts, and social-science research, exploring aspects such as affect, fun, pleasure, flow, and creativity. They present a set of principles for user-driven evaluation of new musical interfaces involving: (1) *validating the basis*, (2) *investigating suitable alternatives to "usability"*, and (3) *tailoring evaluation techniques*. El-Shimy and Cooperstock argue that qualitative and mixed research methods are particularly suited for studies of non-utilitarian systems and propose to use qualitative experiments to develop hypotheses that later can be verified through quantitative studies (as opposed to the traditional approach in which hypotheses are formed before an experiment). Qualitative research methods mentioned include interviews, discussions, case studies, and diaries. Analysis methods include, for example, content analysis, which operates on the principle of grounded theory. El-Shimy and Cooperstock encourage designers to tailor existing evaluation techniques to their own needs or to devise new ones if necessary.

Based on the above discussion, as well as what was briefly mentioned in Section 12.3, we can conclude that there has been a shift from task-based and usability-driven design to more experience-based design and evaluation (so called third wave) HCI, especially within creative and artistic contexts [42]. Third wave HCI is said to be particularly suited to the design and evaluation of novel interactive musical interfaces [42]. Building on the work by Rodger et al. [127], as well as Waters' notion of *"performance ecosystems"*, in which music activities can be understood as a dynamical complex of interacting situated embodied behaviors [161], Jack et al. [69] propose to view DMIs as situated, ecologically valid artefacts, which should be evaluated using qualitative and reflective processes focusing on sociocultural phenomena, rather than first wave HCI techniques.

Finally, when it comes to accessibility of musical expression, it should be noted that the field of Accessible Digital Musical Instruments (ADMIs) still appears to lack a formal framework for evaluation [48], although attempts have been made to formulate design principles and classification methods (see [32, 49, 57]). A set of design guidelines that could be used for the purpose of evaluation were proposed by Frid [49], including *expressiveness, playability, longevity, customizability, pleasure, sonic quality, robustness, multimodality*, and *causality*. A dimension space for evaluation of ADMIs was proposed by Davanzo and Avanzini in [32], in which eight axes are grouped into two subsets: *target users and use contexts* (use context, cognitive impairment, sensory impairment, physical impairment), and *design choices* (simplification, adaptability, design novelty, physical channels). Moreover, Lucas et al. explored ecological perspectives of human activity in the use of DMIs and assistive technology in [99]. The authors used the Human Activity Assistive Technology (HAAT) [27] and the Matching Person and Technology (MPT) [134] frameworks to design and evaluate bespoke ADMIs, concluding that a shortcoming of these tools is that they are biased toward describing persons with disabilities from an external perspective.

12.7 Evaluation in Computational Creativity

Computational Creativity (CC) can be considered a sub-field of Artificial Intelligence (AI) [118]. Computational Creativity is the philosophy, science, and engineering of computational systems which, by taking on particular responsibilities, exhibit behaviors that unbiased observers would deem to be creative [25]. Evaluation of CC systems focuses on determining whether a system is acting creatively or not [118]. However, evaluation attempts in this domain have been found to lack rigor; there is no consensus on how to evaluate creative systems, and the reliability and validity of the proposed methods are in question [90]. Musical Metacreation (MuMe), a subarea of CC which aims to automate aspects of musical creativity with the aim of creating systems or artifacts deemed creative by unbiased observers, has also been found to be characterized by little systematic evaluation [1]. A historical perspective on how CC researchers have evaluated, or not evaluated, the creativity of their systems was presented by Anna Jordanous in [77]. In this work, Jordanous also address the question of how to choose an evaluation method and how to judge its quality via five meta-evaluation standards for comparison of evaluation methods in creativity: *correctness, usefulness, faithfulness as a model of creativity, usability of the methodology,* and *generality.*

Several different theories of creativity evaluation have been proposed throughout the years. Lamb et al. [90] suggest to group these based on their theoretical perspective, building on the taxonomy known as the four Ps: *Person, Process, Product,* and *Press* (see [123]). The four Ps were introduced to Computational Creativity by Jordanous. *Person* or *Producer*[11] is the human or non-human agent that is judged as being creative. Person theories aim to discover which traits (personality, emotional, cognitive) that distinguishes a more creative person from a less creative one. *Process,* on the other hand, refers to a set of internal and external actions that the agent may take when producing creative artifacts. Process theories study the actions that are undertaken in such contexts. This perspective focuses on how creative products are made, i.e. the cognitive steps that must be taken for an activity to be creative. *Product* is an artifact, for example a musical piece, which is seen as creative

---

[11] Jordanous suggested to use the term Producer instead of Person to emphasize that the agent does not need to be a human.

or as having been produced by creativity. Product theories study what it is about a certain product that makes it creative. Finally, *Press* refers to the surrounding culture that influences the other Ps in the model. Press theories study what it is in a culture that leads to the view that something is creative, and what kind of social effect a product needs to have to be called creative.

Person methods include psychometric tests or the study of famous creative people. Some have also attempted to measure the personal traits of computers. *Process* theories are often useful when modelling human creativity. Process evaluations tend to either place the system in a category or use a qualitative analysis of the system's process strengths and weaknesses. They are often somewhat descriptive in their nature, i.e. not always easy to apply to evaluation. Examples include the FACE and IDEA models (see [24]). The FACE[12] model describes creative acts performed by a software, whereas the IDEA model describes how such acts can have an impact on an audience. Another example is the SPECS model, which evaluates systems based on 14 factors that were identified through studies of how humans define creativity [76].[13] The SPECS model is divided into three steps [77]. Step 1 focuses on identifying a definition of creativity that your system should satisfy to be considered creative. Step 2 uses step 1 and focuses on clearly stating what standards you use to evaluate the creativity of the system. Step 3 focuses on testing the creative system against the standards stated in step 2 and reporting on these results. Moving on to the *Product* perspective, the focus lies on the artifact itself (e.g. a music piece or performance) as creative or as having been produced by creativity [90]. Common criteria are "novelty" and "value". Lamb et al. [90] suggest that if using such terms, one should define the specific audience for whom the system's products should be valuable. Ritchie suggested that "typicality" should be used rather than novelty, since creative systems should reliably generate both typical and valuable output [124]. *Product* methods include the Consensual Assessment Technique (CAT), in which a team of human experts evaluates a product [2], and the modified Turing Test (see [90]). The latter focuses on a test in which human subjects are challenged to figure out which products that are human versus computer-created in a set; if they cannot do it, then the computer system is considered creative.[14] Finally, *Press* methods include the Creative Tripod [23] and strategies to measure audience impact (see [90]). The Creative Tripod focuses on whether a system demonstrates skill, imagination, and appreciation, three qualities that are required to be deemed creative.

A review of Creativity Supporting Tools (CSTs) was presented in [121]. In

---

[12]The FACE model has been found to rank musical improvisation systems in the opposite order of other evaluation methods, and its validity has therefore been questioned [75].

[13]These factors include: active involvement and persistence; dealing with uncertainty; domain competence; general intellect; generation of results; independence and freedom; intention and emotional involvement; originality; progression and development; social interaction and communication; spontaneity/subconscious processing; thinking and evaluation; value; and variety, divergence, and experimentation.

[14]However, the modified Turing Test has been criticized on a number of points, see e.g. [118].

this work, the authors discuss six major points that researchers developing CSTs should consider in their evaluation: (1) *clearly define the goal of the CST*; (2) *link to theory to further the understanding of the CST's use and how to evaluate it*; (3) *recruit domain experts, if applicable and feasible*; (4) *consider longitudinal, in-situ studies*; (5) *distinguish and decide whether to evaluate usability or creativity*; and (6) *as a community, help develop a toolbox for CST evaluation* [121]. Related to this, Karimi et al. [82] provided a framework for evaluating creativity in co-creative systems, mentioning four questions that could guide such evaluation: (1) *Who is evaluating the creativity?* (2) *What is being evaluated?* (3) *When does the evaluation occur?* (4) *How is the evaluation performed?*

Agres et al. provided a theoretical motivation for more systematic evaluation of Musical Meta-Creation and computationally creative systems in [1]. The authors present an overview of methods to assess human and machine creativity, dividing creative systems into three categories: (1) *those that have a purely generative purpose*, (2) *those that contain internal or external feedback*, and (3) *those that are capable of reflection and self-reflection*. They present examples of methods to help researchers evaluate their creative systems, test their impact on an audience, and build mechanisms for reflection into creative systems. Other relevant references in the context of Musical Meta-Creation include [41], which describes an evaluation study of several musical meta-creations in live performance settings, and [155], which presents a discussion on evaluation of musical agents. The latter divides evaluations of MuMe systems into informal evaluations and formal evaluations. Informal evaluations do not involve formalized research methodologies (they usually take place as part of the software development), whereas formal evaluations use formalized methodologies to assess the success of systems.

Finally, it should be noted that some question the mere idea of creativity evaluation, and whether this is possible at all. For example, Baer [5] suggested that there are many creative skills, but no underlying process which informs them all. To be creative in one domain does not necessarily imply that you are creative in other domains [6]. Calling a person or process creative without specifying the domain is therefore not considered scientific. Others have suggested that creativity cannot be quantified. For example, Nake [108] suggests that quantification of creativity is an American invention and that there are risks commodifying creativity by framing it as an object one must have a certain amount of, as opposed to considering it a quality that emerges in a social context. Some argue that computational creativity should not be measured by human standards and that it is more interesting to investigate what computers produce according to their own non-human standards [98]. On the other hand, others claim that creativity is inherently human and thus never can be present in computers (although many counterarguments have been presented through the years, see e.g. [14, 106]).

12.8 Evaluation of Musical Robots

To explore the range of different methods used for evaluation of musical robots, a comprehensive search for the keyword "robot" in the title and abstract fields of papers published in the Computer Music Journal, the Journal of New Music Research, and the Leonardo Music Journal was performed. The proceedings of the International Conference on New Interfaces for Musical Expression (NIME) were searched using the same strategy. Chapters from the books *"Musical Robots and Multimodal Interactive Systems"* [144] and *"Robotic Musicianship - Embodied Artificial Creativity and Mechatronic Musical Expression"* [163] as well as the PhD thesis *"Expressive Musical Robots: Building, Evaluating, and Interfacing with an Ensemble of Mechatronic Instruments"* [107] were also skimmed to identify texts that could be included in the review.

In total, 14 journal articles and 50 papers from the NIME proceedings were identified. A total of 7 chapters were selected from the first book, 5 from the second, and 7 from the thesis. From this initial dataset, studies were selected to be included in the review by comparing the information presented in the publication against an inclusion criterion. In order to be included, the studies had to fulfill the following requirements: the publication had to describe a musical robot (the authors had to explicitly define their interface as a robot, and the robot should produce sounds), and an evaluation method must have been used. For this review, I adopted a wide definition of the term *"evaluation"*, including for example performances and installations displayed at public venues as methods. The review focused primarily on summative evaluations performed at the end of a study. The application of the inclusion criteria reduced the dataset from 83 to 62 publications. Information about the employed evaluation strategies was subsequently summarized per publication to identify reoccurring themes.

The publications were initially searched for the keywords "evaluation" or "evaluate", to identify sections describing evaluation methodologies. A total of 38 publications explicitly made use of these terms, corresponding to 61%.[15] Although not explicitly mentioning the term, the remaining publications did indeed describe methods that could be considered evaluation strategies. For example, case studies, measurements, performances, composition processes, and different types of empirical experiments were mentioned, without referring to "evaluation" explicitly.

Severeal evaluation strategies for musical robots could identified. First of all, the authors mentioned both so called "objective" and "subjective" evaluations. The objective evaluations usually focused on technical and mechanical aspects, e.g. sound quality, through measurements and analysis of audio or sensor data. Reoccurring variables were dynamic range, pitch, timbre, speed, repetition,

---

[15]Instances in which the word "evaluate" was used more generally, e.g. to evaluate a mathematical expression, were excluded.

and latency. Some explicitly mentioned that they measured variables relating to *musical expressivity*, e.g. timbre control, peak loudness, decay control, pitch control, etc. When it comes to evaluation of acoustic quality, this was often computationally done, using software and algorithms. However, in some cases humans also analyzed the sonic output through inspection. Building on the methodology proposed by Wanderley and Orio [160], objective evaluations were often based on programming the robot to perform simple musical tasks. For example, robots were instructed to play different polyrhythms or at different dynamics.

It has been suggested that musical robotic systems need to be tested in performance and installation settings for their functionality to be properly understood [107]. Many of the publications in the explored dataset described processes involving composing pieces specifically for the robot, performing with the robot in front of an audience, organising concerts with the robot, or showcasing the robotic system as a music installation open to the public. Such practices were generally described without framing them as evaluation methods. For performances and installations, accounts of methods going beyond the generic description of collecting "informal feedback" from musicians or exhibition visitors were rare. Many papers included general statements describing that the robot had been used in "various applications/contexts". The methodology for analysis of these processes was not clearly defined (one exception was the suggestion to use structured observation in [16]).

In general, there was an overall tendency to focus on "performance analysis" of the technical systems. Relatively few publications included descriptions of subjective evaluations. When it comes to the use of subjective – or qualitative – methods, a rather common strategy was to use questionnaires. Interviews, think-aloud, and observational methods seemed to be less common. Several authors described using different types of listening tests, followed by questionnaires, often involving ratings on Likert scales. When it comes to the use of standardized tools for questionnaires, the vast majority developed their own questions (one exception was the *"Quebec User Evaluation of Satisfaction with Assistive Technology (QUEST)"* questionnaire (see [35]) used in [164]). Questions focused, for example, on rating aspects related to musical performance (e.g. gesture expressiveness) or agreement with statements about robot musicianship (e.g. "the robot played well", see [132]). Although aspects such as "(user) experience" and "usability" were mentioned by some authors, no standardized methods were used to evaluate such dimensions. However, characteristics such as strengths, weaknesses, and frustrations of a system, were sometimes discussed. Overall, relatively little attention was given to aspects concerning the interaction with and creative possibilities of the systems, such as co-creation and agency.[16]

Another common theme was to compare robot-produced sounds and humanly created ones, using methods reminiscent of the modified Turing tests.

---

[16]This might, however, be a direct result of the selection of publication venues.

Such procedures involved, for example, listening tests in which participants listened to a piece generated by an expert performer who played on an acoustic instrument and compared this to the performance of a robot performer playing the same piece. Participants were either asked which performance they thought was generated by a human or rated the music on a set of scales (or both).

Finally, apart from public installations and performances, most of the evaluation experiments were performed in lab settings. Few evaluations involved many participants; often less than 10 subjects took part. This was usually motivated by the fact that experts were invited as participants, or that there was a need for subjects with very specific skill sets (e.g. they needed to not only be able to improvise freely on an instrument but also have good computer skills).

12.9   Prospects for Future Research

To inform the selection of evaluation strategies for musical robots, this chapter has provided an overview of evaluation methods used in the fields of Human–Computer Interaction, Human–Robot Interaction, Human–AI Interaction, New Interfaces for Musical Expression, and Computational Creativity. The chapter has highlighted not only the breadth of systems that can be considered *musical robots* but also the heterogeneous methods employed to evaluate such systems, as well as the sometimes conflicting views on what constitutes an appropriate evaluation method. Based on this heterogeneity, it seems somewhat naive to suggest that it would be possible to develop a general evaluation framework that could be applied to all musical robots. There is no unified reply to the question "what to evaluate" and how to conduct such an evaluation, nor an undivided view of what the term "evaluation" actually means.

Evaluation criteria that are important for one system might not be relevant for another one, and different stakeholders will have different perspectives on what is relevant to explore. For example, the requirements for a physical performance robot playing on stage together with musicians or other robots might be significantly different from the requirements for a virtual agent involved in a collaborative composition task. Different robot designs also pose specific challenges that might not necessarily be relevant to other categories of musical robots. For example, musical robots that act as wearable devices or prosthetic devices raise questions that relate to the notion of cyborgs (see e.g. [56]), aspects that are perhaps less relevant for an intelligent improvising Disklavier piano. To conclude, the employed evaluation frameworks should not be considered "off the shelf" tools that can be readily applied to all settings. The methods should be adapted to different situations, and perhaps also be modified, to make sense for a specific musical context and stakeholder.

Despite the heterogeneous views on evaluation discussed above, there are some themes that re-occur across different research domains. For example, terms such as functionality, usability, and user experience were mentioned in literature from numerous fields. Several authors from various backgrounds also voiced the need for more holistic approaches that go beyond traditional HCI methods focused on standard usability metrics such as task completion rates. Moreover, several mentioned the need for qualitative methods to explore the complex nature of musical interactions and their situated nature, stressing the importance of focusing on emotions and social interactions, as well as cultural contexts. Other topics that reoccurred in the literature were attempts to more clearly define when to evaluate, since this might affect the choice of methods.

Considering the above, a reasonable suggestion would be to propose a workflow for evaluation of musical robots, rather than a set of evaluation metrics. Such a workflow could consist of the following steps:

1. *Classify the robot using existing taxonomies.* For example, is the robot a humanoid, is it physical or virtual, is it a wearable device, and how much AI and autonomous agency is involved? The purpose of this initial step is to place the musical robot in a context and to inform the subsequent steps. Taxonomies and strategies for classification described in Sections 12.1 and 12.2 can be used to help situate the robot in a historical context, and to better understand the breadth of previous work that has explored similar topics.

2. *Specify the context (and goal, if there is such a thing) of the musical interaction.* For example, is the musical robot intended to be used in a solo performance context or in ensemble play, is it performing in front of a large audience, is it composing music (without an audience), and so on. This step can also involve identifying high-level objectives driven by the representative use of the envisioned robot system (this might or might not involve UX goals, see [159]).

3. *Identify stakeholders.* This includes exploring what is the role of the human versus robot, as well as the level of co-creation and collaboration. It also involves situating the work in a socio-cultural context.

4. *Informed by steps 1–3, identify evaluation methods from the literature and adapt them, if necessary.* This involves identifying both objective and subjective measures, as well as formative and summative methods. Once a set of methods have been identified, they can be tuned to fit the specific musical setting, the socio-cultural context, and the stakeholders involved.

To conclude, it is worth noting that the review of evaluation methods presented in Section 12.8 is far from a full systematic review and that it should be expanded to include additional publication venues to be able to draw generalized conclusions about the entire field of musical robots. The review puts a strong emphasize on robotic musicianship in Computer Music and New

Interfaces for Musical Expression (NIME) research. Different tendencies would perhaps be identified if reviewing literature published in HRI journals. Finally, it is likely that the dataset to be reviewed would have become much larger if the search had been expanded to include the term "robot" in the main text (not only in the title and abstract).

Bibliography

[1] AGRES, K., FORTH, J., AND WIGGINS, G. A. Evaluation of Musical Creativity and Musical Metacreation Systems. *Computers in Entertainment (CIE) 14*, 3 (2016), 1–33.

[2] AMABILE, T. *Componential Theory of Creativity.* Harvard Business School Boston, MA, 2011.

[3] AMERSHI, S., WELD, D., VORVOREANU, M., FOURNEY, A., NUSHI, B., COLLISSON, P., SUH, J., IQBAL, S., BENNETT, P. N., INKPEN, K., ET AL. Guidelines for human-AI interaction. In *Proceedings of the ACM SIGCHI Conference on Human Factors in Computing Systems* (2019), pp. 1–13.

[4] ANJOMSHOAE, S., NAJJAR, A., CALVARESI, D., AND FRÄMLING, K. Explainable agents and robots: Results from a systematic literature review. In *Proceedings of the International Conference on Autonomous Agents and Multiagent Systems (AAMAS)* (2019), pp. 1078–1088.

[5] BAER, J. Domain specificity and the limits of creativity theory. *The Journal of Creative Behavior 46*, 1 (2012), 16–29.

[6] BAER, J. The importance of domain-specific expertise in creativity. *Roeper Review 37*, 3 (2015), 165–178.

[7] BAGINSKY, N.A. The three Sirens: A self learning robotic rock band. http://www.baginsky.de/agl/

[8] BARAKA, K., ALVES-OLIVEIRA, P., AND RIBEIRO, T. An extended framework for characterizing social robots. In *Human-Robot Interaction* (2020), Springer, pp. 21–64.

[9] BARBOSA, J., MALLOCH, J., WANDERLEY, M. M., AND HUOT, S. What does "evaluation" mean for the NIME community? In *Proceedings of the International Conference on New Interfaces for Musical Expression (NIME)* (2015), pp. 156–161.

[10] BARKHUUS, L., AND RODE, J. A. From mice to men – 24 years of evaluation in CHI. In *Proceedings of the ACM SIGCHI Conference on Human Factors in Computing Systems* (2007), pp. 1–16.

[11] BATES, J. The role of emotion in believable agents. *Communications of the ACM 37*, 7 (1994), 122–125.

[12] BIRNBAUM, D., FIEBRINK, R., MALLOCH, J., AND WANDERLEY, M. M. Towards a dimension space for musical devices. In *Proceedings of the International Conference on New Interfaces for Musical Expression (NIME)* (2005), pp. 192–195.

[13] BLYTHE, M. A., OVERBEEKE, K., MONK, A. F., AND WRIGHT, P. C. *Funology: From Usability to Enjoyment.* Springer, 2004.

[14] BODEN, M. A. *The Creative Mind: Myths and Mechanisms.* Routledge, 2004.

[15] BOEHNER, K., DEPAULA, R., DOURISH, P., AND SENGERS, P. How emotion is made and measured. *International Journal of Human-Computer Studies 65*, 4 (2007), 275–291.

[16] BUCH, B., COUSSEMENT, P., AND SCHMIDT, L. "Playing Robot": An interactive sound installation in human-robot interaction design for new media art. In *Proceedings of the International Conference on New Interfaces for Musical Expression (NIME)* (2010), pp. 411–414.

[17] BUSCHEK, D., MECKE, L., LEHMANN, F., AND DANG, H. Nine potential pitfalls when designing human-AI co-creative systems. *arXiv preprint arXiv:2104.00358* (2021).

[18] CAMPBELL, D. M. Evaluating musical instruments. *Physics Today 67*, 4 (2014).

[19] CAPTURED! BY ROBOTS. http://www.capturedbyrobots.com/.

[20] CHADEFAUX, D., LE CARROU, J.-L., VITRANI, M.-A., BILLOUT, S., AND QUARTIER, L. Harp plucking robotic finger. In *Proceedings of the IEEE/RSJ International Conference on Intelligent Robots and Systems* (2012), IEEE, pp. 4886–4891.

[21] CHEN, R., DANNENBERG, R. B., RAJ, B., AND SINGH, R. Artificial creative intelligence: Breaking the imitation barrier. In *Proceedings of International Conference on Computational Creativity* (2020), pp. 319–325.

[22] COLLINS, N. Trading faures: Virtual musicians and machine ethics. *Leonardo Music Journal 21* (2011), 35–39.

[23] COLTON, S. Creativity versus the perception of creativity in computational systems. In *Proceedings of the AAAI Spring Symposium: Creative Intelligent Systems* (2008), p. 7–14.

[24] COLTON, S., CHARNLEY, J. W., AND PEASE, A. Computational creativity theory: The FACE and IDEA descriptive models. In *Proceedings of the International Conference on Innovative Computing and Cloud Computing* (2011), pp. 90–95.

[25] COLTON, S., AND WIGGINS, G. A. Computational creativity: The final frontier? In *Proceedings of the European Conference on Artificial Intelligence (ECAI)* (2012), pp. 21–26.

[26] CONNELL, B. R., JONES, M., MACE, R., MUELLER, J., MULLICK, A., OSTROFF, E., SANFORD, J., STEINFELD, E., STORY, M., AND VANDERHEIDEN, G. The Principles of Universal Design, 1997.

[27] COOK, A. M., AND POLGAR, J. M., Mosby (Elsevier), 2015. Assistive technologies: Principles and Practice (2002).

[28] CORBIN, J., AND STRAUSS, A. *Basics of Qualitative Research: Techniques and Procedures for Developing Grounded Theory.* Sage Publications, 2014.

[29] CRYPTON FUTURE MEDIA. Hatsune Miku https://ec.crypton.co.jp/pages/prod/virtualsinger/cv01_us

[30] DANNENBERG, R. B., BROWN, H. B., AND LUPISH, R. McBlare: A robotic bagpipe player. In *Musical Robots and Interactive Multimodal Systems*. Springer, 2011, pp. 165–178.

[31] DAUTENHAHN, K. Some brief thoughts on the past and future of human-robot interaction, *Proceedings of the ACM Transactions on Human-Robot Interaction*, Vol. 7, No. 1, Article 4. 2018.

[32] DAVANZO, N., AND AVANZINI, F. A dimension space for the evaluation of accessible digital musical instruments. In *Proceedings of the International Conference on New Interfaces for Musical Expression (NIME)* (2020), pp. 214–220.

[33] DAVIES, A., AND CROSBY, A. Compressorhead: the robot band and its transmedia storyworld. In *Proceedings of the International Workshop on Cultural Robotics* (2015), pp. 175–189.

[34] DE VAUCANSON, J. Le méchanisme du fluteur automate, 1738.

[35] DEMERS, L., WEISS-LAMBROU, R., AND SKA, B. The quebec user evaluation of satisfaction with assistive technology (QUEST 2.0): An overview and recent progress. *Technology and Disability 14*, 3 (2002), 101–105.

[36] DETERDING, S., HOOK, J., FIEBRINK, R., GILLIES, M., GOW, J., AKTEN, M., SMITH, G., LIAPIS, A., AND COMPTON, K. Mixed-initiative creative interfaces. In *Proceedings of the ACM SIGCHI Conference Extended Abstracts on Human Factors in Computing Systems* (2017), pp. 628–635.

[37] DEWEY, J. Art as experience. In *The Richness of Art Education*. Brill, 2008, pp. 33–48.

[38] DÍAZ-OREIRO, I., LÓPEZ, G., QUESADA, L., AND GUERRERO, L. A. Standardized questionnaires for user experience evaluation: A systematic literature review. In *Proceedings of the International Conference on Ubiquitous Computing and Ambient Intelligence (UCAmI)* (2019), 14, p. 1–12.

[39] DUMAS, J. S. User-based evaluations. In *The Human-Computer Interaction Handbook: Fundamentals, Evolving Technologies and Emerging Applications*. Lawrence Erlbaum, 2002, pp. 1093–1117.

[40] EIDD. The EIDD Stockholm Declaration. https://dfaeurope.eu/what-is-dfa/dfa-documents/the-eidd-stockholm-declaration-2004/, 2004.

[41] EIGENFELDT, A., BURNETT, A., AND PASQUIER, P. Evaluating musical metacreation in a live performance context. In *Proceedings of the International Conference on Computational Creativity* (2012), pp. 140–144.

[42] EL-SHIMY, D., AND COOPERSTOCK, J. R. User-driven techniques for the design and evaluation of new musical interfaces. *Computer Music Journal 40*, 2 (2016), 35–46.

[43] FARMER, H. G. The Organ Of The Ancients From Eastern Sources (Hebrew, Syriac, And Arabic), *The New Temple Press*, 1931.

[44] FERNAEUS, Y., LJUNGBLAD, S., JACOBSSON, M., AND TAYLOR, A. Where third wave HCI meets HRI: Report from a workshop on user-centred design of robots. In *Proceedings of the ACM/IEEE International Conference on Human-Robot Interaction (HRI)* (2009), pp. 293–294.

[45] FORLIZZI, J., AND BATTARBEE, K. Understanding experience in interactive systems. In *Proceedings of the Conference on Designing Interactive Systems: Processes, Practices, Methods, and Techniques* (2004), pp. 261–268.

[46] FORLIZZI, J., AND DISALVO, C. Service robots in the domestic environment: A study of the Roomba vacuum in the home. In *Proceedings of the ACM SIGCHI/SIGART Conference on Human-Robot Interaction* (2006), pp. 258–265.

[47] FRANKLIN, M., AND LAGNADO, D. Human-AI interaction paradigm for evaluating explainable artificial intelligence. In *Proceedings of the International Conference on Human-Computer Interaction* (2022), pp. 404–411.

[48] FRID, E. Accessible digital musical instruments – A review of musical interfaces in inclusive music practice. *Multimodal Technologies and Interaction 3*, 3 (2019), 57.

[49] FRID, E. *Diverse Sounds: Enabling Inclusive Sonic Interaction*. PhD thesis, KTH Royal Institute of Technology, 2019.

[50] FRID, E., GOMES, C., AND JIN, Z. Music creation by example. In *Proceedings of the ACM SIGCHI Conference on Human Factors in Computing Systems* (2020), pp. 1–13.

[51] GATEBOX. INC https://www.gatebox.ai/.

[52] GRAN, F. Cello Concerto No. 1. https://fredrikgran.com/works/cello-suite-no-1/.

[53] GREENBERG, S., AND BUXTON, B. Usability evaluation considered harmful (some of the time). In *Proceedings of the ACM SIGCHI Conference on Human Factors in Computing Systems* (2008), pp. 111–120.

[54] GUNNING, D. Explainable artificial intelligence (XAI). Tech. rep., Defense Advanced Research Projects Agency (DARPA), 2017.

[55] GUREVICH, M. Distributed control in a mechatronic musical instrument. In *Proceedings of the International Conference on New Interfaces for Musical Expression (NIME)* (2014), pp. 487–490.

[56] HARAWAY, D. *A Cyborg Manifesto*. Socialist Review, 1985.

[57] HARRISON, J. *Instruments and access: The role of instruments in music and disability*. PhD thesis, Queen Mary University of London, 2020.

[58] HASSENZAHL, M. The thing and I: Understanding the relationship between user and product. In *Funology*. Springer, 2003, pp. 31–42.

[59] HASSENZAHL, M., LAW, E. L.-C., AND HVANNBERG, E. T. User Experience – Towards a unified view. In *Proceedings of the NordiCHI: COST294-MAUSE Workshop* (2006), pp. 1–3.

[60] HATAKEYAMA, K., SARAIJI, M. Y., AND MINAMIZAWA, K. MusiArm: Extending prosthesis to musical expression. In *Proceedings of the Augmented Human International Conference* (2019), pp. 1–8.

[61] HOFFMANN, L., BOCK, N., AND ROESENTHAL V.D. PÜTTEN, A. M. The peculiarities of robot embodiment (EmCorp-Scale): Development, validation and initial test of the embodiment and corporeality of artificial

agents scale. In *Proceedings of the ACM/IEEE International Conference on Human-Robot Interaction (HRI)* (2018), pp. 370–378.

[62] HÖÖK, K. User-centred design and evaluation of affective interfaces. In *From Brows to Trust.* Springer, 2004, pp. 127–160.

[63] HORVITZ, E. Principles of mixed-initiative user interfaces. In *Proceedings of the ACM SIGCHI Conference on Human Factors in Computing Systems* (1999), pp. 159–166.

[64] IBM. IBM Design for AI. https://www.ibm.com/design/ai/, 2019.

[65] INTERNATIONAL ORGANIZATION FOR STANDARDIZATION. ISO 9241-11:2018 - Ergonomics of human-system interaction — Part 11: Usability: Definitions and concepts. https://www.iso.org/standard/63500.html.

[66] INTERNATIONAL ORGANIZATION FOR STANDARDIZATION. ISO 8373:2021 Robotics – Vocabulary. https://www.iso.org/standard/75539.html, 2021.

[67] ISBISTER, K., HÖÖK, K., SHARP, M., AND LAAKSOLAHTI, J. The sensual evaluation instrument: Developing an affective evaluation tool. In *Proceedings of the ACM SIGCHI Conference on Human Factors in Computing Systems* (2006), pp. 1163–1172.

[68] IVORY, M. Y., AND HEARST, M. A. The state of the art in automating usability evaluation of user interfaces. *ACM Computing Surveys (CSUR) 33*, 4 (2001), 470–516.

[69] JACK, R., HARRISON, J., AND MCPHERSON, A. Digital musical instruments as research products. In *Proceedings of the International Conference on New Interfaces for Musical Expression (NIME)* (2020), pp. 446–451.

[70] JIANG, S., AND ARKIN, R. C. Mixed-initiative human-robot interaction: Definition, taxonomy, and survey. In *Proceedings of the IEEE International Conference on Systems, Man, and Cybernetics* (2015), pp. 954–961.

[71] JOHNSTON, A. J. Beyond evaluation: Linking practice and theory in new musical interface design. In *Proceedings of the International New Interfaces for Musical Expression Conference (NIME)* (2011).

[72] JONES, R. Archaic man meets a marvellous automaton: Posthumanism, social robots, archetypes. *Journal of Analytical Psychology 62*, 3 (2017), 338–355.

[73] JORDÀ, S. Afasia: The ultimate homeric one-man-multimedia-band. In *Proceedings of the International Conference on New Interfaces for Musical Expression (NIME)* (2002), pp. 132–137.

[74] JORDÀ, S. Digital instruments and players: Part II-Diversity, freedom, and control. In *Proceedings of the International Computer Music Conference (ICMC)* (2004).

[75] JORDANOUS, A. *Evaluating Computational Creativity: a Standardised Procedure for Evaluating Creative Systems and its Application.* University of Kent (United Kingdom), 2012.

[76] JORDANOUS, A. A standardised procedure for evaluating creative systems: Computational creativity evaluation based on what it is to be creative. *Cognitive Computation 4*, 3 (2012), 246–279.

[77] JORDANOUS, A. Evaluating evaluation: Assessing progress and practices in computational creativity research. In *Computational Creativity.* Springer, 2019, pp. 211–236.

[78] JOST, C., LE PÉVÉDIC, B., BELPAEME, T., BETHEL, C., CHRYSOSTOMOU, D., CROOK, N., GRANDGEORGE, M., AND MIRNIG, N., Eds. *Human-Robot Interaction: Evaluation Methods and Their Standardization.* Springer, Germany, 2020.

[79] KAJITANI, M. Development of musician robots. *Journal of Robotics and Mechatronics 1*, 1 (1989), 254–255.

[80] KAJITANI, M. Simulation of musical performances. *Journal of Robotics and Mechatronics 4*, 6 (1992), 462–465.

[81] KAPUR, A. A history of robotic musical instruments. In *Proceedings of the International Computer Music Conference (ICMC)* (2005), p. 4599.

[82] KARIMI, P., GRACE, K., MAHER, M. L., AND DAVIS, N. Evaluating creativity in computational co-creative systems. *arXiv preprint arXiv:1807.09886* (2018).

[83] KAYE, J. J. Evaluating experience–focused HCI. In *Proceedings of the ACM SIGCHI Extended Abstracts on Human Factors in Computing Systems* (2007), pp. 1661–1664.

[84] KEMPER, S. Locating creativity in differing approaches to musical robotics. *Frontiers in Robotics and AI 8* (2021), 647028.

[85] KIERAS, D. Model-based evaluation. In *Human-Computer Interaction.* CRC Press, 2009, pp. 309–326.

[86] KIESLER, S., POWERS, A., FUSSELL, S. R., AND TORREY, C. Anthropomorphic interactions with a robot and robot–like agent. *Social Cognition 26*, 2 (2008), 169–181.

[87] KOETSIER, T. On the prehistory of programmable machines: Musical automata, looms, calculators. *Mechanism and Machine Theory 36*, 5 (2001), 589–603.

[88] KRZYZANIAK, M. Prehistory of musical robots. *Journal of Human-Robot Interaction 1*, 1 (2012), 78–95.

[89] KVIFTE, T., AND JENSENIUS, A. R. Towards a coherent terminology and model of instrument description and design. In *Proceedings of the International Conference on New Interfaces for Musical Expression (NIME)* (2006), pp. 220–225.

[90] LAMB, C., BROWN, D. G., AND CLARKE, C. L. A. Evaluating computational creativity: An interdisciplinary tutorial. *ACM Computing Surveys (CSUR) 51*, 2 (2018), 1–34.

[91] LANGLEY, P., MEADOWS, B., SRIDHARAN, M., AND CHOI, D. Explainable agency for intelligent autonomous systems. In *Proceedings of the Association for the Advancement of Artificial Intelligence* (2017).

[92] LAUGWITZ, B., HELD, T., AND SCHREPP, M. Construction and evaluation of a user experience questionnaire. In *Proceedings of the Symposium of the Austrian HCI and Usability Engineering Group* (2008), pp. 63–76.

[93] LAW, E., ROTO, V., VERMEEREN, A. P., KORT, J., AND HASSENZAHL, M. Towards a shared definition of user experience. In *Proceedings of the ACM SIGCHI Extended Abstracts on Human Factors in Computing Systems* (2008), pp. 2395–2398.

[94] LEWIS, C., AND WHARTON, C. Cognitive walkthroughs. In *Handbook of Human–Computer Interaction*. Elsevier, 1997, pp. 717–732.

[95] LI, T., VORVOREANU, M., DEBELLIS, D., AND AMERSHI, S. Assessing human-AI interaction early through factorial surveys: A study on the guidelines for human-AI interaction. *ACM Transactions on Computer-Human Interaction* (2022).

[96] LINDBLOM, J., ALENLJUNG, B., AND BILLING, E. Evaluating the user experience of human–robot interaction. In *Human–Robot Interaction*. Springer Nature Switzerland, 2020, pp. 231–256.

[97] LONG, J., MURPHY, J., CARNEGIE, D., AND KAPUR, A. Loudspeakers optional: A history of non-loudspeaker-based electroacoustic music. *Organised Sound 22*, 2 (2017), 195–205.

[98] LOUGHRAN, R., AND O'NEILL, M. Generative music evaluation: Why do we limit to 'human'. In *Proceedings of the Conference on Computer Simulation of Musical Creativity (CSMC)* (2016).

[99] LUCAS, A., HARRISON, J., SCHROEDER, F., AND ORTIZ, M. Cross-pollinating ecological perspectives in ADMI design and evaluation. In *Proceedings of the International Conference on New Interfaces for Musical Expression (NIME)* (2021).

[100] MACDONALD, C. M., AND ATWOOD, M. E. Changing perspectives on evaluation in HCI: Past, present, and future. In *Proceedings of the ACM SIGCHI Extended Abstracts on Human Factors in Computing Systems* (2013), pp. 1969–1978.

[101] MAGNUSSON, T. An epistemic dimension space for musical devices. In *Proceedings of the International Conference on New Interfaces for Musical Expression (NIME)* (2010), pp. 43–46.

[102] MCCARTHY, J., AND WRIGHT, P. Technology as experience. *Interactions 11*, 5 (2004), 42–43.

[103] MCLEOD, K. Living in the immaterial world: Holograms and spirituality in recent popular music. *Popular Music and Society 39*, 5 (2016), 501–515.

[104] MCNAMARA, N., AND KIRAKOWSKI, J. Defining usability: Quality of use or quality of experience? In *Proceedings of the International Professional Communication Conference (IPCC)* (2005), pp. 200–204.

[105] MCNAMARA, N., AND KIRAKOWSKI, J. Functionality, usability, and user experience: Three areas of concern. *Interactions 13*, 6 (2006), 26–28.

[106] MINSKY, M. L. Why people think computers can't. *AI Magazine 3*, 4 (1982), 3–3.

[107] MURPHY, J., KAPUR, A., AND CARNEGIE, D. Musical robotics in a loudspeaker world: Developments in alternative approaches to localization and spatialization. *Leonardo Music Journal 22* (12 2012), 41–48.

[108] NAKE, F. Construction and intuition: Creativity in early computer art. In *Computers and Creativity*. Springer, 2012, pp. 61–94.

[109] NESS, S. R., TRAIL, S., DRIESSEN, P. F., SCHLOSS, W. A., AND TZANETAKIS, G. Music information robotics: Coping strategies for musically challenged robots. In *Proceedings of International Society for Music Information Retrieval Conference (ISMIR)* (2011), pp. 567–572.

[110] NIELSEN, J. Usability inspection methods. In *Conference Companion on Human Factors in Computing systems* (1994), pp. 413–414.

[111] NIELSEN, J., AND MOLICH, R. Heuristic evaluation of user interfaces. In *Proceedings of the SIGCHI Conference on Human Factors in Computing Systems* (1990), pp. 249–256.

[112] NORMAN, D. A. *The Design of Everyday Things*. Basic Books, 1988.

[113] NORMAN, D. A. *Emotional Design: Why We Love (or Hate) Everyday Things*. Basic, 2004.

[114] O'MODHRAIN, S. A framework for the evaluation of digital musical instruments. *Computer Music Journal 35*, 1 (2011), 28–42.

[115] ONNASCH, L., AND ROESLER, E. A taxonomy to structure and analyze human–robot interaction. *International Journal of Social Robotics 13*, 4 (2021), 833–849.

[116] GOOGLE PAIR, G. People + AI Guidebook. https://pair.withgoogle.com/guidebook/, 2019.

[117] PAWEL ZADROZNIAK. Floppy music DUO – Imperial March. https://youtu.be/yHJOz_y9rZE. See also http://silent.org.pl/home/.

[118] PEASE, A., AND COLTON, S. On impact and evaluation in computational creativity: A discussion of the Turing test and an alternative proposal. In *Proceedings of the AISB Symposium on AI and Philosophy* (2011), pp. 15–22.

[119] PETRIE, H., AND BEVAN, N. The evaluation of accessibility, usability, and user experience. In *The Universal Access Handbook* (2009), 20:1–16. See https://www.routledge.com/The-Universal-Access-Handbook/Stephanidis/p/book/9780805862805.

[120] REIMER, P. C., AND WANDERLEY, M. M. Embracing less common evaluation strategies for studying user experience in NIME. In *Proceedings of the International Conference on New Interfaces for Musical Expression (NIME)* (2021).

[121] REMY, C., MACDONALD VERMEULEN, L., FRICH, J., BISKJAER, M. M., AND DALSGAARD, P. Evaluating creativity support tools in HCI research. In *Proceedings of the 2020 ACM Designing Interactive Systems Conference* (2020), pp. 457–476.

[122] REZWANA, J., AND MAHER, M. L. Designing creative AI partners with COFI: A framework for modeling interaction in human-AI co-creative systems. *ACM Transactions on Computer-Human Interaction* (2022).

[123] RHODES, M. An analysis of creativity. *The Phi Delta Kappan 42*, 7 (1961), 305–310.

[124] RITCHIE, G. Some empirical criteria for attributing creativity to a computer program. *Minds and Machines 17*, 1 (2007), 67–99.

[125] ROADS, C. The Tsukuba musical robot. *Computer Music Journal 10*, 2 (1986), 39–43.

[126] ROBERT, J.-M., AND LESAGE, A. From usability to user experience with interactive systems. In *The Handbook of Human-Machine Interaction*. CRC Press, 2017, pp. 303–320.

[127] RODGER, M., STAPLETON, P., VAN WALSTIJN, M., ORTIZ, M., AND PARDUE, L. What makes a good musical instrument? A matter of

processes, ecologies and specificities. In *Proceedings of the International Conference on New Interfaces for Musical Expression (NIME)* (2020), pp. 405–410.

[128] ROGERS, T., KEMPER, S., AND BARTON, S. MARIE: Monochord - aerophone robotic instrument ensemble. In *Proceedings of the International Conference on New Interfaces for Musical Expression (NIME)* (2015), pp. 408–411.

[129] ROWE, R. *Machine Musicianship.* MIT Press, 2004.

[130] SAVERY, R., AND WEINBERG, G. Shimon the robot film composer and DeepScore. In *Computer Simulation of Musical Creativity (2018)* (2018), pp. 1–14.

[131] SAVERY, R., AND WEINBERG, G. Robotics - Fast and curious: A CNN for ethical deep learning musical generation. In *Artificial Intelligence and Music Ecosystem.* Focal Press, 2022, pp. 52–67.

[132] SAVERY, R., ZAHRAY, L., AND WEINBERG, G. Shimon the Rapper: A real-time system for human-robot interactive rap battles. In *International Conference on Computational Creativity* (2020), pp. 212–219.

[133] SAVERY, R., ZAHRAY, L., AND WEINBERG, G. Shimon sings – Robotic musicianship finds its voice. In *Handbook of Artificial Intelligence for Music.* Springer, Cham, 2021, pp. 823–847.

[134] SCHERER, M. J. Matching person & technology: Model and assessment process, 2007.

[135] SCHMID, G.-M. *Evaluating the Experiential Quality of Musical Instruments.* Springer, 2017.

[136] SCRIVEN, M. Beyond formative and summative evaluation. In *Evaluation and Education: At Quarter Century*, M. W. M. D. C. Phillips, Ed. University of Chicago Press, 1991, pp. 18–64.

[137] SENGERS, P., AND GAVER, B. Staying open to interpretation: Engaging multiple meanings in design and evaluation. In *Proceedings of the Conference on Designing Interactive Systems* (2006), pp. 99–108.

[138] SINGER, E., FEDDERSEN, J., REDMON, C., AND BOWEN, B. LEMUR's musical robots. In *Proceedings of the International Conference on New Interfaces for Musical Expression (NIME)* (2004), pp. 181–184.

[139] SINGER, E., LARKE, K., AND BIANCIARDI, D. LEMUR GuitarBot: MIDI robotic string instrument. In *Proceedings of the International Conference on New Interfaces for Musical Expression (NIME)* (2003), pp. 188–191.

[140] SMALL, C. *Musicking: The Meanings of Performing and Listening.* Wesleyan University Press, 1998.

[141] SNAKE-BEINGS, E. The Do-it-Yourself (DiY) craft aesthetic of The Trons – Robot garage band. *Craft Research 8*, 1 (2017), 55–77.

[142] SOLIS, J., BERGAMASCO, M. ISODA, S., CHIDA, K., AND TAKANISHI, A. "Learning to Play the Flute with an Anthropomorphic Robot", Proceedings of the International Computer Music Conference, Miami, Florida, 2004.

[143] SOLIS, J., CHIDA, K., TANIGUCHI, K., HASHIMOTO, S. M., SUEFUJI, K., AND TAKANISHI, A. The Waseda flutist robot WF-4RII in comparison with a professional flutist. *Computer Music Journal* (2006), 12–27.

[144] SOLIS, J., AND NG, K., Eds. *Musical Robots and Interactive Multimodal Systems.* Springer, 2011.

[145] SOLIS, J., AND NG, K. Musical robots and interactive multimodal systems: An introduction. In *Musical Robots and Interactive Multimodal Systems.* Springer, 2011, pp. 1–12.

[146] SOLIS, J., AND TAKANISHI, A. An overview of the research approaches on musical performance robots. In *Proceedings of the International Conference on Computer Music (ICMC)* (2007), pp. 356–359.

[147] SOLIS, J., AND TAKANISHI, A. Wind instrument playing humanoid robots. In *Musical Robots and Interactive Multimodal Systems.* Springer, 2011, pp. 195–213.

[148] SOLIS, J., TAKANISHI, A., AND HASHIMOTO, K. Development of an anthropomorphic saxophone-playing robot. In *Brain, Body and Machine.* Springer, 2010, pp. 175–186.

[149] STEPHENS, E., AND HEFFERNAN, T. We have always been robots: The history of robots and art. In *Robots and Art.* Springer, 2016, pp. 29–45.

[150] STOWELL, D., ROBERTSON, A., BRYAN-KINNS, N., AND PLUMBLEY, M. D. Evaluation of live human–computer music-making: Quantitative and qualitative approaches. *International Journal of Human-Computer Studies 67*, 11 (2009), 960–975.

[151] STRIEBE, D. The Prayer. https://theprayer.diemutstrebe.com/.

[152] SUNG, J.-Y., GUO, L., GRINTER, R. E., AND CHRISTENSEN, H. I. "My Roomba is Rambo": Intimate home appliances. In *Proceedings of the International Conference on Ubiquitous Computing* (2007), pp. 145–162.

[153] SZOLLOSY, M. Freud, Frankenstein and our fear of robots: Projection in our cultural perception of technology. *AI & Society 32*, 3 (2017), 433–439.

[154] TAKAGI, S. Toyota partner robots. *Journal of the Robotics Society of Japan 24*, 2 (2006), 208–210.

[155] TATAR, K., AND PASQUIER, P. Musical agents: A typology and state of the art towards musical metacreation. *Journal of New Music Research 48*, 1 (2019), 56–105.

[156] TEAM UEQ. User Experience Questionnaire. https://www.ueq-online. org/.

[157] TRIMPIN. Wikipedia. https://en.wikipedia.org/wiki/Trimpin.

[158] YAMAHA CORPORATION. Vocaloid. https://www.vocaloid.com/en/.

[159] WALLSTRÖM, J., AND LINDBLOM, J. Design and development of the USUS goals evaluation framework. In *Human-Robot Interaction* (2020), Springer, pp. 177–201.

[160] WANDERLEY, M. M., AND ORIO, N. Evaluation of input devices for musical expression: Borrowing tools from HCI. *Computer Music Journal 26*, 3 (2002), 62–76.

[161] WATERS, S. Performance ecosystems: Ecological approaches to musical interaction. *EMS: Electroacoustic Music Studies Network* (2007), 1–20.

[162] WEINBERG, G., BRETAN, M., HOFFMAN, G., AND DRISCOLL, S. Introduction. In *Robotic Musicianship: Embodied Artificial Creativity and Mechatronic Musical Expression*, vol. 8. Springer Nature, 2020, pp. 1–24.

[163] WEINBERG, G., BRETAN, M., HOFFMAN, G., AND DRISCOLL, S. *Robotic Musicianship: Embodied Artificial Creativity and Mechatronic Musical Expression*, vol. 8. Springer Nature, 2020.

[164] WEINBERG, G., BRETAN, M., HOFFMAN, G., AND DRISCOLL, S. "Wear it"—Wearable robotic musicians. In *Robotic Musicianship*. Springer, 2020, pp. 213–254.

[165] WEINBERG, G., AND DRISCOLL, S. Toward robotic musicianship. *Computer Music Journal* (2006), 28–45.

[166] WEISS, A., BERNHAUPT, R., LANKES, M., AND TSCHELIGI, M. The USUS evaluation framework for human-robot interaction. In *Proceedings of the Symposium on New Frontiers in Human-Robot Interaction* (2009), pp. 11–26.

[167] WILLIAMSON, M. M. *Robot arm control exploiting natural dynamics*. PhD thesis, Massachusetts Institute of Technology, 1999.

[168] WOOD, G. *Edison's Eve: A Magical History of the Quest for Mechanical Life.* Alfred A. Knopf, 2002.

[169] YANCO, H. A., AND DRURY, J. Classifying Human-Robot Interaction: An updated taxonomy. In *Proceedings of the IEEE International Conference on Systems, Man and Cybernetics* (2004), vol. 3, pp. 2841–2846.

[170] YANCO, H. A., AND DRURY, J. L. A taxonomy for human-robot interaction. In *Proceedings of the Association for the Advancement of Artificial Intelligence (AAAI) Fall Symposium on Human-Robot Interaction* (2002), pp. 111–119.

[171] YANG, N., SAVERY, R., SANKARANARAYANAN, R., ZAHRAY, L., AND WEINBERG, G. Mechatronics-driven musical expressivity for robotic percussionists. In *Proceedings of the International Conference on New Interfaces for Musical Expression (NIME)* (2020).

[172] YANNAKAKIS, G. N., LIAPIS, A., AND ALEXOPOULOS, C. Mixed-initiative co-creativity. In *Proceedings of the International Conference on Foundations of Digital Games* (2014), pp. 1–8.

[173] YOUNG, G. W., AND MURPHY, D. HCI models for digital musical instruments: Methodologies for rigorous testing of digital mmsical instruments. *arXiv preprint arXiv:2010.01328* (2020).

[174] YOUNG, J. E., SUNG, J., VOIDA, A., SHARLIN, E., IGARASHI, T., CHRISTENSEN, H. I., AND GRINTER, R. E. Evaluating human-robot interaction. *International Journal of Social Robotics 3*, 1 (2011), 53–67.

[175] ZIMOUN. Zimoun selected works 4.2 (video). https://www.zimoun.net/ and https://vimeo.com/7235817.

13

Robotic Dancing, Emotional Gestures and Prosody: A Framework for Gestures of Three Robotic Platforms

Richard Savery, Amit Rogel and Gil Weinberg

DOI: 10.1201/9781003320470-13

13.1   Introduction

Robots have the potential to be effective music-interpretive dancers that entertain, foster trust, and provide new ways to interact with humans. A robot's movement (including dance) can change the way a person perceives and interacts with a robot [31]. In many cultures, music and dance co-evolved and serve as important elements in social behavior [23]. This is because humans relate to both music and dance on physical and psychological levels. Music not only induces movement in people [18,37] but also can elicit empathy. According to Leman, this empathy is the necessary connection between the emotions hidden in music and the expression in body movements [46].

When a listener feels an emotion in music, they often reflect that emotion in their gestures. Similarly, a good dancer understands how to portray emotions in their movements, and therefore, is able to reflect the emotions from music in their movements. Taking this one step further, we hoped to create a robot dancer that can give people a way to experience emotion, as well as using this shared emotion to help establish a positive connections with robots in general, especially non-anthropomorphic ones. This chapter explores two fundamental issues, firstly how can emotional gestures be portrayed across different robot platforms while incorporating sound. The second issues address re-purposing the developed gesture framework for longer form dance to music.

We first developed a comparative approach for depicting robotic emotions using sound and gesture in three different platforms: Stretch by Hello Robot[1], Panda by Franka Emica [2], and SeekerBot, an internally developed social robot. These robotic platforms represent unique current trends in robotics, and each comes with its own Degrees of freedom (DoF), size, design, and utility. Stretch is a human size mobile manipulator with a wheel base and a telescoping arm; Panda, is a 7 DoF robotic arm designed to function as an industrial arm or cobot. The SeekerBot is a 2 DoF tabletop social robot with a LED display for a face. Based on the similarities to human morphology, posture recreation and manipulation, we regarded the SeekerBot as the most anthropomorphic platform in our study, followed by the Panda. While the Stretch is a human size mobile robot, it's morphology, speed, and non-humanoid telescopic arm, rendered it as the least anthropomorphic platform. We believed that each

---

[1] https://hello-robot.com/product
[2] https://www.franka.de/robot-system

of these platforms represent an important category in robotic design, which can help provide a comprehensive comparative evaluation of robotic emotion depiction through sound and gesture.

To conduct our study, we used an interactive design methodology based on [84], [58], and tested the perception of emotion for each gesture in a user study. We generalized this methodology to map human gestures to robots of similar physiology. We used emotional musical prosody to generate sounds, as described in Chapters 10 and 11. We evaluated the impact of our emotional gesture and prosody generators across the three robotic platforms for animacy, anthropomorphism, likeability, and perceived intelligence. The contribution of this research is a comprehensive evaluation of the manner in which different robotic platforms and approaches for gesture generation alter common HRI metrics and the portrayal of emotion, and a guideline for mapping human gestures to non-humanoid robots. This guide can help future robotic platforms incorporate emotional movements in their trajectory/movement planning for various non-humanoid robots.

The second research area of this chapter, uses this framework to create robotic dance based on musical features from a song. A seven degree of freedom (7DoF) robotic arm was used as the non-humanoid robot. The gestures, or dance movements, were designed to respond to music in a way that corresponds to a human's perception of an expressed emotion. As a result, the robot will appear to be improvising and designing it's own dance based on the music. The robot can be used as another bridge between music and dance. We leveraged Burger's research on correlations between different musical elements and various body motions [18] to create a mapping between musical features and a single robot's gestures as well as accompanying gestures of additional robots. Video samples of all gestures can be viewed at www.soundandrobotics.com/ch13

13.2 Related Work

13.2.1 Emotional Gestures in Robotics

Conveying emotions in robotic behaviors is an important tool for facilitating social interactions with humans [6]. Robotic researchers have been using a wide variety of modalities to convey emotions in robots including speech [14, 47], sound [24, 77], body posture [29, 80], facial expression [11, 13] [19], and body gestures [36, 54, 67]. These projects have commonly utilized either discrete [5, 66], continuous [53, 56] or integrative approaches [70] for emotion classification. Emotion conveyance in robots has been helpful in improving the perception of robots by human subjects in human–robot interaction scenarios. Bartneck [13] has shown that emotional facial expression in the social robot iCat significantly increased its likeability and anthropomorphism. Monceaux showed how emotion-driven robotic posture on the robot NAO enhanced its perception of animacy [54].

Gestures are often used to signal communication and display meaning or ideas [51]. A common approach of gesture design simulates human motion [20]. Cha discusses the challenges of mimicing human motion with non-humanoid robots and suggests an approach to increase animation of robots. Motion re-targeting looks at emulating a series of human postures. Kaushik looked at the skeletal model of a human and digitally reduced the profile to smaller non-anthropomorphic agents. Kaushik's research showed reduction of human movements, but did not focus on emotions [41]. In emotional gestures, motion re-targeting can be used to adapt human body gestures to robotic joints. Novikova and Watts created five flowcharts mapping emotional gestural expressions for a lego mindstorm non-anthropomorphic robot [59].

A few researchers such as Read, Dautenhahn, and Braeazeal [63] [25] [15] focused on sound generation as a tool to help convey emotions in robots such as Aibo, PeopleBot, and Kismet. Read [63], for example, has shown how embedding prosody over robotic speech can help its perception of intelligence. Embedding emotions in robots has also been shown to contribute to improving robotic performance in tasks such as museum tours, assistance for the elderly, and healthcare. Vasquez [83] has shown that using emotions in their guide robot Doris improved it's effectiveness and entertainment as a guide, while Ferreira [48] has shown that the NAO robot can reduce loneliness with the inclusion of emotions.

One of the most prominent current challenges in robotic research is establishing trust in human–robot interaction. Researchers have explored multiple approaches to enhance trust including emotion conveyance through gesture and sound. Araiz-Bekket showed that unpredictable robot movements lowers a human's trust in them and also increases that persons discomfort [1]. Other prior work showed gestures and music improved trust with Shimi robot [69]. Additionally, it has been shown that musical prosody improved trust when interacting with a virtual robotic arm [71]. While conveying emotions in robots has proven to be useful for a wide variety of purposes, no comparative and integrative research has been done to our knowledge, that embed both sound and gesture generation in multiple robotic platforms to generate and assess emotions and their effect on subjects.

13.2.2   Robotic Dance Generation

There are different approaches to automatic dance generation in robots. Alemi and Pasquier listed a survey of machine learning techniques that generate human motion based on various features of human motion [4]. This was similar to their work on GrooveNet, a real time animation of humans dancing to music [3]. Joshi and Chakrabarty created a notation style to analyze and generate dance trajectories. Similar systems of feature extraction and notation can be used for robotic dancers [39]. Another, possibly more popular approach, to robot dancing is manual coding. Boston Dynamics released videos of their humanoid robots dancing to various songs [33]. Merrit Moore also released a set of dances with a robotic arm (7 degree of freedom) [3]. Both of these projects involve manually programming the robot's dance moves. Alternatively, Shimon and Shimi use the beat of the music to bob their head and follow different musicians [16, 73]. Xia and Shiratori used multiple musical qualities to generate dances for a humanoid robot [75, 86]. This work is expanded in a variety of humanoid robots using primarily the music and rhythmic as input. Most of the systems sync the robot trajectories with a beat gathered from music [52, 60, 61, 65, 68, 78].

Other attempts at robotic dance look to adapt their trajectories based on human motion. Augello and Jochem explored different machine learning approaches to humanoid dancing [8, 38]. Both Hagendoorn's and Wallis' work center around human-inspired dance for humanoid robots [35, 85]. LaViers and Alcubilla looked at rule-based systems that are inspired from dance theory. Alcubilla created a mapping from Laban Movement analysis, and expanded Forsythe's tools for improvisation on robots [2].

13.3 Approach

We created two research questions to guide the design and improve movement for non-humanoid robot dancers:

RQ 1 In what ways can a non-humanoid robot express the emotion of human gestures?

RQ 2 In what ways can a non-humanoid robot express emotion and fluency of human gestures as a dance response to music?

We designed emotion-driven gestures and in combination with emotional musical prosody (EMP) for three robotic platforms: a social robot (Seekerbot), a collaborative robotic arm (Panda), and a mobile manipulator (Stretch). To generate EMP we used the same process described in Chapters 10 and 11. Emotional musical prosody (EMP) was used instead of speech as it better matches the morphology of a range of robots and has shown significant results for improving trust and likeability [72]. EMP consisted of short, emotionally tagged musical phrases designed to convey meaning to human collaborators using non verbals means.

---

[3] https://www.universal-robots.com/blog/dancing-through-the-pandemic-how-a-quantum-physicist-taught-a-cobot-to-dance/

13.4   Gesture and Emotional Musical Prosody Generation

Research question 1 started with the development of robotic movement inspired by human emotion. We first developed a framework for designing gestures that can appear emotional based on our generated emotional musical prosody phrases.

13.4.1   Gesture Design and Generation

Our gesture design approach for each robotic platform considered the similarities and differences in affordances in relation to a human's physical methods of conveying emotions. To generate emotion gestures, we related motions from each robots degrees of freedom to similar human gestures. The robotic gestures were designed in an effort to mimic emotional human gestures as described by [22, 40, 58, 74, 81, 84]. Tables 1-3 present an overview of our design approach for mapping emotional human gestures to the DoFs available in the three robotic platforms. Column 3 in the table presents a the human gesture for an emotion described in Column 2. Each emotional gesture guideline was generalized to address the focused movement directions for non-humanoid robots, as presented in Column 4. Columns 5 generalizes the speed interpretation. While the guideline tables do not suggest specific mapping instructions, they are designed as a reference for other researchers looking to create emotional gestures. Each table is intended to be general enough so that it could be applied to a wide range of non-anthropomorphic robots. These tables are designed to act as a baseline which could be used to create more specific rules to create emotional gestures for different robotic platforms.

13.4.1.1   Stretch Robot

The mobile manipulator robot Stretch provides a unique combination of movements and DoFs with its telescopic arm that can move up and down, camera movement with tilt-pan mount, and spatial movement through two wheels as seen in Figure 13.1. The robot's emotional gestures were designed by Mohammad Jafari based on the guideline tables. This was done by matching each of the robot's DoF to a combination of human joints. The gripper mechanism was mapped to all human arm movements. The camera was mapped to the human head, rotating upward and downward for positive and negative emotions respectively. The camera also panned to simulate human eye contact and avoidance. Since leaning can symbolize the distance from the stimulus, we decided to map these gestures to the Stretch's wheel base. Raising and lowering the robotic arm represented erect and collapsed shoulder positions.

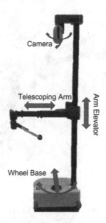

FIGURE 13.1
Model of the stretch robot.

13.4.1.2 Panda

The co-bot Panda arm we used has seven DoFs, allowing it to rotate in unique ways that can support human motion recreation and improved object manipulation. While the Panda is not a humanoid robot, it's degrees of freedom can be used to create various postures and gestures. Figure 13.2 shows the joint labels and movements. Gestures were created by assigning each joint specific angular movements with various velocities. To convey emotions, we matched the robot's DoFs to a profile of human posture and gesture as can be seen in Figure 13.3. Each gesture was mapped to a series of robot poses to match these profiles. The time it would take to switch between poses was determined by velocities indicated in the guideline table.

In general, joint 4 was used for posture changes, joint 2 for leaning positions. Joint 1 focused on adding sideways motions such as shaking body for fear or avoidance for shame. Since joint 6 moves the end effector of the robot, it was mapped to human head motions and would rotate to tuck one's head in/out. Joint 5 also acted as side movement for head shaking or any extra motions that would be needed for additional expression of a gesture. For example, an erect posture commonly used in joyous gestures were mapped to move joints 2 and 4 to 30 and 150 degrees. This actuated the robot to move from slightly to fully erect position. The up and down movement simulated jumping up and down or moving hands up and down; both common human gestures for joy. Forward leaning posture in emotions such as anger and love had joint 2 positioned at 210 degrees. A collapsed position, seen in sadness, set joint 4 to 215 degrees and joint 6 tucked in.

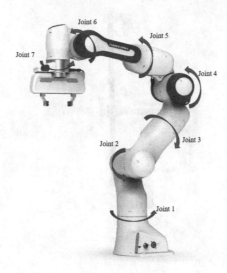

FIGURE 13.2
Panda robot with labeled joints and movements.

13.4.1.3 SeekerBot

The SeekerBot (see Figure 13.4) is a biped robot designed and built in-house by Rishikesh Daoo to portray emotions through gestures and facial expressions. The design of the robot is based on OttoBot [4], an open source platform for free education in the field of robotics. Gestures are designed by Rishikesh Daoo and used as comparison to the other robotic platforms.

Given the limited mechanical abilities of the SeekerBot, most mappings focused around expressions from the LEDs and leg movements. The legs were used for side to side motion of human bodies or head. The robot's second actuator could move itself toward and away from a stimulus. Human motions that had higher movements such as anger would be linked to the legs, as well as any forward or backward leaning posturing.Supplemental gestures were embellished with movements of the LED screen to mimic basic facial movements observed by studied works.

---

[4]https://www.ottodiy.com/

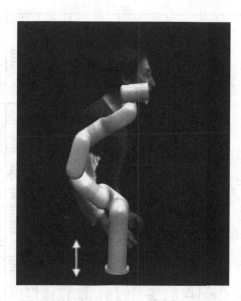

FIGURE 13.3
Example of robotic arm creating a linear profile of Dael's [58] fear position.

FIGURE 13.4
The SeekerBot.

	Co-Bot Arm (Panda)			
DoF	Human Movement	Emotion	Position Adaptation	Speed Adaptation
6	Head bent down [22]	Sadness, Shame	Joint tilts end of robot downwards	Slow
	Head bent up [22]	Joy	Joint tilts end of robot upwards	Fast
	Up-down repetitive arm motion [58]	Pride	Joint tilts end of robot upward (rest of joints give up and down appearance)	Fast
	Backwards shoulders [84]	Disgust	Joint tilts end of robot away from stimulus	Fast
5	High arm movement [58]	Fear	Twists front part of robot moving head in and out of body	Fast
	Looking away from the interactor (toward the right) [40]	Guilt	Rotates top half of robot to avoid stimulus	Slow
	Higher knee movement [58]	Pride	Joint rotates top half of robot side to side	Slow
	"Collapsed Upper Body" [58]	Sadness	Joint collapses top half of robot toward itself	Slow
4	Opening/Closing many self manipulators [7]	Fear	Collapses robot in on itself	Fast
	Collapsed position [40]	Guilt	Collapses top half of robot down	Slow
	Up-down repetitive arm motion [58]	Joy	Raises and lowers top half of robot	Fast
	Collapsed Shoulders [84]	Shame	Collapses top half of robot down	Slow
	High arm movement [81]	Pride	Joint moves top half of robot up and down	Medium
	Arms at rest [84]	Relief	Joint collapses top half of robot into itself	Fast to slow
3	Looking away from the interactor (toward the right) [40]	Guilt	Rotates top half of robot to avoid stimulus	Slow
	Higher knee movement [58]	Pride	Joint rotates top half of robot side to side	Slow
2	Weight transfer backwards [74]	Fear	Leans whole robot back	Fast
	Backwards leaning [40]	Shame	Leans robot away from stimulus	Slow
	Weight transfer backwards [22]	Disgust	Leans robot away from stimulus	Fast

Mobile Manipulator (Stretch)				
DoF	Human Movement	Emotion	Position Adaptation	Speed Adaptation
Camera	Head Bent Down [22]	Sadness	Camera looks at floor	Slow
	Looking away from the interactor (toward the right) [40]	Guilt	Camera tilts up then pans side to side	Medium speed
	Head bent up [22]	Joy	Camera tilts up	Medium speed
	Head facing down [22]	Shame	Camera tilts down	Slow
	Arms at rest [32]	Relief	Camera tilts up	Slow
Telescoping Manipulator	Low Movement Dynamics [74]	Sadness	Gripper telescopes inward	Slow
	High arm movement [74]	Fear	Gripper telescopes inward	Fast
	High arm movement [81]	Pride	Gripper telescopes slightly	Slow
	Smooth falling hands [84]	Sadness	Arm slides down to floor	Slow
	High arm movement [58]	Fear	Arm slides upward	Fast
Manipulating Elevator	Collapsed Shoulders [40]	Shame	Arm slides down to floor	Slow
	Up-down repetitive arm motion [58]	Pride	Arms slides up and down	Slow
	Arms at rest [84]	Relief	Arm slides down a bit	Slow
	High shoulder swings [22]	Disgust	Arm slides up and down	Fast
Wheel Base	Weight transfer backwards [58]	Fear	Wheels drive robot away from stimulus	Fast
	Looking away from the interactor [40]	Guilt	Wheels rotate robot away from stimulus	Medium speed
	Backwards leaning [40]	Shame	Wheels rotate robot away from stimulus	Slow
	Higher knee movement [58]	Pride	Wheels move toward stimulus	Slow
	Weight transfer backwards [22]	Disgust	Drives away from stimulus then rotates side to side	Fast

Social Robot (Seekerbot)				
DoF	**Human Movement**	**Emotion**	**Position Adaptation**	**Speed Adaptation**
Eyes	Head bent down [22]	Sadness	Eyes look down at floor	Fast
	Opening/Closing many self manipulators [7]	Fear	Eyes open wide	
	Brows lowered [40]	Guilt	Eyes look down at floor	Slow
	Cheek Raiser [21]	Joy	Eyes squint	Fast
	Head looking down [84]	Shame	Eyes look down	Slow
	Closed Eyes [81]	Pride	Narrow Eyes	Slow
	Brows are lowered [34]	Disgust	Eyes narrow toward each other	Fast
Eyelids	Inner corners of eyebrows are drawn up [34]	Sadness	Inner corners of eyebrows move upwards	Slow
	Brows are lowered [40]	Guilt	Eyelids squint inwards	Slow
Eyebrows	Inner corners of eyebrows are drawn up [34]	Sadness	Inner corners of eyebrows move upwards	Slow
	Raising of inner brows [28]	Shame	Inner corners of eyebrows move upwards	Slow
	Brows are lowered [34]	Disgust	Eyebrows squint inward	Fast
Mouth	Corners of the lips are drawn downwards [34]	Sadness	Mouth Frowns	Slow
	Fear	Fear		
	Frowning, lips stretched [40]	Guilt	Frowning	Slow
	corners of lips are drawn back and up [27,84]	Joy	Mouth smiles widely	Sudden Change
	Small [81]	Pride	Mouth smiles	Slow
	Lower lip is raised and pushed up to upper lip [34]	Disgust	Mouth frowns	

Social Robot Cont'd				
Legs	Collapsed Upper Body [58]	Sadness	Legs bend to collapse robot	Slow
	Weight transfer backwards [22]	Fear	Robot moves away from stimulus	Fast
	Looking away from the interactor (toward the right) [40]	Guilt	Legs move robot away from stimulus	Slow
	Body action: Jumping, Shape change: Expansion [74]	Joy	Legs tilt robot side to side	Rapid speed changes
	Collapsed Shoulders [84]	Shame	Legs tilt robot	Slow
	Fully visible, expanded posture [81]	Pride	Legs stand robot up and slightly tilts side to side	Slow
	Shoulders lean down [32]	Relief	Legs tilt robot side to side	Slow
	Backwards shoulders [84]	Disgust	Legs move robot away from stimulus	Fast

13.4.2   User Study

After developing a new framework for gesture creation we conducted a study to validate how humans perceive both gestural and EMP on each one of the robotic platforms. We further break down the research question into sub-questions to evaluate the system:

RQ1A *Can non-anthropomorphic robots of various structures express emotions within each emotion quadrant through gestures and emotional musical prosody?*

We hypothesize that for each platform, people will be able to interpret the simulated robotic emotions. The ability for humans to recognize emotions of other humans has been measured to be around 72% [10, 64], giving us a baseline recognition goal. We also hypothesize that the more anthropomorphic a robot is, the better it will perform in depiction of emotions .

RQ1B *How does the use of emotional musical prosody influence the recognition of emotional gestures on all platforms?*

Our hypothesis is that when combined across platforms the emotion driven prosodic voice will significantly outperform robotic performance with no audio. We also hypothesize that adding voice will increase variation between the platforms.

RQ1C *How will embedding emotional musical prosody and gesture design alter the perception of anthropomorphism, animacy, likeability perceived intelligence for different robotic platforms ?*

We expected to have no significant results for this research question as we were comparing six conditions (three robots, each with two audio types) using a between group study design. We thus posed the third question as an exploratory comparison using the widely used Godspeed metrics [12].

13.4.3   Experiments

We designed a between subjects experiment for participants to validate emotions on each robotic platform. The first part of the experiment aimed to address Research Questions 1-2. The second part analyzed the Godspeed metrics for Animacy, Anthropomorphism, Likeability and Perceived Intelligence.

As part of the experiment we asked participants to identify the emotion of different stimuli based on the GEW. We chose the GEW to match the models used for our EMP generation and validation method. The GEW provides both discreet and continuous selection of emotions. The stimuli consisted of combinations of robot EMP or no audio with each platform for a total of 9 different sets of stimuli. The stimuli consisted of the robotic platform displaying an emotion in sync with audio using the same gestures used for each group.

Due to social distancing requirements, we used videos of all three robots in front of a white background as our stimuli. Studies have shown that levels of a robot's presence affects some variables in human–robot interactions [9, 79]. We believe having all the robotic platforms at the same level of presence, through videos, would mitigate this effect.

After completing a consent form, participants were introduced to the GEW and were given a test question to teach them the layout of a GEW. The test question required selecting a specific emotion without which participants were not allowed to continued. Participants were then shown one of the stimuli with audio, such as Stretch with robot EMP or SeekerBot with no EMP. This was followed by completing the Godspeed test. Participants then viewed and rated on the GEW a second set of stimuli of a different platform without audio. No part of the stimuli was repeated for any participant. The average study length was 12 minutes.

For the experiment, 150 participants were recruited, with 11 rejected for incorrect answers on attention checks, leaving a total of 139 participants. From the participants used in the study 89 identified as male, while the other 50 identified as female. 101 were currently in the United States, 30 in India, with the remaining 8 spread between Ireland, Brazil, and Thailand. For both studies the mean age was 41 with a standard deviation of 11 and a range of 18–77. We found no statistically significant difference between demographics. The study was 15 minutes long and paid USD\$2.00 per study. The study was approved by the university Institutional Review Board.

13.5   Results

13.5.1   RQ1A: Emotion Validation

We found high ratings for accuracy for emotion detection in three of the four quadrants for gestures and audio. For the Panda arm and SeekerBot participants consistently score over 85% accuracy, achieving higher accuracy than the 72% achieved by humans recognizing other human faces. For the high valence / low arousal quadrant emotion detection accuracy was significantly lower. However participants still scored consistently above random. Across the majority of categories the Stretch performed worse than the other two robots. Figure 13.5 shows the percentage of accurate results for EMP and gesture combined.

We first conducted a one-way Anova comparing between the three robots, with the results *(p <0.001, f = 14.335)*. A summary of related statistics is presented in Table 13.1. We then conducted a post-hoc test, using Tukey Honestly Significant Difference. We found significant difference in how clearly the emotions were perceived between Arm and the Stretch and the Arm and the Seekerbot, implying the audio and gesture design was more effective on these platforms.

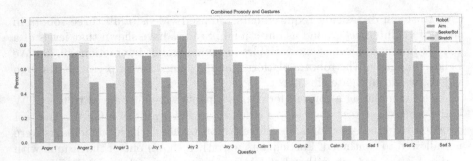

FIGURE 13.5
Combined EMP and gesture emotion recognition (dashed line indicates 72%).

TABLE 13.1
One-way ANOVA statistic summary.

Robot	N	Mean	SD	SE	95% Conf.	Interval
Arm	36	0.7326	0.1668	0.0278	0.6762	0.7890
SeekerBot	36	0.7426	0.2248	0.0375	0.6665	0.8187
Stretch	36	0.5163	0.2116	0.0353	0.4447	0.5879

13.5.2 RQ1B: EMP and Gesture Comparison

We found no significant results between EMP and gesture for the SeekerBot or the Panda arm (p >0.05). For the Stretch, EMP improved the recognition of emotion across all 12 questions. A pair-wise t-test for EMP and gesture returned $p = 0.035$, $f = 0.04$, indicating a significant result. Figure 13.6 shows the accuracy for gesture alone, and Figure 13.7 show the results for EMP. Figure 13.8 shows a box-plot comparing between platforms.

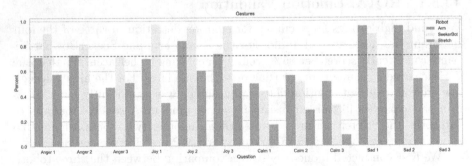

FIGURE 13.6
Gesture emotion recognition (dashed line indicates 72%).

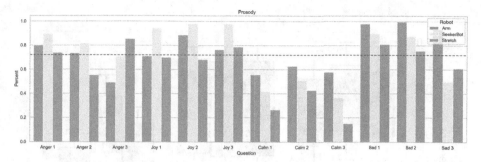

FIGURE 13.7

EMP emotion recognition (dashed line indicates 72%).

13.5.3 RQ1C: HRI Metrics

We first calculated Cronbach's alpha for each question, with each metric over >0.85, indicating high internal reliability. After Holm–Bonferroni correction we found no significance between any audio and gesture comparison. Nevertheless, between platforms there was wide variability while EMP did consistently score higher than gesture alone.

13.6 Discussion

13.6.1 Emotion Recognition

For every quadrant except for Calm (low valance / low arousal), participants correctly identified the quadrant. The SeekerBot performed best for quadrants 1 and 3 (Joy and Anger), while the Panda arm had the best recognition for quadrants 2 and 4 (Calm and Sadness). The Stretch performed significantly worse in all quadrants; matching our hypothesis. The Seekerbot and Panda arm received closer ratings than the Panda arm and Stretch. We theorize that this can be explained by the fact that the SeekerBot and Panda arm have a similar number of degrees of freedom. Anger and Joy quadrants contain emotions that are highly correlated to facial movements. The Seekerbot, being the only robot with facial display, improved its expression in those quadrants.

The fourth quadrant (Calm) had the least accurate emotion identification results. While EMP helped improve emotion identification even in this quadrant, all three robotic adaptations struggled to effectively express their emotions. One possible explanation may be due to the lesser amount of gestures specifically related to emotions of this quadrant.Emotions in the fourth quadrant such as love and relief can have more personal interpenetrations, which might indicate that other researchers had similar trouble correlating specific gestures and rules for emotions within the fourth quadrant.

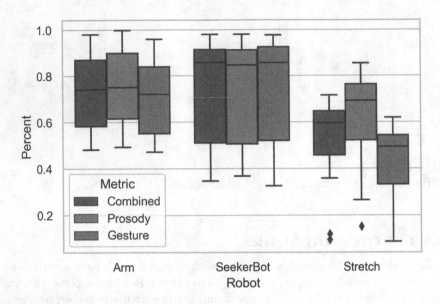

FIGURE 13.8
Comparison between accuracy for EMP and gesture on each platform.

13.6.2   Emotional Musical Prosody

While insignificant when tested across all gestures, there was a slight improvement with the Panda arm when EMP was added. The Panda performed worse than SeekerBot for no audio but was not significantly different from the SeekerBot for either EMP group. Because some gestures performed worse than others, as evidenced by Figure 13.9 we speculate that EMP was helpful for detecting emotions that are more difficult to identify from gestures alone. The accuracy for SeekerBot did not improve when EMP was added. This might imply that its face, which the other robots lacked, provided enough information that significant improvement was not achievable.

The Stretch robot was the only platform that showed significant improvement in emotion detection when EMP was added. One possible explanation for this could be that the Stretch design has the least emotion conveying anthropomorphic affordances, leading to more challenges in producing subtle differences between emotions. Our gesture design guidelines generalize the movements for each emotion, where subtleties between emotions can be more easily tuned with more degrees of freedom. However, without enough degrees of freedom, some emotions may have appeared too similar to each other. We hypothesize that the addition of EMP to robots that are otherwise less anthropomorphic and expressive, such as the Stretch, can improve and enrich subtleties in emotion conveyance.

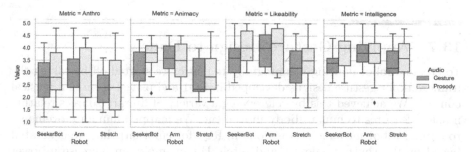

FIGURE 13.9
Godspeed metrics between platforms.

13.6.3 Godspeed Metrics

For every category except perceived intelligence, the Panda arm performed the best. While understanding motivation for godspeed metrics can be challenged, we believe that The Panda performed best due to the increased variety in degrees of freedom. The Panda Arm has 7 high ranged Degrees of freedom (more than any other the other bots). We suggest that this allows for increased variety of expression and can enable a robot to more accurately simulate a specific gesture. Increased expressivity would correlate well with a perception of animacy, where a robot with the least degrees of freedom had the lowest ratings out of the three. The difference in degrees of freedom between SeekerBot and the Panda arm are small, and therefore, they received similar ratings. We theorize that the Panda arm has more variety in its uses of each degree of freedom where some of the SeekerBot movements are more limited.

13.6.4 Limitations

In order to comply with social distancing, all studies were conducted online. This limited our opportunity to evaluate the effects of physical interactions with each robot. Future work could look at how the sizes of each robot can affect the users perception. While using the GEW as our emotion model gave users a variety of options to choose from, we suspect it gave the participants a potentially overwhelming amount of options. For example, some participants could have choosen the term pleasure instead of joy. While these terms are next to each other in the GEW it shows that their recognition was accurate, but not exact.

After the development of new movements, we looked to apply them to more intricate dances. We combined these gestures to create a series of dance steps, that we then combined to make complete dances.

13.7   Applying Gesture Design to Dance

To create a robot dancing meaningfully to music, we started off by taking the framework for gesture designs developed in RQ1 and played them in a musical context. We achieved this by referencing a variety of literary sources relating musical features to human body movement. We mapped human movements to similar robot movements (as done in RQ1 for emotion) that we generated based on the rhythm, energy, and pitch. For example, an increase in lower frequency spectral flux results in an increase in head movements. We used this as a basis for programming a robotic arm to react in specific ways to music. We created dances in three systems:

- Pick a dance that best matches the music playing

- Generate new dances in real-time

- A combination of both systems

A successful system would show an improvement in the Performance Competence Evaluation Measure [43], and HRI godspeed metric. In addition, a successful system would show an observer that the robots can improvise and react artistically to music. We first developed a system to pick a dance from a gesture database. After we performed basic evaluations, we use this feedback to help improve the development of the next system. After performing another set of evaluations, we combined both systems and conducted a final study to compare all the three systems.

13.7.1   System 1: Dance Selection Based on Music

We used Burger's work mapping human movement to musical features to develop a mappings table that links musical features to human and robot movements. The robot movements were mapped based on our emotional gesture mappings in Research Question 1 and a literary review of publications relating musical features to movement [17,26,30,49,50,57,62,76,82]. Table 13.2 shows the correlations between musical features and human body movements.

We used a set of music information retrieval libraries to extract song information. We used Madmom's recursive neural net to detect onset and rhythmic information, Numpy to detect frequency features as well as RMS, and Msaf to section the piece. Librosa performed source separation for instrument changes. The Msaf library determined the sections for each piece. We chose these libraries for their accuracy throughout a variety of songs and genres.

TABLE 13.2

Table showing mappings of musical features with human and robotic motion.

Musical Feature	Human Movement	Robot equivalent Movement	Relationship
Low frequency spectral flux	Head Movement [17,45]	Joint 5, 6	Speed of movement increases with higher flux
High frequency spectral flux	Head Movement [17]	Joint 2, 4	Speed of movement increases with higher flux
	Hand Movement [17,26]	Joint 3, 4, 5	Hand distance and amount of movement increases with higher flux
Onset strength	Center of Mass [62]	Joint 2 and 4	Speed and distance travel increases with higher onset strength
	Shoulder movement [17]	Joint 4	Speed of motion and amount of shoulder wiggle increases with higher onset strength
Percusiveness (envelope slope)	Center of mass [50,62]	Joint 2, 4	Speed of body motion and distance increases with more percusiveness
	Shoulder [17,82]	Joint 4, 5	Speed of motion and amount of shoulder wiggle increases with higher percusiveness
	Hand movement [17,76,82]	Joint 3, 4, 5	Hand distance and amount of movement increases with higher percusiveness
	Head movement [17,62]	Joint 5, 6	Speed of movement and amount of head bobs increases
Energy (RMS)	Head movement [62]	Joint 5, 6	Distance of movement increases with higher RMS
	Body movement [62]	Joint 2, 3, 4	Distance of movement increases with higher RMS

13.7.2 Musical Feature Recognition

The musical features were detected using a variety of machine learning algorithms. To analyze the musical features (for any system), a song was first blocked, and FFT's were performed. The audio information was sent to each respective library to determine sections and rhythmic information. We proceeded to generate values for all desired musical features every five milliseconds. We then normalized these values and calculate the differential of each musical feature. Each gesture was selected based on the values for each section.

Music analysis was performed offline before generating a dance to get more accurate data than real-time music information retrieval. After we analyze a song once, we can generate multiple dances with (the same or) different systems quickly in real time. Analyzing music offline gave us the opportunity to look ahead of the song and time section changes accordingly with robot dances.

13.7.3 Robot Gesture Database

A dance is composed of a series of gestures, which can be further broken down into joint movements. To create a gesture, we combined joint movements that can be designed by inputting the change in degrees of each joint, the time to start this motion, and the time to finish this joint movement. Each robot gesture was then scaled based on the calculated BPM of a song. A script was designed to convert desired joint movements into smooth robot trajectory plots that also included follow through.

We used the emotional gestures developed in RQ1 as well as a variety of dance gestures that had been derived from the emotional gestures by members of the music-technology department at Georgia Tech. Such gestures were used in previous performances and developed by dancers. Each dance had a starting position, number of robots involved, duration, and max speed. When a dance is selected, the gesture speed was modified to match the calculated BPM of the song. The trajectory of each dance used a 5th order polynomial (as determined in RQ1) and included follow through. Dances were sorted based on their max speed, maximum movement of each joint, and amount of robot total movement (total displacement of movement in each joint).

13.7.4   Decision Tree for Dance Selection

Dance selection was performed iteratively over each section of the input song. We first looked at the total duration. If a section was longer than our longest gesture in the database, we selected an additional robot gestures to increase diversity of the dance. We then looked at the following analyses for each musical feature: Average value, maximum change in value, range of value, time difference between max change.

We then selected a robot gesture from the gesture database that was most similar to the values of music features. We did this by going through each musical features and selecting a group that best matches the features. For example, the section with the highest RMS selects one third of the gestures with the biggest amplitude changes in joints. We then checked the next musical feature and took one third of this new group based on the features values. The decision tree below (Figure 13.10) shows the process for selecting gestures in a piece.

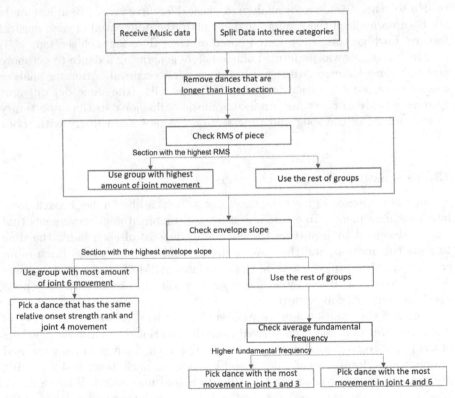

FIGURE 13.10
Flowchart of selecting a dance to best match the musical features.

Every iteration of a section took a third of the dance gestures from the database that best matched the music feature desired. We prioritized features with the most correlations to human movement as the first set of filtering. This ensured that the strongest links between movement and music are most accurately depicted in the selected gesture. We repeated this process for each of the sections in a song.

13.8 System 2: Dance Generation Based on Music

13.8.1 Generate New Dances in Real-time

The original inspiration of linking music to movement started with human movements that are inspired by music. However, these correlations did not guarantee mappings that audience members can easily identify. Using qualitative feedback received from informal interviews, we improved the choice of musical features and and created a system that generated and manipulated live trajectories for each section of a song. We modified the robot section of all mappings to focus on the frequency and amplitude of continuous movements, rather than discreet motions.

13.8.2 Mapping Musical Features to Robot Movement

We first looked to improve our dance generator by modifying the mappings from musical features to robot movements. Krumhan's work showed that humans can best correlate simple musical features to body movements [44]. In our new dance generation system, we chose to only use RMS, tempo, and song section as they were the most recognizable features for non-musicians [42]. For trajectory design, we set joint angles to follow a sine wave that changed in amplitude and frequency based on musical features. The robotic arm had a set of unique poses that it could go to after each section. Then, a weighted probability table determined which joint would start moving next, at a given amplitude and frequency based on a musical feature. The robots design allows different aesthetics based on which joint is moving. The joint moved was chosen based on the fundamental frequency; where a lower fundamental frequencies would move a joint closer to the head joint 5 and 6. RMS value increased the amplitude of a joints movement and the frequency of the sine wave was determined by the tempo.

13.9 System 3: Combination of System 1 and 2

13.9.1 Dance Generation

A third system of dance generation was created in an attempt to combine the positive feedback from both dance generators. This version utilized the whole group of robots by having a number of robots play visually aesthetic gestures(from system 1), and having at least one robot moving with simple mappings from system 2. Each section used the average RMS to determine how many robots would play choreographed gestures, and how many were expressing musical features. Based on the layout and number of robots selected, there were set patterns for which specific robot to play.

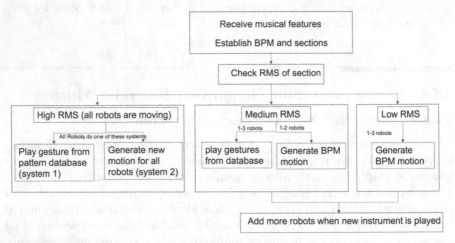

FIGURE 13.11
Dance metrics for full body and body part separation.

As shown in Figure 13.11, the decision tree for the third version made sure that there was always one robot to keep a connection to music playing. The other robots had more complex gestures to keep the choreography engaging.

13.9.2   Discussion

Throughout the design of each system, there was a common trade-off between having intricate gestures and establishing an evident connection from robot movement and music playing. Simple robot movements required a user to make less mental connections in order to see what the robot is doing. However, the simple gestures were not very interesting to the viewer. On the other hand, too many complex gestures tended to be less discreet and more distracting, thereby decreasing the connection between the gestures and the music. Group robot dances were helpful in this aspect because different robots could be doing different gestures. Having at least one robot do simple gestures establishes the connection with the music, while the rest of the robots can do gestures with more engaging and elaborate gestures. According to the analysis, most dances that used the third model had similarly high ratings for HRI and dance metrics, despite working very differently. This could potentially be a result of the impressiveness of a dancing robot arm. Many participants saw dancing robots for the first time in this study, and therefore, may not have picked up on all of the subtle changes. When comparing the second and the third studies, there was increased variety in opinions of the robot. This showed that participants can pick up on subtle differences in dance design.

13.10 Conclusion and Future Work

We presented a comparison between three robotic platforms using EMP and gestures to convey emotions. The studies we presented can suggest broad guidelines and a framework for improving emotion conveyance for non-anthropomorphic robots. Our use of EMP increased the accuracy of identifying a correct gestural emotion and lowered the overall variance. Our results showed that using EMP and gestures gave a recognition rate better than human face detection for three of the four valance-arousal quadrants. Overall, adding EMP was most significant for improving emotion conveyance for the least anthropomorphic robot – The Stretch, followed by the Panda arm, while only causing minor changes for the more social SeekerBot. This reinforces the idea that that robots should have a matching artificial sounding voice to increase their overall perceptions [55]. Each of these gestures was put in a database, for the robots to generate dances based on music

In future work we will continue to improve the emotional gesture portrayal on all platforms through an iterative design experiment methodology, revising and testing our gestures in consecutive user studies. We also plan to implement our gesture design to evaluate specific emotions, and explore the effects of gestures as lube reactions to human input. We also want to further explore new gesture generations methods with a human dancer in hopes of developing a human–robot interactive system. We will use this framework to start developing other types of gestures, and apply these guidelines to implicit gestures. We believe that for future developments in emotion and robotics it is crucial to consider studies with multiple robotic platforms.

Bibliography

[1] AÉRAÏZ-BEKKIS, D., GANESH, G., YOSHIDA, E., AND YAMANOBE, N. Robot movement uncertainty determines human discomfort in co-worker scenarios. In *2020 6th International Conference on Control, Automation and Robotics (ICCAR)* (2020), IEEE, pp. 59–66.

[2] ALCUBILLA TROUGHTON, I., BARAKA, K., HINDRIKS, K., AND BLEEKER, M. Robotic improvisers: Rule-based improvisation and emergent behaviour in hri. In *Proceedings of the 2022 ACM/IEEE International Conference on Human-Robot Interaction* (2022), pp. 561–569.

[3] ALEMI, O., FRANÇOISE, J., AND PASQUIER, P. Groovenet: Real-time music-driven dance movement generation using artificial neural networks. *Networks 8*, 17 (2017), 26.

[4] ALEMI, O., AND PASQUIER, P. Machine learning for data-driven movement generation: a review of the state of the art. *arXiv preprint arXiv:1903.08356* (2019).

[5] ANDO, T., AND KANOH, M. A self-sufficiency model using urge system. In *International Conference on Fuzzy Systems* (2010), IEEE, pp. 1–6.

[6] ARKIN, R., AND ULAM, P. An ethical adaptor: Behavioral modification derived from moral emotions. pp. 381–387.

[7] ATKINSON, A. P., DITTRICH, W. H., GEMMELL, A. J., AND YOUNG, A. W. Emotion perception from dynamic and static body expressions in point-light and full-light displays. *Perception 33*, 6 (2004), 717–746. PMID: 15330366.

[8] AUGELLO, A., CIPOLLA, E., INFANTINO, I., MANFRE, A., PILATO, G., AND VELLA, F. Creative robot dance with variational encoder. *arXiv preprint arXiv:1707.01489* (2017).

[9] BAINBRIDGE, W., HART, J., KIM, E., AND SCASSELLATI, B. The effect of presence on human-robot interaction. pp. 701–706.

[10] BÄNZIGER, T., MORTILLARO, M., AND SCHERER, K. R. Introducing the geneva multimodal expression corpus for experimental research on emotion perception. *Emotion 12*, 5 (2012), 1161.

[11] BARTNECK, C. How convincing is mr. data's smile: Affective expressions of machines. *User Modeling and User-Adapted Interaction 11*, 4 (2001), 279–295.

[12] BARTNECK, C., KULIĆ, D., CROFT, E., AND ZOGHBI, S. Measurement instruments for the anthropomorphism, animacy, likeability, perceived intelligence, and perceived safety of robots. *International Journal of Social Robotics 1*, 1 (2009), 71–81.

[13] BARTNECK, C., REICHENBACH, J., AND BREEMEN, V. A. In your face, robot! the influence of a character's embodiment on how users perceive its emotional expressions.

[14] BREAZEAL, C. Emotive qualities in robot speech. vol. 3, pp. 1388–1394 vol.3.

[15] BREAZEAL, C., AND ARYANANDA, L. Recognition of affective communicative intent in robot-directed speech. *Autonomous Robots 12*, 1 (2002), 83–104.

[16] BRETAN, M., HOFFMAN, G., AND WEINBERG, G. Emotionally expressive dynamic physical behaviors in robots. *International Journal of Human-Computer Studies 78* (2015), 1–16.

[17] BURGER, B., SAARIKALLIO, S., LUCK, G., THOMPSON, M. R., AND TOIVIAINEN, P. Relationships between perceived emotions in music and music-induced movement. *Music Perception: An Interdisciplinary Journal 30*, 5 (June 2013), 517–533.

[18] BURGER, B., THOMPSON, M., LUCK, G., SAARIKALLIO, S., AND TOIVIAINEN, P. Influences of rhythm- and timbre-related musical features on characteristics of music-induced movement. *Frontiers in Psychology 4* (2013), 183.

[19] CANAMERO, L. D., AND FREDSLUND, J. How does it feel? emotional interaction with a humanoid lego robot. In *Proceedings of American Association for Artificial Intelligence Fall Symposium, FS-00-04* (2000), pp. 7–16.

[20] CHA, E., KIM, Y., FONG, T., MATARIC, M. J., ET AL. A survey of nonverbal signaling methods for non-humanoid robots. *Foundations and Trends® in Robotics 6*, 4 (2018), 211–323.

[21] COHN, J. F., AND EKMAN, P. Measuring facial action.

[22] COULSON, M. Attributing emotion to static body postures: Recognition accuracy, confusions, and viewpoint dependence. *Journal of Nonverbal Behavior 28*, 2 (2004), 117–139.

[23] CROSS, I. Music, cognition, culture, and evolution. *Annals of the New York Academy of Sciences 930*, 1 (2001), 28–42.

[24] CRUMPTON, J., AND BETHEL, C. L. A survey of using vocal prosody to convey emotion in robot speech. *International Journal of Social Robotics 8*, 2 (2016), 271–285.

[25] DAUTENHAHN, K., WOODS, S., KAOURI, C., WALTERS, M. L., KHENG LEE KOAY, AND WERRY, I. What is a robot companion - friend, assistant or butler? In *2005 IEEE/RSJ International Conference on Intelligent Robots and Systems* (2005), pp. 1192–1197.

[26] DAVIDSON, J. W. Bodily movement and facial actions in expressive musical performance by solo and duo instrumentalists: Two distinctive case studies. *Psychology of Music 40*, 5 (2012), 595–633.

[27] EKMAN, P. Facial expression and emotion. *American Psychologist 48*, 4 (1993), 384.

[28] EKMAN, P. *Approaches to Emotion*. Psychology Press, 2014.

[29] ERDEN, M. S. Emotional postures for the humanoid-robot nao. *International Journal of Social Robotics 5*, 4 (2013), 441–456.

[30] GEMEINBOECK, P., AND SAUNDERS, R. Movement matters: How a robot becomes body. In *Proceedings of the 4th International Conference on Movement Computing* (2017), pp. 1–8.

[31] GOETZ, J., KIESLER, S., AND POWERS, A. Matching robot appearance and behavior to tasks to improve human-robot cooperation. In *The 12th IEEE International Workshop on Robot and Human Interactive Communication, 2003. Proceedings. ROMAN 2003.* (2003), pp. 55–60.

[32] GROSS, M. M., CRANE, E. A., AND FREDRICKSON, B. L. Effort-shape and kinematic assessment of bodily expression of emotion during gait. *Human Movement Science 31*, 1 (2012), 202–221.

[33] GUIZZO, E. By leaps and bounds: An exclusive look at how boston dynamics is redefining robot agility. *IEEE Spectrum 56*, 12 (2019), 34–39.

[34] GUNES, H., AND PICCARDI, M. Bi-modal emotion recognition from expressive face and body gestures. *Journal of Network and Computer Applications 30*, 4 (2007), 1334–1345.

[35] HAGENDOORN, I. Cognitive dance improvisation: How study of the motor system can inspire dance (and vice versa). *Leonardo 36*, 3 (2003), 221–228.

[36] HOFFMAN, G., AND BREAZEAL, C. Anticipatory perceptual simulation for human-robot joint practice: Theory and application study. In *AAAI* (2008), pp. 1357–1362.

[37] JANATA, P., TOMIC, S. T., AND HABERMAN, J. M. Sensorimotor coupling in music and the psychology of the groove. *Journal of Experimental Psychology: General 141*, 1 (2012), 54.

[38] JOCHUM, E., AND DERKS, J. Tonight we improvise! real-time tracking for human-robot improvisational dance. In *Proceedings of the 6th International Conference on Movement and Computing* (2019), pp. 1–11.

[39] JOSHI, M., AND CHAKRABARTY, S. An extensive review of computational dance automation techniques and applications. *Proceedings of the Royal Society A 477*, 2251 (2021), 20210071.

[40] JULLE-DANIÈRE, E., WHITEHOUSE, J., MIELKE, A., VRIJ, A., GUSTAFSSON, E., MICHELETTA, J., AND WALLER, B. M. Are there non-verbal signals of guilt? *PloS One 15*, 4 (2020), e0231756.

[41] KAUSHIK, R., AND LAVIERS, A. Imitating human movement using a measure of verticality to animate low degree-of-freedom non-humanoid virtual characters. In *International Conference on Social Robotics* (2018), Springer, pp. 588–598.

[42] KIM, J.-Y., AND BELKIN, N. J. Categories of music description and search terms and phrases used by non-music experts. In *ISMIR* (2002), vol. 2, pp. 209–214.

[43] KRASNOW, D., AND CHATFIELD, S. J. Development of the "performance competence evaluation measure": assessing qualitative aspects of dance performance. *Journal of Dance Medicine & Science 13*, 4 (2009), 101–107.

[44] KRUMHANSL, C. Musical tension: Cognitive, motional and emotional aspects. In *Proceedings of the 3rd Triennial ESCOM Conference* (1997).

[45] KRUMHUBER, E., MANSTEAD, A. S., AND KAPPAS, A. Temporal aspects of facial displays in person and expression perception: The effects of smile dynamics, head-tilt, and gender. *Journal of Nonverbal Behavior 31*, 1 (2007), 39–56.

[46] LEMAN, M., ET AL. *Embodied Music Cognition and Mediation Technology*. MIT Press, 2008.

[47] LI, Y., ISHI, C., WARD, N., INOUE, K., NAKAMURA, S., TAKANASHI, K., AND KAWAHARA, T. Emotion recognition by combining prosody and sentiment analysis for expressing reactive emotion by humanoid robot. pp. 1356–1359.

[48] LÓPEZ RECIO, D., MÁRQUEZ SEGURA, L., MÁRQUEZ SEGURA, E., AND WAERN, A. The nao models for the elderly. In *2013 8th ACM/IEEE International Conference on Human-Robot Interaction (HRI)* (2013), pp. 187–188.

[49] MADISON, G., GOUYON, F., ULLÉN, F., AND HÖRNSTRÖM, K. Modeling the tendency for music to induce movement in humans: first correlations with low-level audio descriptors across music genres. *Journal of Experimental Psychology: Human Perception and Performance 37*, 5 (2011), 1578.

[50] MARTÍNEZ, I. C., AND EPELE, J. Embodiment in dance-relationships between expert intentional movement and music in ballet. In *ESCOM 2009: 7th Triennial Conference of European Society for the Cognitive Sciences of Music* (2009).

[51] MCNEILL, D. *Gesture and Thought*. University of Chicago Press, 2008.

[52] MICHALOWSKI, M. P., SABANOVIC, S., AND KOZIMA, H. A dancing robot for rhythmic social interaction. In *Human-Robot Interaction (HRI), 2007 2nd ACM/IEEE International Conference on* (2007), IEEE, pp. 89–96.

[53] MICHAUD, F., AUDET, J., LÉTOURNEAU, D., LUSSIER, L., THÉBERGE-TURMEL, C., AND CARON, S. Experiences with an autonomous robot attending aaai. *IEEE Intelligent Systems 16*, 5 (2001), 23–29.

[54] MONCEAUX, J., BECKER, J., BOUDIER, C., AND MAZEL, A. Demonstration: first steps in emotional expression of the humanoid robot nao. In *Proceedings of the 2009 International Conference on Multimodal Interfaces* (2009), ACM, pp. 235–236.

[55] MOORE, R. K. Is spoken language all-or-nothing? implications for future speech-based human-machine interaction. In *Dialogues with Social Robots.* Springer, 2017, pp. 281–291.

[56] NANTY, A., AND GELIN, R. Fuzzy controlled pad emotional state of a nao robot. In *2013 Conference on Technologies and Applications of Artificial Intelligence* (2013), IEEE, pp. 90–96.

[57] NAVEDA, L., AND LEMAN, M. The spatiotemporal representation of dance and music gestures using topological gesture analysis (tga). *Music Perception 28*, 1 (2010), 93–111.

[58] NELE DAEL, M. M., AND SCHERER, K. R. The body action and posture coding system (bap): Development and reliability. *Journal of Nonverbal Behavior 36* (2012), 97–121.

[59] NOVIKOVA, J., AND WATTS, L. A design model of emotional body expressions in non-humanoid robots. In *Proceedings of the Second International Conference on Human-Agent Interaction* (New York, NY, USA, 2014), HAI '14, Association for Computing Machinery, p. 353–360.

[60] OLIVEIRA, J. L., INCE, G., NAKAMURA, K., NAKADAI, K., OKUNO, H. G., REIS, L. P., AND GOUYON, F. An active audition framework for auditory-driven hri: Application to interactive robot dancing. In *2012 IEEE RO-MAN: The 21st IEEE International Symposium on Robot and Human Interactive Communication* (2012), IEEE, pp. 1078–1085.

[61] OLIVEIRA, J. L., REIS, L. P., FARIA, B. M., AND GOUYON, F. An empiric evaluation of a real-time robot dancing framework based on multi-modal events. *TELKOMNIKA Indonesian Journal of Electrical Engineering 10*, 8 (2012), 1917–1928.

[62] PHILLIPS-SILVER, J., AND TRAINOR, L. J. Vestibular influence on auditory metrical interpretation. *Brain and Cognition 67*, 1 (2008), 94–102.

[63] READ, R. G., AND BELPAEME, T. Interpreting non-linguistic utterances by robots: studying the influence of physical appearance. In *Proceedings of the 3rd international workshop on Affective Interaction in Natural Environments* (2010), ACM, pp. 65–70.

[64] RECIO, G., SCHACHT, A., AND SOMMER, W. Recognizing dynamic facial expressions of emotion: Specificity and intensity effects in event-related brain potentials. *Biological Psychology 96* (2014), 111–125.

[65] ROGEL, A., SAVERY, R., YANG, N., AND WEINBERG, G. Robogroove: Creating fluid motion for dancing robotic arms. In *Proceedings of the 8th International Conference on Movement and Computing* (2022), pp. 1–9.

[66] SAINT-AIMÉ, S., LE PÉVÉDIC, B., AND DUHAUT, D. Emotirob: an emotional interaction model. In *RO-MAN 2008-The 17th IEEE International Symposium on Robot and Human Interactive Communication* (2008), IEEE, pp. 89–94.

[67] SALEM, M., KOPP, S., WACHSMUTH, I., ROHLFING, K., AND JOUBLIN, F. Generation and evaluation of communicative robot gesture. *International Journal of Social Robotics 4*, 2 (2012), 201–217.

[68] SANTIAGO, C. B., OLIVEIRA, J. L., REIS, L. P., AND SOUSA, A. Autonomous robot dancing synchronized to musical rhythmic stimuli. In *6th Iberian Conference on Information Systems and Technologies (CISTI 2011)* (2011), IEEE, pp. 1–6.

[69] SAVERY, R., ROSE, R., AND WEINBERG, G. Establishing human-robot trust through music-driven robotic emotion prosody and gesture. In *2019 28th IEEE International Conference on Robot and Human Interactive Communication (RO-MAN)* (2019), IEEE, pp. 1–7.

[70] SAVERY, R., AND WEINBERG, G. A survey of robotics and emotion: Classifications and models of emotional interaction. In *Proceedings of the 29th International Conference on Robot and Human Interactive Communication* (2020).

[71] SAVERY, R., ZAHRAY, L., AND WEINBERG, G. Emotional musical prosody for the enhancement of trust in robotic arm communication. In *Trust, Acceptance and Social Cues in Human-Robot Interaction, RO-MAN 2020* (2020).

[72] SAVERY, R., ZAHRAY, L., AND WEINBERG, G. Emotional musical prosody for the enhancement of trust in robotic arm communication. In *Trust, Acceptance and Social Cues in Human-Robot Interaction, RO-MAN 2020* (2020).

[73] SAVERY, R., ZAHRAY, L., AND WEINBERG, G. Shimon the rapper: A real-time system for human-robot interactive rap battles. *arXiv preprint arXiv:2009.09234* (2020).

[74] SHAFIR, T., TSACHOR, R. P., AND WELCH, K. B. Emotion regulation through movement: Unique sets of movement characteristics are associated with and enhance basic emotions. *Frontiers in Psychology 6* (2016), 2030.

[75] SHIRATORI, T., AND IKEUCHI, K. Synthesis of dance performance based on analyses of human motion and music. *Information and Media Technologies 3*, 4 (2008), 834–847.

[76] SIEVERS, B., POLANSKY, L., CASEY, M., AND WHEATLEY, T. Music and movement share a dynamic structure that supports universal expressions of emotion. *Proceedings of the National Academy of Sciences 110*, 1 (jan 2013), 70–75.

[77] SLOBODA, J. Music: Where cognition and emotion meet. In *Conference Proceedings: Opening the Umbrella; an Encompassing View of Music Education; Australian Society for Music Education, XII National Conference, University of Sydney, NSW, Australia, 09-13 July 1999* (1999), Australian Society for Music Education, p. 175.

[78] SOUSA, P., OLIVEIRA, J. L., REIS, L. P., AND GOUYON, F. Humanized robot dancing: humanoid motion retargeting based in a metrical representation of human dance styles. In *Portuguese Conference on Artificial Intelligence* (2011), Springer, pp. 392–406.

[79] THIMMESCH-GILL, Z., HARDER, K., AND KOUTSTAAL, W. Perceiving emotions in robot body language: Acute stress heightens sensitivity to negativity while attenuating sensitivity to arousal. *Computers in Human Behavior 76* (06 2017).

[80] THIMMESCH-GILL, Z., HARDER, K. A., AND KOUTSTAAL, W. Perceiving emotions in robot body language: Acute stress heightens sensitivity to negativity while attenuating sensitivity to arousal. *Computers in Human Behavior 76* (2017), 59–67.

[81] TRACY, J. L., AND ROBINS, R. W. Show your pride: Evidence for a discrete emotion expression. *Psychological Science 15*, 3 (2004), 194–197.

[82] VAN DYCK, E., MOELANTS, D., DEMEY, M., COUSSEMENT, P., DEWEPPE, A., AND LEMAN, M. The impact of the bass drum on body movement in spontaneous dance. In *Proceedings of the 11th International Conference in Music Perception and Cognition* (2010), pp. 429–434.

[83] VÁSQUEZ, B. P. E. A., AND MATÍA, F. A tour-guide robot: moving towards interaction with humans. *Engineering Applications of Artificial Intelligence 88* (2020), 103356.

[84] WALBOTT, H. G. Bodily expression of emotion. *European Journal of Social Psychology 28*, 6 (1998), 879–896.

[85] WALLIS, M., POPAT, S., MCKINNEY, J., BRYDEN, J., AND HOGG, D. C. Embodied conversations: performance and the design of a robotic dancing partner. *Design Studies 31*, 2 (2010), 99–117.

[86] XIA, G., TAY, J., DANNENBERG, R., AND VELOSO, M. Autonomous robot dancing driven by beats and emotions of music. In *Proceedings of the 11th International Conference on Autonomous Agents and Multiagent Systems-Volume 1* (2012), pp. 205–212.

14

Dead Stars and "Live" Singers: Posthumous "Holographic" Performances in the US and Japan

Yuji Sone

14.1   Introduction

This essay examines the recent controversy concerning the modelling of performing "holograms" (a concept I will qualify shortly) on deceased singers in the USA and in Japan. Since the 2012 digital recreation of hip-hop artist Tupac Shakur, who died in 1996, "live" concerts starring the holographic doubles of late, well-known singers such as Michael Jackson (2014) and Whitney Houston (2020) have been organized in the USA. In Japan, the public broadcaster NHK collaborated with Yamaha to produce a concert in 2019 that featured "AI Misora Hibari", a synthetic double of the late Japanese singer Misora Hibari. Misora, who rose to fame in the period following World War II, is regarded as one of Japan's greatest singers of the 20th century.[1] In this essay, I examine how

---

[1] I use Japanese names in this essay in the Japanese manner: family name first, followed by given name. Long vowel sounds are indicated by macrons, unless the word is in common usage in Romanised form.

DOI: 10.1201/9781003320470-14

the reception of AI Misora Hibari's performance paralleled and diverged from the reception of some of its Western counterparts, referring to the debates that sprang from live performances featuring the digital double of Tupac Shakur.

Existing discussions on holographic performances of deceased musicians who were famous in Western popular music generally focus on ethical concerns and the "inauthenticity" of such performances, and attend to a visual paradigm. Debates around these performances have tended to focus on the visual representations of artists. As well, such performances' critics in the Western context have seemed preoccupied with the question of future holographic technologies, and have raised concerns for technological encroachment on artists' rights, and emerging commercial interests. In contrast, I will discuss the positive reception in Japan for AI Misora Hibari by the people who were closest to the late singer and her fans, highlighting culturally-specific, yet potentially hidden, characteristics of the posthumous holographic performance of the song *"Arekara* (Ever Since Then)" (2019). *"Arekara"* is an *enka* song, a type of popular ballad that emphasizes emotion and, importantly, nostalgic sentiment. I will explain in what ways I see AI Misora Hibari as a metonymical sign for Japan's post-war recovery, rapid economic growth in the mid-1950s to the early '70s, and its relative economic decline since the 1990s. I will argue how such sentiment meshes with a Japanese quasi-spirituality that characterizes custom and tradition.

This essay contributes to *Sound and Robotics* by highlighting the significance of voice, and the implications of culturally specific contexts for robotics research.

Before I examine AI Misora Hibrari, I will first consider some "live" concerts of the holographic doubles of deceased pop singers in the USA in order to set the context and the basis for my comparative discussion.

14.2 Posthumous "Holographic" Performances of Deceased Singers

In the mainstream popular music scene of the global new millennium, "live" music performances through the use of the "holographic" image are presented as an entertainment spectacle. These spectacular shows deploy a modern version of "Pepper's Ghost", the 19th century optical trick in which an image of an actor below the stage is projected onto a pane of glass that is placed between the stage and the audience. With clever lighting effects, the glass becomes invisible, allowing a projected image to appear 3D-like (Stout 2021). It is not a true hologram, as there is no recording of 3D images that is played back for the audience, but the Pepper's Ghost visual trick creates a startlingly believable 3D image through what we might call "holographic effects" (Nick Smith, president of AV Concepts, audio-visual solutions provider, quoted in Dodson 2017). (In the rest of this essay, I will refer to these effects as holograms.) Contemporary holographic music performances have reinvigorated this technique with high-definition CGI projection and advanced AV systems.

Iconic singer Madonna's collaborative performance with Gorillaz, a UK virtual band who presented themselves as animated characters in the 2006 Grammy Awards ceremony would be one notable example of the contemporary utilization of this holographic technique. Madonna performed through a human-sized holographic image of herself with the virtual band. The moving image of Madonna sang and moved between the animated images of Gorillaz members, successfully presenting this display as a "live" performance, which was followed by a performance by the "real" Madonna, who wore the same costume as her image (Kiwi 2017). The most recent example of a holographic "live" show appeared in ABBA's comeback concert *Voyage* in 2022. On stage, images of ABBA's members are turned into 'eerily de-aged digital avatars' through CGI, or what have been dubbed "Abbatars" (Petridis 2022). For producer Ludvig Andersson, *Voyage* presented these avatars as real, "young" ABBA members rather than "four people pretending to be ABBA" because they had clearly aged (Stout 2021).

Significantly, instead of presenting holographic doubles of living artists, there have also been "live" concerts starring the digital doubles of *deceased* singers and musicians, an approach that quickly became highly controversial. Such concerts include those featuring holograms of well-known singers and musicians from diverse genres of American popular music: hip-hop artists, such as Tupac Shakur (2012), Eazy-E (2013), and Ol' Dirty Bastard (2013); pop singers, including Michael Jackson (2014) and Whitney Houston (2020); the heavy metal singer Ronnie James Dio (2017); rock'n'roll musicians of the 1950s Roy Orbison and Buddy Holly (2019); jazz rock musician Frank Zappa (2019); and opera singer Maria Callas (2018).

The very first case of such holographic performances was a duet by the rapper and media personality Snoop Dogg and a digitally recreated, animation of the late Tupac Shakur in the 2012 Coachella Valley Music and Arts Festival. This event gave rise to critical debates highlighting ethical concerns. Critics saw it as an abuse of the dead artist's legacy. The digital performance of Tupac Shakur is worth discussing in further detail.

At the end of Dr. Dre and Snoop Dogg's set at the Coachella festival, the very realistic, shirtless figure of Tupac Shakur appeared to rise from the stage. The digital double shouted "What the f— is up, Coachellaaaaa!" in front of the crowd of 80,000 (Dodson 2017). The digital Tupac Shakur then performed *Hail Mary* (1997), one of his posthumously released singles. The second song was *2 Of Amerikaz Most Wanted* (1996) with Snoop Dogg, a re-enactment of a hit duo performance that the two had earlier sung in concert at the House of Blues in Los Angeles in July, 1996. Notably, this was Shakur's last live performance before he was murdered in September of the same year (Dodson 2017). At the end of the second song, the digital Shakur stood in the center stage with his head down, then disappeared with a lighting burst that splintered into fragments. A stunned audience was re-energised by the appearance of Eminem, who began his act at Coachella.

The creation of a holographic double of the late Tupac Shakur for the Coachella event was an entrepreneurial idea that was conceived by rapper, producer, entrepreneur Dr. Dre, who organized the show along with Snoop Dogg for performance on two consecutive Sunday nights in April 2012. Tupac Shakur, along with Dr. Dre and Snoop Dogg, was a key artist produced by the American record label Death Row Records in the 1990s. Dr. Dre was also the co-founder of Death Row Records. Leaving Death Row Records in 1996, he started his own record label, Aftermath Entertainment, and later signed prominent rappers such as Eminem and 50 Cent to the label. According to journalist Aaron Dodson, by creating the AI double, Dr. Dre appropriated Tupac Shakur's view that Machiavelli had faked his own death, such and that "Tupac Shakur faked his death and is still alive" (2017). Fans and music journalists also speculated that Dr. Dre would organize a tour with the virtual Tupac Shakur (Dykes 2012b). The rumour was denied by Dr. Dre during the Coachella festival (Dodson 2017).

From a business point of view, Coachella was a highly successful event, and the holographic participation of Shakur contributed significantly, resulting in 15 million views on YouTube, 50 million Google search for the digital Tupac Shakur, a "500%" increase in sales of Shakur's album, 1,500% increase in downloads of Shakur's song *Hail Mary* (Wolfe 2012). Fans were commenting on Twitter that, to see the virtual performance of Tupac Shakur, they would make sure to attend the Coachella festival in the following year, and would attend a tour of it if there ever were one (Dykes 2012a). As an indication of the success of the holographic Shakur within the industry, Digital Domain, a Hollywood visual effects and digital production firm that created the CGI images of Tupac Shakur, won a prestigious Cannes Lions Titanium Award at the 59th annual Cannes Lions International Festival of Creativity in 2012 (Wolfe 2012).

However, the performance of digital Tupac Shakur received a mixed reception from his fans. There was a great degree of confusion: i.e. "many people didn't know what to make of it" (Dodson 2017). The holographic effects that created the figure of Tupac Shakur were so effective that they may have generated "something authentic and visceral about the projection of Tupac that Coachella attendees experienced" (Dodson 2017). While fans were amazed to see the holographic performance, many of them expressed mixed feelings, using terms such as "freaky", "weird", "scary", and "troubling" (Jones, Bennett, and Cross 2015, 130). This kind of feeling toward non-human entities and representations is often described as uncanny, related to the well-known concept of "the uncanny valley" coined by Masahiko Mori in a landmark 1970 paper. Mori is a pioneering figure in the history of Japanese robotics. In this paper, Mori sets forth a hypothesis that "as robots appear more humanlike, our sense of their familiarity increases until we come to a valley", at which point robots become eerie (Mori 2012, 98). Digital Tupac Shakur fell into the "uncanny valley", as it was presented not only as a resurrection of a deceased star, but one who died violently, as signaled by the inclusion of *Hail Mary.*

Significantly, the event was also seen as an exploitation of the dead. Journalist Tony Wong described the event as "[c]reepy maybe, but profitable": a succinct characterization that captures the ambivalence of fans and critics concerning the digital recreation of Shakur (Wong 2012). The ethical issues around "the resurrection of Tupac 'from the grave'", are more precisely to do with appropriateness and legitimacy (Jones, Bennett, and Cross 2015, 129). According to Nick Smith of AV Concepts, the visual technologies company that projected and staged the digital performance, Dr. Dre had a vision to "bring [the digital image of Shakur] back to life" and "to utilize the technology to make it come to life" (Kaufman 2012). For journalist Jason Lipshutz, however, the re-creation of Shakur's stage performance was felt as "incorrect" in some way, and it looked like an "imitation" (Lipshutz 2012). The digital performance was also seen as an act of "profiting off of live re-creations of dead artists' music" (Lipshutz 2012). While Dr. Dre had the approval of Tupac Shakur's estate and his mother Afeni Shakur (Ugwu 2012), it is impossible to determine if the performer himself would have agreed to perform forever in digital form. The religious tone of the "resurrection" image of Shakur in the performance – his emergence from the bottom of the stage, the gold crucifix necklace, and the song 'Hail Mary'- sent the audience the clear message that they were invited "to accept the presence on stage 'as Tupac', not a holographic tribute to him" (Jones, Bennett, and Cross 2015, 129). Under the influence of drugs and alcohol, some audience members were "confused and thought [Tupac Shakur] might be alive now", and they questioned "why are [Dr Dre and Snoop Dogg] doing this [?]" (Dykes 2012a). The digital performance of Tupac Shakur triggered mixed reactions from the producers, the audience, and the critics, dividing opinions.[2]

The ambivalence felt by Tupac Shakur's fans in 2012 regarding his digital double did not deter music promoters, record labels, technology-driven creative entrepreneurs, and the estates of deceased singers to create "holographic" doubles of other late pop music stars. The business opportunities arising through this new technologically enhanced representation sought to take advantage of "people's desire for reliving an experience with their favorite artists from the past" (Marinkovic 2021). In 2013, Rock the Bells, an annual hip-hop festival, included holographic performances of Eazy-E and Ol' Dirty Bastard. The 2014 Billboard music awards presented Michael Jackson's holographic double. As concerts generate "the most profitable revenue" (Marinkovic 2021), music producers also organized touring concerts involving holographic performances, such as those featuring the late heavy metal rock vocalist Ronnie James Dio in 2017 and, in 2019, rock musician Frank Zappa. To avoid the expensive and complex staging requirements for the Pepper's Ghost method, advanced

---

[2]Literary critic John Freeman discusses Tupac Shakur's holographic image in terms of slavery, the ownership and control of the Black subject in the USA (2016). A deeper investigation of race and ownership in relation to the deployment of representations of Tupac Shakur would be useful, but it falls outside the parameters of this essay.

computer rendering and high-resolution projection were used for the digital performances in the 2018 tours of opera singer Maria Callas and singer and songwriter Roy Orbison (with Buddy Holly in 2019). The touring performances of "Holographic" Roy Orbison (2018) and Frank Zappa (2019) were highly successful at the box office, re-emphasising the demand for such shows (Binelli 2020). A show featuring Whitney Houston's digital double was organized in the UK in 2020, and in Las Vegas in 2022, with a plan for it to tour in the USA in 2023 (Zeitchik 2021).

There is a tacit acknowledgment in the music industry that there is (and will continue to be) a market for holographic performances of deceased music celebrities. Demand will continue because the current music touring market is dominated by "artists who [are] at least 60 years old, among them Cher, Kiss, Fleetwood Mac, Paul McCartney, Dead & Company and Billy Joel" and the top three musicians are "The Rolling Stones, Elton John and Bob Seger" (Binelli 2020). Understanding the commercial opportunities of this situation, promoters like Ahmet Zappa, the son of Frank Zappa, representative of the Zappa estate, and the Executive Vice President of Eyellusion, the hologram company that created the digital doubles of Ronnie James Dio and Frank Zappa, argue for technological intervention. As Kory Grow indicates, Ahmet Zappa has stated that "Other artists are going to pass away, and if we want to keep having these magical experiences, technology is going to be the way to keep people engaged and hearing the music" (2019).

The technological extension of deceased superstars' musical careers raises serious concerns for some, not least because the practice reflects the music industry's insatiable appetite for monetization. As well, established stars could continue "their market domination after death", potentially limiting opportunities for younger artists (Myers 2019). Critics describe these digital performances with phrases such as "morbid cash grab" (Moran 2019) and "ghost slavery" (Myers 2019). There is no consent, as such, from the deceased artists (Moran 2019; Marinkovic 2021). In relation to such ethical concerns, some critics point out that there are emotional risks for fans. Because the technologies have not yet been perfected, performances can become flat and repetitive, as they cannot include spontaneous improvisation, and so they cannot sustain excitement for long (Moran 2019; Grow 2019; Marinkovic 2021). While some fans may feel that their memories of the original performances of a pop icon become tarnished when the market is made up of holographic performances, but on the other hand "the novelty factor and affordability will be a worthwhile tradeoff" for others (Myers 2019). As long as the fans in the second category do not mind the potentially repetitive spectacles of holographic doubles of their beloved musical icons, the business of holographic concert production is likely to continue.

It is therefore not surprising that some critics in the Japanese and US contexts refer to Hatsune Miku as evidencing an acceptable style of holographic performance. Hatsune Miku is a commercial software product that synthesizes a female singing voice. It was released in 2007 by Japanese company Crypton

Future Media, a media company founded by Itō Hiroyuki (Crypton Future Media 2007). Essentially, a user inputs lyrics and a melody to create a song that is sung by this synthesized female voice. The voice can suggest that there is a human being "inside" the vocal synthesizer. The product uses Yamaha's computer music software engine called VOCALOID. Crypton Future Media launched Hatsune Miku with the animated image of a slender teenaged girl with long, turquoise ponytails. Hatsune Miku has a typical anime face that features large eyes. The use of this kind of a cartoon image was a promotional strategy to make the character more appealing.

Itō is an advocate of "consumer generated media", a phrase that refers to media products that consumers themselves create (Kubo 2011). Itō allows the generation of music pieces and images featuring Hatsune Miku by non-commercial users. Hatsune Miku became a popular image and video meme proliferating on the Internet. The staging of a "live" concert with projected holographic animation of the character was organized by Crypton Future Media in collaboration with SEGA in 2009. "Live" concerts of Hatsune Miku became annual events. These concerts have attracted thousands of Hatsune Miku fans online. Lady GaGa used a singing animation of Hatsune Miku to open her concert tour ArtRave: The Artpop Ball in 2014 (Baseel 2014). More recently, Hatsune Miku was due to appear at the 2020 Coachella Valley Music and Arts Festival, which was canceled due to the global Covid pandemic (Cirone 2020).

The popularity of Hatsune Miku can be attributed to its openness to fans' engagement. Jones, Bennett, and Cross observe that "Hatsune Miku fans are able to participate generatively in the writing of the hologram's 'personality'" (Jones, Bennett, and Cross 2015, 129). Fans' own creation of Hatsune Miku songs, images, and video footage is "the virtual embodiment of a collective idea, a meme in human form" (McLeod 2016, 505). Critics of the use of holographic singing avatars argue that people consuming such products may feel comfortable with Hatsune Miku because they are more accepting of animated images that "are *brought to life*, rather than actual human bodies that are brought *back* to life" (Jones, Bennett, and Cross 2015, 129, original emphasis). The artificial existence of Hatsune Miku is understood as such, and is appreciated by fans (Michaud 2022). It is a very different aim to "resurrect" a deceased artist for business reasons.

It is therefore useful to consider an alternate conceptual framework for a holographic performance of a dead singer, one that, I suggest, is concerned with communality, a context that partly explains Hatsune Miku's popularity.

As a case study that also emerges from the Japanese cultural context, I will now discuss AI Misora Hibari. AI Misora Hibari did not escape from criticism that was very similar to the criticism that holographic performances by Western singers were subject to. However, it is my contention that AI Misora Hibari presents a unique instance of the melding of the fan community and culturally specific nostalgia that gave rise to a context in which AI Misora Hibari would be appreciated and accepted.

14.3 "Resurrection" of Misora Hibari

Japan's public broadcasting station NHK (Nippon Hōsō Kyōkai, meaning the Japan Broadcasting Corporation), collaborating with Yamaha Corporation, organized the "AI (artificial intelligence) Misora Hibari" project in 2019, which aimed to reproduce the singing voice of the late Misora Hibari. Misora was a legendary, highly popular singer of the post-war era in Japan. AI Misora Hibari used Yamaha's VOCALOID AI singing synthesis technology, which was a more developed version of the VOLCALOID technology that was deployed for Hatsune Miku, combined with artificial intelligence and 4K 3D video projection.

Unlike the Western counterparts of holographic performances, in which existing recordings of the deceased singers were used, VOCALOID AI analysed Misora's voice and was able to "sing" "*Arekara*", a song that was written for the project, by putting the singer's phonemes together. "*Arekara*" was written by Akimoto Yasushi, a well-known lyricist and record producer who wrote the lyrics for "*Kawa no nagare no yō ni* (Like The Flow of The River)" (1989), Misora's last song, released as a single, before her death in 1989. In September 2019, AI Misora Hibari was presented to a live audience, including a family member of Misora's, people who had had close professional relationships with her, and the singer's devoted fans. A film documentary of the project titled *AI de Yomigaeru Misora Hibari (The resurrection of Misora Hibari by AI)* (2019) was aired on television in the same month. AI Misora Hibari was also presented in Kōhaku Uta Gassen, an annual New Year's Eve television special broadcast by NHK.

AI Misora Hibari was well received at its first showing by its assembled audience members, who were closely involved in the production, as well as a group of long-term fans who provided feedback during the process of creating AI Misora Hibari. However, the public's views on AI Misora Hibari have been divided. As may have been anticipated, the main criticism concerning AI Misora Hibari was whether it was appropriate to create a digital double of Misora Hibari at all. Singer-songwriter Yamashita Tatsurō criticized AI Misora Hibari as a "blasphemy" in January 2020 (Kimura 2020, my translation). Yamashita's underlying message that was communicated by the use of the term is that the essential qualities of Misora's singing should not and cannot be reproduced by artificial intelligence (Kimura 2020). A similar point is also raised by actor Nakamura Meiko, who was a close friend of Misora's: "I want the actual person [Misora Hibari] to be here and sing again for us" (quoted in Nishioka 2019, my translation). For Yamashita and Nakamura, the uniqueness of Misora's live presence simply cannot be replaced or synthesized. It is also possible to speculate that perhaps Yamashita would not wish his *own* singing voice to be similarly turned into a synthetic voice through the use of artificial intelligence (Suzukake 2020). Even if Japanese fans felt that a synthesization of her voice would be acceptable, there is also the question of what visual representation (or avatar) would truly reflect who the famous person really is (Koyama 2020). The "wrong" representation could taint the appropriately reflective individuality of a dead celebrity.

These criticisms express an anxiety that the uniqueness of an individual is being violated in order to be commodified, as for their Western counterparts. Fukui Kensaku, a lawyer specialising in copyright law in Japan, warns of the perils of the commodification of AI doubles of celebrities, given the lack of clarity in Japanese copyright law concerning the rights over such products. He refers to "deepfake" media technologies that can, for example, replace a person's face with that of another person on video (quoted in Nishioka 2019). The creation of AI digital doubles would be legal in Japan as long as it does not violate the "honor" and "privacy" of a dead celebrity (Fukui in Nishioka 2019).

On a technical level, the appearance and gestural movement of the projected image of Misora Hibari seemed somewhat crude in contrast to the level of sophistication of the holographic image's vocal component. The holographic image did not match the high standards for sound quality generated by Yamaha's VOCALOID AI. AI Misora Hibari was criticized for being "too mechanical" or "creepy" (Harada 2020, my translation). Some audience members expressed the view that, as Misora passed away 30 years ago, people still remember her singing, and therefore, differences between the actual and the copy may be noticeable (Suzukake 2020). In contrast to the strong visual presence of Tupac Shakur's digital double, AI Misora Hibari's visual representation was mediocre, which seemed to degrade or demean Misora Hibari, according to critics, who dismissed it as "a gimmick" (Michel 2020).

Despite such concerns, for the audience attending AI Misora Hibari's first performance, the AI double created a special occasion that transcended ethical questions and technological weaknesses. I would like to discuss this transformative moment, as I see it as a culturally specific instance in which, paradoxically, it would not have been appropriate for a human, that is, a live Misora Hibari impersonator, to perform. For reasons that I will now discuss, only a *non-human performer,* like a holographic double of the dead singer, could have succeeded as thoroughly as AI Misora Hibari did. The success of the digital performer can be almost entirely attributed to its particular voice performance.

Misora Hibari was famous for her unusually talented singing abilities. She was able to shift gracefully from "natural voice (*jigoe*)" to "falsetto voice (*uragoe*)" and "dramatic melismas and throbbing vibratos (*yuri*)" (Tansman 1996, 127). Yamaha's team of sound engineers and audio scientists stored recordings of Hibari Misora's singing and speech in VOCALOID AI, using the audio synthesis technology to carry out "machine learning" (automated learning) and "deep learning" (replicating a learning capability that mimics the human brain) to duplicate the singer's voice at the level of phonemes, as I mentioned earlier. The feedback the team received from the deceased singer's longtime fans was critical. Interestingly, they advised that the singing of the vocal double produced in VOCALOID AI did not entirely sound like the real Misora. It sounded flat to them (NHK 2019). The Yamaha sound design team later discovered that Misora Hibari's very unique vocalizations included "overtone singing" and "derivate misalignment of singing timing and interval". VOCALOID AI was reprogrammed to replicate this vocal styling in its singing of the newly composed song "*Arekara*" (NHK 2019).

The performance coupled the replication of such subtleties with effects described by voice theorist Steven Connor as "the embodying power of the voice" in his study of ventriloquism. Significantly, ventriloquism is highly successful at generating what Connor calls "the vocalic body" (Connor 2000, 35–36). This means that a dummy or even a non-anthropomorphic object can become a speaking "body", as it is animated by the ventriloquist's voice. For AI Misora Hibari, the vocalic body that can be said to have emerged from its "dummy", that is, the avatar of the visual representation, however simplistic, was supported to a great extent by some registers of cultural meaning that are particularly Japanese: i.e. the use of *enka* music, which is a type of ballad in Japanese popular music, and Japanese cultural traditions in dealing with the dead. I will now discuss each of these in turn, and how they may have influenced the positive reception of the voice of AI Misora Hibari.

14.4   *Enka* Singer Misora Hibari

Misora Hibari is regarded as a quintessential *enka* singer. Lyrics in *Enka* are sentimental and melodramatic, and full of pathos concerning lost love, loneliness, and pain (Martin 2008); yearning for "a lost past" or "an unattained future" (Tansman 1996, 116); and longing for one's place, one's home town (Yano 2002, 18). Sharing similarities with Blues melodies and often utilising a vibrato vocal technique, *Enka* songs actively reflect the hardships of ordinary people, and especially those whom the Blues might refer to as "down and out". The "hard luck" narratives of *enka* express the nostalgia of such people for an idealised past situation or life circumstance (Martin 2008). *Enka* became very popular in the late 1960s, corresponding with the anti-establishment movements of Japan's disenfranchised youth and the working class during that decade (Yano 2002; Wajima 2010). *Enka* was seen as "an alternative to imported [Western] musical styles", and it was meant to provide "'authentic' Japanese experience" to a wide audience (Tansman 1996, 111).

Misora was seen as the embodiment of the Japanese spirit of "postwar forbearance" and "stoicism and endurance" who could bind people together with the "honesty and sincerity" she projected in her stage persona (Tansman 1996, 105 and 107). While the late singer has been described as the "Queen of *Enka*", she has also been labeled the "Queen of Showa", as her death in 1989 occurred at the same time as, and, for many Japanese people, has signified the symbolic closure of, the Showa period (Martin 2008). There is an important correlation therefore between *enka* and the Showa era. *Enka* represents songs that deeply symbolise the Shōwa-era, the period of the 20th century between 1926–1986, corresponding with the reign of Emperor Shōwa (Hirohito), during which Japan experienced significant cultural and economic change. Misora is further described as the "queen of tears" because the singer's performance style demonstrated crying that is "profuse, unchecked and unwiped" (Yano 2002, 121). Misora expressed "emotion and pain" that reflected the central concerns of *enka*. *Enka* singer Misora Hibari became an idealised image of the "Showa woman", which has been mythologised since the deaths of the singer and the emperor (Tansman 1996, 107). Misora emblematised the Showa woman's traditional sensibilities and presentation of herself as a unifying figure of diligence through adversity.

The *enka* song that AI Misora Hibari performed in its first public show, "*Arekara*", also moved the audience. It was written by Akimoto Yasushi, a well-known television producer and lyricist, and accompanied by a highly emotive *enka* melody written by Yoshinori Satō especially for this event, as I have mentioned. Akimoto had written *Kawa no nagare no yō ni* (*Like the Flow of the River*), the last single produced by Hibari Misora in the same year as her death in 1989. These choices were selected carefully, and would have clearly associated this *enka* performance with those of the real Misora Hibari.

"*Arekara*" fully embodies the nostalgic ethos of *enka*. Its lyrics first describe a sunset as a metaphor signalling farewell to a loved one, leaving behind a sense of regret. The second part of the song describes a night illuminated by stars that cleanse any negative feelings and offer hope for tomorrow. Songwriter Akimoto's lyrics are written as if the singer, the voice of Hibari Misora, were talking to the audience, addressing them directly. There are two key set lines that repeat. The first is: "How have you been since then? I too have become old. I'm still humming an old song without realising it. If I look back, it was a happy time, don't you agree". The second reprises these themes of aging, memory, and yearning: "How have you been since then? The long time has passed. I want to sing an old song, which I had sung many times. Yes, memory is precious to one's life" (NHK 2019, my translation). The lyrics are written in the first person singular, as if Misora is gently talking to the audience in a caring and familiar way. The song also includes *serifu*, an oral address to the audience without singing. The Misora double states: "It has been a long time since I saw you last. I have been watching you. You've been working hard. Please keep up the hard work for my sake" (NHK 2019, my translation). It is easy to imagine that these comments must have been delivered as comforting messages for devoted Misora fans, reflecting Misora's perceived persona as a "giver of succour", as Tansman indicates (1996, 117).

Indeed, as shown in the documentary *The Resurrection of Misora Hibari by AI*, the first performance of AI Misora Hibari in 2019 moved the audience members, including those who had had close relationships with the late singer or were devoted to her memory for decades, to tears (NHK 2019). These fans, many of whom were in their 70s, had lived through the Showa era, which, in effect, Misora Hibari had emotively narrated for them. Fans expressed their excitement when they entered the venue to observe the first public live performance (NHK 2019). Kazuya Kato, the son of Misora, commented that of course he was aware that his mother would not literally be brought back to life, but he appreciated that the most advanced AI technologies would be able to fill the "time gap" for him since her death (NHK 2019). I suggest that the emotional responses by Misora's son, close colleagues, and fans are testimonies to the effectiveness of AI Misora Hibari in terms of its framing with regard to the real Misora's emblematising of both the deep emotion typical of *enka* and of sentiments tied to a nostalgic view of the Showa era. Significantly, these factors also resonated with longstanding Japanese cultural traditions regarding the dead, i.e. the *kuyō* (commemoration service) and the involvement of *Itako* (blind female shamans). In these cultural practices, which persist in present-day Japan, people are allowed to be demonstrably emotional, and are even expected to be so.

14.5   Japanese Cultural Traditions and the Dead

At an event involving the Buddhist practice of *kuyō*, family members come to a temple for what is essentially a memorial service for a deceased person, laying flowers and incense at the altar. In Japan's indigenous belief system, when a person dies, the spirit leaves the human body and flies off to another world. The spirits of the dead return at these events. It is therefore not an uncommon practice for family members to "talk" to the dead at the altar as if they are listening. The theatrical precondition of "as if", coupled with the particular backdrop of temple and graveyard, is essential to the emotive part of *kuyō*. I suggest that the first performance of AI Misora Hibari sought to facilitate a flow of emotion in relation to the dead in ways that are very similar to the tradition of *kuyō*.

I'd also like to suggest that cultural knowledge regarding *Itako*, blind female shamans, is also relevant here. *Itako* are traditionally understood to act as mediums who call upon the spirits of the dead, allowing relatives to communicate with them. Religion scholar Omichi Haruka indicates that although the actual *Itako* practice was originally site-specific, localised to a northern region of Japan, the history of the *Itako* in the 20th and in the 21st centuries has demonstrated a process of simplification and delocalization (Oomichi 2017, 246). According to Omichi, in the 1960s in Japan, the notion of the *Itako* became popularised along with the concept of a "rediscovered" older Japan, which was regarded as the antithesis of the Westernised, modern Japan (2017, 90–91). The *enka* received renewed attention in the same period, the 1960s. While AI Misora Hibari was not explicitly referenced in relation to *Itako*, I suggest that the same cultural logic is apparent, that is, AI Misora Hibari can conjure up past relationships, and even the sensibility of a past era, by essentially channelling the dead for its audience.

Views regarding the dead, as expressed in the practices of *kuyō* and *Itako*, are cultural knowledge or internalized, customary thinking that Japanese people are not conscious of in everyday life. In this sense, *kuyō* and *Itako* can be discussed in terms of religion scholar Inaba Keishin's notion of "unconscious religiosity" that describes a vague sense of connection beyond the boundaries of the self to ancestors, gods, Buddha, and the community (2011, 15). These ideas become meaningful for Japanese individuals through recitation and other forms of repetition at cultural events such as *kuyō* and *Itako*. In other words, the cultural practices of *kuyō* and *Itako* permits a "feeling" for and a "performance" of dialog with the dead. I argue that AI Misora Hibari's performance capitalised on this same logic, even to the extent of emotionally signifying a past era that would have complex, loaded meanings for an audience that lived through it, and may desire to view it nostalgically.

The Showa era was indirectly referenced in *"Arekara"*. The song's lyrics "If I look back, it was a happy time" could be taken to refer to Misora's career from the 1950s until her death in 1989, which was also a prosperous period for Japan, until the economic crash of the late 1980s/early 1990s. For anthropologist Christine Yano, *enka* is simultaneously "a technology for creating national and cultural memory and it is an archive of the nation's collective past" (Yano 2002, 17). Because the audience for the first performance of AI Misora Hibari consisted primarily of the generations that lived through the Showa period, the song conveys a nostalgia for this idealised, shared past. Essentially, remembrance of the Showa era has been conflated with remembrance of Misora Hibari. The audience's longing for Misora is integrally associated with memories of this past period, as both describe a "better" past that would have seemed, in the 1960s and 1970s, out of reach. The unsurmountable distance of the complicated present from a mythologised past fuels a yearning desire for a "lost" object, as literary scholar Susan Stewart discusses as "the desire for desire" (1993, 23), which she sees as a fundamental principle for nostalgic feeling to flourish.

Significantly, *enka*'s calling up of nostalgic feeling, the desire for a simplified past that never actually existed, is *instrumentalised* in relation to AI Misora Hibari. I will explain this point further after discussing the particular ways that *enka* codifies emotion and memory.

14.6   Codes of *enka*

Enka generates nostalgic feeling effectively through the Japanese concept of "*kata*", meaning a repertoire of patterned forms. In his book *Empire of Signs*,

Roland Barthes (1983) examines Japanese life as composed of social rules and manners embodied in multiple modes of performance (corporeal, verbal, visual, aural, kinetic, and spatial) in the forms of writing, gestures, architecture, or theatrical forms such as Kabuki. For example, as Barthes explains it, Japanese traditional arts function through abstracted expressions that are specific to each genre: i.e. the *"kata"* for that genre. These are passed on and repeated from one generation to another by the masters of each genre. This tradition of *kata* explains, for instance, why the audience for Kabuki can enjoy and take seriously an old, male *onnagata* actor playing a young female role without his performance seeming grotesque or comical. It is the legibility of the *kata* in Kabuki for an audience that understands its codes as icons of distilled gestures that allows the expressions of the *onnagata* actor to be understood and appreciated. In Barthes's words, famously interpreting Japanese culture as a system of signs, "Femininity is presented to read, not to see" (Barthes 1983, 53). The audience for Kabuki appreciates the skilled delivery of this unique arrangement of referents within a familiar framework – that is, the expert audience can enjoy this particular pattern. Importantly, the emphasis here is on internalization and embodiment, rather than superficial imitation. Literal visual resemblance is not important for this traditional art. What is important is establishing a mutual understanding between the actor and the audience through a kind of abstraction that requires an audience that understands the its culturally encoded viewing perspective.

Enka is appreciated through a similar cultural, and *historical*, codification. As Yano pints out, *enka* exhibits "varying degrees of 'patternedness' that can include the smallest nuance of breathiness, the lifting of a heel, and the streaming of tears, as well as the sounds, sights, and situations that evoke those tears" (Yano 2002, 25). In the sign system of *enka*, emotion and pain are presented to be read in terms of a repertoire of codified expressions that signal gender as well as Showa-era attitudes. As customary responses that are typical in Japanese traditional performing arts, such as Kabuki, fans respond to well executed singing by *enka* performers with applause or shouts at appropriate moments in the performance.

My contention is that AI Misora Hibari, seen as enacting the particular interplay of signs for the genre of *enka*, worked reflexively for the audience at the first showing. AI Misora Hibari was built as an assemblage of signs that were drawn from the late popular singer's signature gesture in terms of voice, singing, image, and movement. The vocal signs of the deceased singer seem to have been extremely convincing. As such, at its inaugural performance, AI Misora Hibari became a medium to trigger the memory of the singer for each audience member, as well as catalysing collective memory for the attendees. A "collective memory" can recreate "an image of the past" through "certain instruments", and this image takes shape in terms of "collective frameworks of memory", as French philosopher and sociologist Maurice Halbwachs puts it (1992, 40). As I have discussed above, *enka* is such an instrument, giving rise to vivid images (for a knowledgeable audience) of the lives of ordinary people, especially those living at the edges of Japanese society. These are precisely the people who were forgotten and disenfranchised during the late postwar period under a lengthy, and still ongoing, Americanization of the country in which entrepreneurialism, capitalism, and class mobility became more important than perceived community.

Japanese studies scholar Alan Tansman stated: "To her fans, Misora Hibari remains a vibration from the past [the 1960s and the 1970s] that echoes into the future" (Tansman 1996, 129). These remarks suggest that this kind of "reliving" can be a positive and forward-thinking event for Japanese society, not merely a closed loop of cultural and historical regression. Tia DeNora, a sociologist specialising in music, argues that "as an identification with or of 'the past', is part of the work of producing one's self as a coherent being over time, part of producing a retrospection that is in turn a resource for projection into the future" (DeNora 2000, 66). In other words, the past evoked through music can provide "a resource for the reflexive movement from present to future" (DeNora 2000, 66). Such deep recollections, which link the past to one's self in the present to a "projected" common future that draws upon the past and present, can help a society to move forward in a stable and coherent way.

Just like the Phoenix Concert, symbolising the rebirth of the mythical, immortal phoenix, the performance of AI Misora Hibari recalled her "comeback" concert in 1988 after her illness in 1985, and was a major event before her death in 1989: it was as if Misora Hibari had returned again. The singer's message of support and empathy for the audience, as emblematised in the song "*Arekara*", encouraged the audience to look the future, just as the real Misora Hibari had done before.

14.7 Conclusion

In this essay, I have examined posthumous holographic performances of deceased singers in the American context and in the Japanese context. The 2012 performance of the late Tupac Shakur's holographic double was the earliest show that was presented in a "live" setting, which is to say with a live audience and other highly effective, which brought the troubling idea of "resurrection" to the fore for its critics, triggering debates on whether or not such a performance was ethically and morally acceptable. Since then, many similar holographic performance events featuring the voice and appearance of deceased pop singers have been organized, although there is no agreement within the industry on fundamental ethical issues concerning these projects. However, in addition to raising similar concerns in the Japanese context, this approach *also* appears to be valued rather differently. This essay has examined the culturally different perspective on these debates that is illustrated by the case of AI Misora Hibari.

It is my contention that the meaning of "resurrection" concerning AI Misora Hibari was deployed inaccurately by some Japanese critics when they discussed the project. Unlike the performance featuring a holographic Tupac Shakur, the "resurrection" of Misora Hibari for the audience at the first show was not the "reincarnation" of the singer. While the holographic image of Tupac Shakur produced a strangely visceral sense of presence for its audience, as is evident in the comments of audience members, the digital image of Misora Hibari was visually imperfect. Instead, it was the highly successful recreation of Misora Hibari's voice by VOCALOID AI that the audience paid the most attention to: it strongly generated the vocalic body of the singer.

The song *"Arekara"* was a newly written *enka* song for VOCALOID AI, not a remastering of an existing recorded song by Misora Hibari. As the audience members were involved in the process of making AI Misora Hibari, their collective frameworks for cultural and historical memory were deliberately activated. This cultural-historical memory was triggered by the enacting of the cultural practices of *enka*, *kuyō* (commemoration service), and *Itako* (the historical, place-specific practice of divination led by female shamans), and, importantly, the positive, nostalgic recalling of the Showa era. These factors were all at play, enhancing the commemorative aspects of this first show in 2019. AI Misora Hibari was, in a strongly symbolic sense, deployed as a "phoenix" that facilitated a revitalization of collective memory of the Showa era. It *also* evoked concerning the economically bright outlook of the 1960s and the 1970s in Japan for the audience at the first recital. The evocation of strong feeling was facilitated and even enhanced by these multiple layers of emotive and historical recollection. It was a unique one-off, culturally specific show that could only be performed by non-human performer AI Misora Hibari, untouched by the vagaries of time.

Bibliography

[1] Barthes, Roland. 1983. *Empire of Signs*. Translated by Richard Howard. London: Jonathan Cape.

[2] Baseel, Casey. 2014. 'Lady Gaga Goes Gaga for Hatsune Miku, Makes Virtual Idol Her Opening Act'. *SoraNews24 -Japan News-* (blog). 16 April 2014. https://soranews24.com/2014/04/17/lady-gaga-goes-gaga-for-hatsune-miku-makes-virtual-idol-her-opening-act/.

[3] Binelli, Mark. 2020. 'Dead Ringers: How Holograms Are Keeping Music Legends Alive'. The Sydney Morning Herald. 17 January 2020. https://www.smh.com.au/culture/music/dead-ringers-how-holograms-are-keeping-music-legends-alive-20200109-p53q0g.html.

[4] Cirone, David. 2020. 'Kyary Pamyu and Pamyu and Hatsune Miku to Turn Coachella "Kawaii" in 2020'. *J-Generation* (blog). 3 January 2020. https://j-generation.com/2020/01/kyary-pamyu-and-pamyu-and-hatsune-miku-to-turn-coachella-kawaii-in-2020/.

[5] Connor, Steven. 2000. *Dumbstruck: A Cultural History of Ventriloquism.* Oxford; New York: Oxford University Press.

[6] Crypton Future Media. 2007. 'Vocaloid 2, Hatsune Miku'. 2007. http://www.crypton.co.jp/mp/pages/prod/vocaloid/cv01.jsp. Accessed on 04-15-2016.

[7] DeNora, Tia. 2000. *Music in Everyday Life.* Cambridge: Cambridge University Press. https://doi.org/10.1017/CBO9780511489433.

[8] Dodson, Aaron. 2017. 'The Strange Legacy of Tupac's "Hologram" Lives on Five Years after Its Historic Coachella Debut'. *Andscape* (blog). 14 April 2017. https://andscape.com/features/the-strange-legacy-of-tupacs-hologram-after-coachella/.

[9] Dykes, Brett Michael. 2012a. 'A First-Hand Account Of Hologram Tupac's Coachella Performance: "I Was Completely Freaked Out" '. Uproxx. 17 April 2012. https://uproxx.com/music/a-first-hand-account-of-hologram-tupacs-coachella-performance-i-was-completely-freaked-out/.

[10] ———. 2012b. 'Hologram Tupac May Be Touring With Dr. Dre Soon'. Uproxx. 17 April 2012. https://uproxx.com/technology/hologram-tupac-may-be-touring-with-dr-dre-soon/.

[11] Freeman, John. 2016. 'Tupac's "Holographic Resurrection": Corporate Takeover or Rage against the Machinic?' *CTHEORY*, Theorizing 21C. https://journals.uvic.ca/index.php/ctheory/article/view/15700.

[12] Grow, Kory. 2019. 'Inside Music's Booming New Hologram Touring Industry'. Rolling Stone. 10 September 2019. https://www.rollingstone.com/music/music-features/hologram-tours-roy-orbison-frank-zappa-whitney-houston-873399/.

[13] Halbwachs, Maurice. 1992. *On Collective Memory.* Translated by Lewis A. Coser. Heritage of Sociology. Chicago: University of Chicago Press.

[14] Harada, Shinya. 2020. ' "AI Misora Hibari" Ni Sanpi: Kojin No "Saigen" Giron No Keiki Ni (Pros and Cons on "AI Misora Hibari": Opportunity to Debate on "reviving" the Dead'. *Tokyo Shimbun/Tokyo Web*, 4 February 2020. https://www.tokyo-np.co.jp/article/7226.

[15] Inaba, Keishin. 2011. 'Unconscious Religiosity and Social Capital'. *Religion and Social Contribution* 1 (1): 3–26.

[16] Jones, Angela, Rebecca Bennett, and Samantha Cross. 2015. 'Keepin'it Real? Life, Death, and Holograms on the Live Music Stage'. In *The Digital Evolution of Live Music*, edited by Angela Jones and Rebecca Bennett, O'Reilly Online Learning: Academic/Public Library Edition. UK: Chandos Publishing. https://doi.org/10.1016/B978-0-08-100067-0.00010-5.

[17] Kaufman, Gil. 2012. 'Exclusive: Tupac Coachella Hologram Source Explains How Rapper Resurrected'. MTV News. 16 April 2012. http://www.mtv.com/news/1683173/tupac-hologram-coachella/.

[18] Kimura Shingo. 2020. 'Yamashita Tatsurō, "Bōtoku hatsugen" ni sandō no koe: AI Misora Hibari ni iwakan ga astimatta wake (Support for Tatsurō Yamashita's "blasphemy comments": The reasons for uneasiness concerning AI Misora Hibari)'. Excite News. 22 January 2020. https://www.excite.co.jp/news/article/Asajo_84512/.

[19] Kiwi, Andy, dir. 2017. *Gorillaz Ft. Madonna - Feel Good Inc & Hung Up Live at Grammy Awards 2006*. https://www.youtube.com/watch?v=CGoSlY2sT04.

[20] Koyama Yasuhiro. 2020. 'Kōhaku de wadai: "AI Misora Hibari" no "naka no hito" no issha ga kataru 'dejotaru hyūman no kanōsei (A topical subject at Kōhaku: A possibility for "digital human" told by an "insider" company who participated in the "AI Misora Hibari" project)'. Business Insider Japan. 3 January 2020. https://www.businessinsider.jp/post-205167.

[21] Kubo, Toshiya. 2011. 'Special Interview'. Inter-x-Cross Creative Center. 2011. http://www.icc-jp.com/special/2011/01/001726.php. Accessed on 04-15-2013.

[22] Lipshutz, Jason. 2012. 'Opinion: The Problem with the Tupac Hologram'. Billboard. 16 April 2012. https://www.billboard.com/music/music-news/opinion-the-problem-with-the-tupac-hologram-494288/.

[23] Marinkovic, Pavle. 2021. 'Hologram Artists — The Future of Live Performance'. *Predict* (blog). 13 December 2021. https://medium.com/predict/hologram-artists-the-future-of-live-performance-9851d2e02ae1.

[24] Martin, Alex. 2008. ' "Enka" Still Strikes Nostalgic Nerve Bluesy Ballads Stand the Test of Time, Hold Their Vibrato Own and Enter Japan's DNA'. *The Japan Times Online*, 18 November 2008. https://www.japantimes.co.jp/news/2008/11/18/reference/enka-still-strikes-nostalgic-nerve/.

[25] McLeod, Ken. 2016. 'Living in the Immaterial World: Holograms and Spirituality in Recent Popular Music'. *Popular Music and Society* 39 (5): 501–515. https://doi.org/10.1080/03007766.2015.1065624.

[26] Michaud, Alyssa. 2022. 'Locating Liveness in Holographic Performances: Technological Anxiety and Participatory Fandom at Vocaloid Concerts'. *Popular Music* 41 (1): 1–19. https://doi.org/10.1017/S0261143021000660.

[27] Michel, Patrick St. 2020. 'NHK Raises the Dead for "Kohaku" to Mixed Reviews'. The Japan Times. 10 January 2020. https://www.japantimes.co.jp/culture/2020/01/10/films/nhk-raises-dead-kohaku-mixed-reviews/.

[28] Moran, Padraig. 2019. 'As Frank Zappa Is Resurrected as a Hologram, Expert Warns Wishes of Dead Artists Should Be Considered'. CBC. 3 May 2019. https://www.cbc.ca/radio/thecurrent/the-current-for-may-3-2019-1.5121329/as-frank-zappa-is-resurrected-as-a-hologram-expert-warns-wishes-of-dead-artists-should-be-considered-1.5121464.

[29] Mori, Masahiro. 2012. 'The Uncanny Valley'. Translated by Karl F. MacDorman and Norri Kageki. *IEEE Robotics Automation Magazine* 19 (2): 98–100. https://doi.org/10.1109/MRA.2012.2192811.

[30] Myers, Owen. 2019. ' "It's Ghost Slavery": The Troubling World of Pop Holograms'. *The Guardian*, 1 June 2019, sec. Television & radio. https://www.theguardian.com/tv-and-radio/2019/jun/01/pop-holograms-miley-cyrus-black-mirror-identity-crisis.

[31] NHK, dir. 2019. 'AI de Youmigaeru Misora Hibari (Revival of Hibari Misora by AI)'. *NHK Special*. NHK.

[32] Nishioka Yukifumi. 2019. 'Kōhaku shutsujō, AI Misora Hibari, "kimochi warusa" no shōtai: Hō kisei ha hitsuyō ka (Appearance in Kōhaku Uta Gassen, AI Misora Hibari, the sauce of creepiness: Is legal regulation necessary?)'. AERA dot. 31 December 2019. https://dot.asahi.com/dot/2019122900003.html.

[33] Oomichi, Haruka. 2017. *Itako No Tanjō: Masu Media to Shukyō Bunka (The Birth of Itako Shaman: Mass Media and Religious Culture)*. Tokyo: Koubundou.

[34] Petridis, Alexis. 2022. 'Abba Voyage Review: Jaw-Dropping Avatar Act That's Destined to Be Copied'. *The Guardian*, 27 May 2022, sec. Music. https://www.theguardian.com/music/2022/may/26/abba-voyage-review-jaw-dropping-avatar-act-thats-destined-to-be-copied.

[35] Stewart, Susan. 1993. *On Longing: Narratives of the Miniature, the Gigantic, the Souvenir, the Collection*. Durham: Duke University Press.

[36] Stout, Andy. 2021. 'How ABBA's Abbatars Rely on a Victorian Magic Trick'. 2021. https://www.redsharknews.com/how-abbas-abbatars-rely-on-a-victorian-magic-trick.

[37] Suzukake Shin. 2020. 'AI Misora Hibari Kōsatsu: "Shisha heno Bōtoku" no kotoba no omomi wo sitte imasuka? (A thought on AI Misora Hibari: Are⁻ you aware of the weight of the phrase "the blasphemy against the dead"?)'. Magazine. FRaU. 26 January 2020. https://gendai.ismedia.jp/articles/-/70013.

[38] Tansman, Alan M. 1996. 'Mournful Tears and Sake: The Postwar Myth of Misora Hibari'. In *Contemporary Japan and Popular Culture*, edited by John Whittier Treat, 103–133. Honolulu: University of Hawai'i Press.

[39] Ugwu, Reggie. 2012. 'Everything You Ever Wanted To Know About The Tupac Hologram | Complex'. Complex. 19 April 2012. https://www.complex.com/pop-culture/2012/04/everything-you-ever-wanted-to-know-about-the-tupac-hologram.

[40] Wajima, Yūsuke. 2010. *Tsukurareta 'Nihon no kokoro' shinwa: 'enka' o meguru sengo taishū ongakushi*. Tokyo: Kōbunsha.

[41] Wolfe, Jennifer. 2012. 'Virtual 2Pac Wins Cannes Lion Award'. Animation World Network. 25 June 2012. https://www.awn.com/news/virtual-2pac-wins-cannes-lion-award.

[42] Wong, Tony. 2012. 'Tupac Hologram Opens up Pandora's Box; Dead Rapper's Performance at Coachella Raises Too Many Ethical Questions'. *The Toronto Star*, 22 April 2012, ONT edition, sec. Entertainment.

[43] Yano, Christine. 2002. *Tears of Longing: Nostalgia and the Nation in Japanese Popular Song*. Harvard East Asian Monographs. Cambridge, MA: Harvard University Asia Center. https://doi.org/10.2307/j.ctt1tfjck1.

[44] Zeitchik, Steven. 2021. 'Nine Years after She Died, Whitney Houston Is Back to Entertain You'. *Washington Post*, 29 October 2021. https://www.washingtonpost.com/technology/2021/10/29/whitney-houston-hologram-concert-innovations/.

Index

Page numbers in **bold** refer to tables and those in *italic* refer to figures.

Printed in the United States
by Baker & Taylor Publisher Services

Printed in the United States
by Baker & Taylor Publisher Services